SPECIAL RELATIVITY

For the Enthusiastic Beginner

David Morin
Harvard University

ISBN-10: 1542323517
ISBN-13: 978-1542323512

Printed by CreateSpace

Cover image: iStock photograph by Andrey Armyagov

Additional resources located at:
www.people.fas.harvard.edu/~djmorin/book.html

Contents

1 Kinematics, Part 1 **1**

1.1 Motivation . . . 2

1.1.1 Galilean transformations, Maxwell's equations . . . 2

1.1.2 Michelson–Morley experiment . . . 5

1.2 The postulates . . . 10

1.3 The fundamental effects . . . 12

1.3.1 Loss of simultaneity . . . 12

1.3.2 Time dilation . . . 17

1.3.3 Length contraction . . . 23

1.3.4 A few other important topics . . . 28

1.4 Four instructive examples . . . 31

1.5 Velocity addition . . . 35

1.6 Summary . . . 43

1.7 Problems . . . 44

1.8 Exercises . . . 49

1.9 Solutions . . . 54

2 Kinematics, Part 2 **71**

2.1 Lorentz transformations . . . 71

2.1.1 First derivation . . . 72

2.1.2 Second derivation . . . 75

2.1.3 The fundamental effects . . . 78

2.2 Velocity addition . . . 79

2.2.1 Longitudinal velocity addition . . . 79

2.2.2 Transverse velocity addition . . . 80

2.3 The invariant interval . . . 82

2.4 Minkowski diagrams . . . 87

2.5 The Doppler effect . . . 92

2.5.1 Longitudinal Doppler effect . . . 92

2.5.2 Transverse Doppler effect . . . 94

2.6 Rapidity . . . 96

2.6.1 Definition . . . 96

2.6.2 Physical meaning . . . 98

2.7 Relativity without c . . . 99

2.8 Summary . . . 102

2.9 Problems . . . 103

2.10 Exercises . . . 106

2.11 Solutions . . . 110

3 Dynamics 123
3.1 Energy and momentum 123
3.1.1 Momentum 124
3.1.2 Energy 125
3.2 Transformations of E and p 131
3.3 Collisions and decays 134
3.4 Particle-physics units 139
3.5 Force 140
3.5.1 Force in one dimension 140
3.5.2 Force in two dimensions 142
3.5.3 Transformation of forces 143
3.6 Rocket motion 145
3.7 Relativistic strings 149
3.8 Summary 150
3.9 Problems 151
3.10 Exercises 155
3.11 Solutions 159

4 4-vectors 177
4.1 Definition of 4-vectors 177
4.2 Examples of 4-vectors 178
4.3 Properties of 4-vectors 181
4.4 Energy, momentum 183
4.4.1 Norm 183
4.4.2 Transformations of E and p 183
4.5 Force and acceleration 184
4.5.1 Transformation of forces 184
4.5.2 Transformation of accelerations 185
4.6 The form of physical laws 186
4.7 Summary 187
4.8 Problems 188
4.9 Exercises 188
4.10 Solutions 189

5 General Relativity 192
5.1 The Equivalence Principle 192
5.2 Time dilation 194
5.3 Uniformly accelerating frame 197
5.3.1 Uniformly accelerating point particle 197
5.3.2 Uniformly accelerating frame 199
5.4 Maximal-proper-time principle 200
5.5 Twin paradox revisited 202
5.6 Summary 204
5.7 Problems 204
5.8 Exercises 208
5.9 Solutions 211

6 Appendices **225**
6.1 Appendix A: Qualitative relativity questions . . . 225
6.2 Appendix B: Derivations of the Lv/c^2 result . . . 235
6.3 Appendix C: Resolutions to the twin paradox . . . 237
6.4 Appendix D: Lorentz transformations . . . 238
6.5 Appendix E: Nonrelativistic dynamics . . . 241
6.6 Appendix F: Problem-solving strategies . . . 245
6.6.1 Solving problems symbolically . . . 245
6.6.2 Checking units/dimensions . . . 246
6.6.3 Checking limiting/special cases . . . 248
6.6.4 Taylor series . . . 249
6.7 Appendix G: Taylor series . . . 251
6.7.1 Basics . . . 251
6.7.2 How many terms to keep? . . . 253
6.7.3 Dimensionless quantities . . . 254
6.8 Appendix H: Useful formulas . . . 255
6.9 References . . . 256

Preface

Einstein's theory of relativity is one of the most exciting and talked-about subjects in physics. It is well known for its "paradoxes," which often lead to intense and animated discussions. There is, however, nothing at all paradoxical about it. The theory is logically and experimentally sound, and the whole subject is actually quite straightforward, provided that one proceeds calmly and keeps a firm hold of one's wits.

This book is a revised and expanded version of the last four chapters in my textbook *Introduction to Classical Mechanics, with Problems and Solutions* (Cambridge University Press, 2008). A great deal of additional commentary has been included in order to make the book accessible to a wider audience. This new version assumes a mild familiarity with standard Newtonian physics (often called "freshman physics"). More precisely, the first two chapters (on kinematics) assume very little prior physics knowledge; there is no mention of force, energy, momentum, etc. But Chapters 3 and onward assume a bit more, although still not a huge amount. (Appendix E gives a quick review of force, etc., for readers with a limited physics background.) The nice thing (or not-so-nice thing, depending on your point of view) about relativity is that it is challenging due to its inherent strangeness, as opposed to a heavy set of physics prerequisites. Likewise for the math prerequisite: calculus is used on a few occasions, but it isn't essential to the overall flow of the book. Generally, all that is required is standard algebra.

If you happen to have a strong background in standard Newtonian physics, then you might be in for a shock, because relativity is where you discover that most of what you know about physics is wrong. Or perhaps "incomplete" would be a better word. The important point to realize is that Newtonian physics is a limiting case of the more correct relativistic theory. Newtonian physics works perfectly fine when the speeds you're dealing with are much less than the speed of light. Indeed, it would be silly to use relativity to solve a problem involving the length of a baseball trajectory. But in problems involving large speeds, or in problems where a high degree of accuracy is required, you must use the relativistic theory.

Of course, relativity isn't the end of the story either. In your future physics studies, you will eventually discover that classical special relativity (the subject of this book) is similarly the limiting case of another theory (quantum field theory). And likewise, quantum field theory is the limiting case of yet another theory (string theory). And likewise . . . well, you get the idea. Who knows, maybe it really *is* turtles all the way down.

There are two main topics in relativity; *special relativity* (which doesn't deal with gravity) and *general relativity* (which does). We'll deal mostly with the former, although Chapter 5 contains a brief introduction to the latter. Special relativity may be divided into two topics, *kinematics* and *dynamics*. Kinematics deals with lengths, times, speeds, etc. It is concerned only with the space and time coordinates of an abstract particle, and not with masses, forces, energy, momentum, etc. Dynamics, on the other hand, does deal with these quantities. The first two chapters cover kinematics. Most of the fun paradoxes fall into this category. Chapter 3 covers dynamics. In Chapter 4 we introduce 4-vectors, which tie together much of the material in Chapters 1–3. After the brief introduction to general relativity in Chapter 5, we conclude with a number of appendices in Chapter 6. Among the topics covered are a collection of qualitative relativity questions and answers (Appendix A), a review of nonrelativistic dynamics (Appendix E), and strategies for solving problems (Appendix F).

I have included a large number of puzzles/paradoxes in the problems and exercises. ("Problems" have solutions included. "Exercises" do not, so that they can be used for homework assignments.) When attacking these, be sure to follow them through to completion, and don't say, "I could finish this one if I wanted to, but all I'd have to do would be such-and-such, so I won't bother," because the essence of the paradox may very well be contained in the such-and-such, and you will have missed out on all the fun. Most of the paradoxes arise because different frames of reference *seem* to give different results. Therefore, in explaining a paradox, you not only need to give the correct reasoning, you also need to say what's wrong with incorrect reasoning.

The difficulty of the problems and exercises is indicated by stars (actually asterisks). One star means fairly straightforward, while four stars mean diabolically difficult. Two stars are the most common. Be sure to give a solid effort when solving a problem, and don't look at the solution too soon. If you can't solve a problem right away, that's perfectly fine. Just set it aside and come back to it later. It's better to solve a problem later than to read the solution now. If you do eventually need to look at a solution, cover it up with a piece of paper and read one line at a time, to get a hint to get started. Then set the book aside and work things out for real. That way, you can still (mostly) solve it on your own. You will learn a great deal this way. If you instead head right to the solution and read it straight through, you will learn very little. Be warned that the four-star and some of the three-star problems are *extremely* difficult. Even if you understand everything in the text, that doesn't mean you should be able to solve every problem right away (as I can attest to!).

For your reading pleasure (I hope!), I have included limericks throughout the text. Although these might be viewed as educational, they certainly don't represent any deep insight I have into the teaching of physics. I have written them for the sole purpose of lightening things up. Some are funny. Some are stupid. But at least they're all physically accurate (give or take).

A few informational odds and ends: This book contains many supplementary remarks that are separated off from the main text; these end with a shamrock, "♣." The letter β denotes the ratio of v to c; that is, $\beta \equiv v/c$. We will often drop the c's in calculations, lest things get too messy. The c's can always be brought back in at the end, by putting them where they need to go in order to make the units correct. Fractions like $L/\gamma v$ are understood to mean $L/(\gamma v)$, and not $(L/\gamma)v$. Although the latter is technically correct, people generally assume the $L/(\gamma v)$ meaning. I will often use an " 's" to indicate the plural of one-letter items (like the m's of particles). I will unabashedly construct unrealistic scenarios where people run at speed $3c/5$, etc. And I will frequently be grammatically sloppy and use the phrase "clocks run slow" instead of "clocks run slowly."

I would like to thank Carey Witkov for meticulously reading through the entire book and offering many valuable suggestions, and Jacob Barandes for numerous illuminating discussions. Other friends and colleagues whose input I am grateful for are Andrzej Czarnecki, Carol Davis, Howard Georgi, Brian Hall, Theresa Morin Hall, Don Page, and Alexia Schulz. I would also like to thank the students in my Special Relativity course in Harvard University's Pre-College Program in the summer of 2016. The lively discussions were very helpful in fine-tuning the presentation of many topics throughout the book.

For instructors using this book as the assigned textbook for a course, a solutions manual for the exercises is available upon request. When sending a request, please point to a syllabus and/or webpage for your course.

Despite careful editing, there is zero probability that this book is error free. If anything looks amiss, please check *www.people.fas.harvard.edu/~djmorin/book.html* for a list of typos, updates, additional material, etc. And please let me know if you discover something that isn't already posted. Suggestions are always welcome.

David Morin
Cambridge, MA

Chapter 1

Kinematics, Part 1

TO THE READER: This book is available as both a paperback and an eBook. I have made the first chapter available on the web, but it is possible (based on past experience) that a pirated version of the complete book will eventually appear on file-sharing sites. In the event that you are reading such a version, I have a request:

If you don't find this book useful (in which case you probably would have returned it, if you had bought it), or if you do find it useful but aren't able to afford it, then no worries; carry on. However, if you do find it useful and are able to afford the Kindle eBook (priced below $10), then please consider purchasing it (available on Amazon). If you don't already have the Kindle reading app for your computer, you can download it free from Amazon. I chose to self-publish this book so that I could keep the cost low. The resulting eBook price of around $10, which is very inexpensive for a 250-page physics book, is less than a movie and a bag of popcorn, with the added bonus that the book lasts for more than two hours and has zero calories (if used properly!).

– David Morin

Special relativity is an extremely counterintuitive subject, and in this chapter we will see how its bizarre features come about. We'll build up the theory from scratch, starting with the postulates of relativity, of which there are only two. A surprisingly large number of strange effects can be derived from these two easily stated postulates.

The postulate that most people find highly counterintuitive is that the speed of light has the same value in any inertial (that is, non-accelerating) reference frame. This speed, which is about $3 \cdot 10^8$ m/s, is much greater than the speed of everyday objects, so most of the consequences of relativity aren't noticeable. If we instead lived in a world identical to ours except for the speed of light being only 50 mph, then the consequences of relativity would be ubiquitous. We wouldn't think twice about time dilation, length contraction, and so on.

As mentioned in the preface, this chapter is the first of two that cover kinematics. (*Kinematics* deals with lengths, times, speeds, etc., whereas *dynamics* deals with masses, forces, energy, momentum, etc.) The outline of this chapter is as follows. In Section 1.1 we discuss the historical motivations that led Einstein to his theory of special relativity. Section 1.2 covers the two postulates of relativity, from which everything in the theory can be obtained. Section 1.3 is the heart of the chapter, where we derive the three main consequences of the postulates (*loss of simultaneity*, *time dilation*, and *length contraction*). In Section 1.4 we present four instructive examples that utilize the three fundamental effects. Section 1.5 covers the *velocity-addition formula*, which gives the proper correction to the naive Newtonian result (simply adding the velocities). In Chapter 2 we will continue our discussion of kinematics, covering more advanced topics.

1.1 Motivation

Although it was certainly a stroke of genius that led Einstein to his theory of relativity, it didn't just come out of the blue. A number of conundrums in 19th-century physics suggested that something was amiss. Many people had made efforts to explain away these conundrums, and at least a few steps had been taken toward the correct theory. But Einstein was the one who finally put everything together, and he did so in a way that had consequences far beyond the realm of the specific issues that people were trying to understand. Indeed, his theory turned our idea of space and time on its head. But before we get to the heart of the theory, let's look at two of the major problems in late 19th-century physics. (A third issue, involving the addition of velocities, is presented in Problem 1.15.) If you can't wait to get to the postulates (and subsequently the results) of special relativity, you can go straight to Section 1.2. The present section can be skipped on a first reading.

1.1.1 Galilean transformations, Maxwell's equations

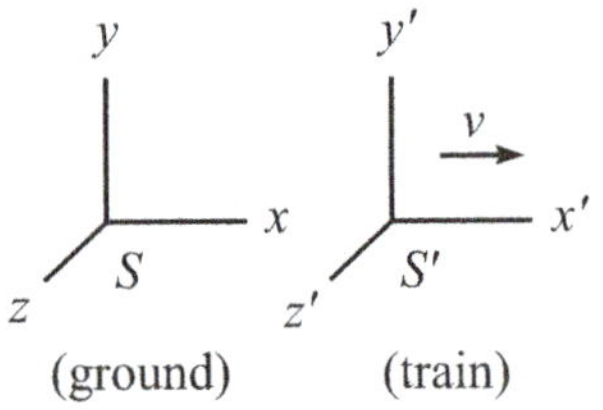

Figure 1.1

Imagine standing on the ground and watching a train travel by with constant speed v in the x direction. Let the reference frame of the ground be labeled S, and let the reference frame of the train be labeled S', as shown in Fig. 1.1. Consider two events that happen on the train. An *event* is defined as something that occurs with definite space and time coordinates (as measured in a given frame). For example, a person might clap her hands; this clap takes place at a definite time and a definite location. Technically, the clap lasts for a nonzero time (a few hundredths of a second), and the hands extend over a nonzero distance (a few inches). But we'll ignore these issues and assume that the clap can be described be unique x, y, z, and t values. Note that a given event isn't associated with one particular frame. The event simply happens, independent of a frame. For any arbitrary frame we then choose to consider, we can describe the event by specifying the coordinates as measured in that frame.

On our train, the two events might be one person clapping her hands and another person stomping his feet. If the space and time separations between these two events in the frame of the train are $\Delta x'$ and $\Delta t'$, what are the space and time separations, Δx and Δt, in the frame of the ground? Ignoring what we'll be learning about relativity in this chapter, the answers are "obvious" (although, as we'll see in Section 2.1 when we derive the Lorentz transformations, obvious things can apparently be incorrect!). The time separation Δt is "clearly" the same as on the train, so we have $\Delta t = \Delta t'$. We know from everyday experience that nothing strange happens with time. When you see people exiting a train station, they're not fiddling with their watches, trying to recalibrate them with a ground-based clock.

The spatial separation is a little more interesting, but still fairly simple. If the train weren't moving, then we would just have $\Delta x = \Delta x'$. This is true because if the train isn't moving, then the only possible difference between the frames is the location of the origin. But the only consequence of this difference is that every x' coordinate in the train is equal to a given fixed number plus the corresponding x coordinate on the ground. This fixed number then cancels when calculating the separation, $\Delta x' \equiv x'_2 - x'_1$.

However, in the general case where the train is moving, everything in the train gets carried along at speed v during the time Δt (which equals $\Delta t'$) between the two events.

So as seen in the ground frame, the person stomping his feet ends up $v\,\Delta t'$ to the right (or left, if v is negative) of where he would be if the train weren't moving. The total spatial separation Δx between the events in the ground frame is therefore the $\Delta x'$ separation that would arise if the train weren't moving, plus the $v\,\Delta t'$ separation due to the motion of the train. That is, $\Delta x = \Delta x' + v\,\Delta t'$, as shown in Fig. 1.2.

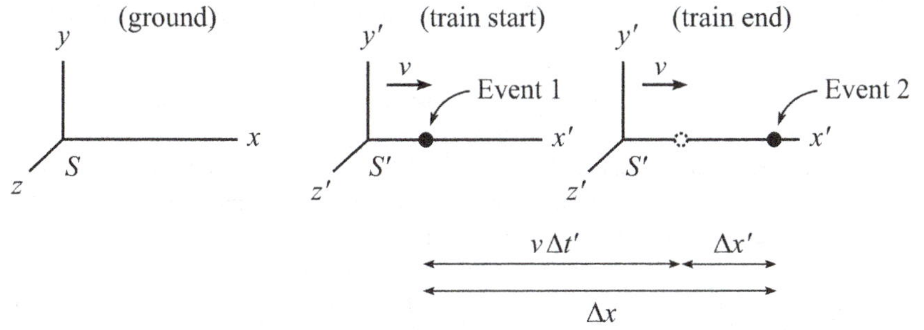

Figure 1.2

The *Galilean transformations* (first written down by Galileo Galilei in 1638) are therefore

$$\boxed{\begin{aligned} \Delta x &= \Delta x' + v\,\Delta t' \\ \Delta t &= \Delta t' \end{aligned}} \tag{1.1}$$

Nothing interesting happens in the y and z directions (assuming the train is traveling in the x direction), so we additionally have $\Delta y = \Delta y'$ and $\Delta z = \Delta z'$. Both of these common values are zero for the events in Fig. 1.2, because the events occur on the x axis. But for general locations of the events, the Δ's will be nonzero.

A special case of Eq. (1.1) arises when the two events occur at the same place on the train, so that $\Delta x' = 0$. In this case we have $\Delta x = v\,\Delta t'$. This makes sense, because the spot on the train where the events occur simply travels a distance $v\,\Delta t$ (which equals $v\,\Delta t'$) by the time the second event occurs.

The principle of *Galilean invariance* says that the laws of physics are invariant under the above Galilean transformations. Alternatively, it says that the laws of physics hold in all inertial (non-accelerating) frames. (It was assumed prior to Einstein that these two statements say the same thing, but we will soon see that they do not. The second statement is the one that remains valid in relativity.) This principle is quite believable. For example, in Galilean (nonrelativistic) physics, Newton's second law, $F = ma$ (or really $F = dp/dt$) holds in all inertial frames, because (1) the force F is the same in all inertial frames, and (2) the constant relative velocity v_{rel} between any two inertial frames implies that the acceleration of a given particle is the same in all inertial frames. Written out explicitly, the velocities v_1 and v_2 in the two frames are related by $v_1 = v_2 + v_{\text{rel}}$, so

$$a_1 \equiv \frac{dv_1}{dt_1} = \frac{d(v_2 + v_{\text{rel}})}{dt_1} = \frac{dv_2}{dt_2} + 0 \equiv a_2, \tag{1.2}$$

where we have used the facts that $t_1 = t_2$ (at least in a Galilean world) and that the derivative of a constant is zero.

Remarks:

1. Note that the Galilean transformations in Eq. (1.1) aren't symmetric in x and t. This isn't necessarily a bad thing, but it turns out that it will in fact be a problem in special relativity, where space and time are treated on a more equal footing. We'll find in Section 2.1 that the Galilean transformations are replaced by the *Lorentz transformations*, and the latter are in fact symmetric in x and t (up to factors of the speed of light, c).

2. Eq. (1.1) deals only with the *differences* in the x and t values between two events, and not with the values of the coordinates themselves of each event. The values of the coordinates of a single event depend on where you pick your origin, which is an arbitrary choice. The coordinate differences between two events, however, are independent of this choice, and this allows us to make the physically meaningful statements in Eq. (1.1). Since it makes no sense for a physical result to depend on your arbitrary choice of origin, the Lorentz transformations we derive in Section 2.1 will also need to involve only differences in coordinates.

3. We've been talking a lot about "events," so just to make sure we're on the same page with the definition of an event, we should give some examples of things that are *not* events. If a train is at rest on the ground (or even if it is moving), and if you look at it at a snapshot in time, then this doesn't describe an event, because the train has spatial extent. There isn't a unique spatial coordinate that describes the train. If you instead consider a specific point on the train at the given instant, then that does describe an event. As another example of a non event, if you look at a pebble on the ground for a minute, then this doesn't describe an event, because you haven't specified the time coordinate. If you instead consider the pebble at a particular instant in time, then that does describe an event. (We'll consider the pebble to be a point object, so that the spatial coordinate is unique.) ♣

We introduced the Galilean transformations above because of their relation (more precisely, their conflict) with *Maxwell's equations*. One of the great triumphs of 19th-century physics was the theory of electromagnetism. In 1864, James Clerk Maxwell wrote down a set of equations that collectively described everything that was known about the subject. These equations involve the electric and magnetic fields through their space and time derivatives. Maxwell's original formulation consisted of a large number of equations, but these were later written more compactly, using vector calculus, as four equations. We won't worry about their specific form here, but it turns out that if you transform the equations from one reference frame to another via the Galilean transformations, they end up taking a different form. That is, if you've written down Maxwell's equations in one frame (where they take their standard nice-looking form), and if you then replace the coordinates in this frame by those in another frame, using Eq. (1.1), then the equations look different (and not so nice).

This different appearance presents a major problem. If Maxwell's equations take a nice form in one frame and a not-so-nice form in every other frame, then why is one frame special? Said in another way, it can be shown that Maxwell's equations imply that light moves with a certain speed c. But which frame is this speed measured with respect to? The Galilean transformations imply that if the speed is c with respect to a given frame, then it is *not* c with respect to any other frame. (You need to add or subtract the relative speed v between the frames.) The proposed special frame where Maxwell's equations are nice and the speed of light is c was called the frame of the *ether*. We'll talk in detail about the ether in the next subsection, but experiments showed that light was surprisingly always measured to move with speed c in every frame, no matter which way the frame was moving through the supposed ether. We say "supposed" because the final conclusion was that the ether simply doesn't exist.

There were thus various possibilities. Something was wrong with either Maxwell's equations, the Galilean transformations, or the way in which measurements of speed were done (see Footnote 2 on page 8). Considering how "obvious" the Galilean transformations are, the natural assumption in the late 19th century was that the problem lay elsewhere. However, after a good deal of effort by many people to make everything else fit with the Galilean transformations, Einstein finally showed that these were in fact the culprit. It was well known that Maxwell's equations were invariant under the Lorentz transformations (in contrast with their non-invariance under the Galilean ones), but Einstein was the first to recognize the full meaning of these transformations. Instead of being relevant only to electromagnetism, the Lorentz transformations replaced the Galilean ones universally.

More precisely, in 1905 Einstein showed why the Galilean transformations are simply a special case of the Lorentz transformations, valid (to a high degree of accuracy) only when the speed involved is much less than the speed of light. As we'll see in Section 2.1, the coefficients in the Lorentz transformations depend on both the relative speed v of the frames and the speed of light c, where the c's appear in various denominators. Since c is quite large (about $3 \cdot 10^8$ m/s) compared with everyday speeds v, the parts of the Lorentz transformations involving c are negligible, for any typical v. The surviving terms are the ones in the Galilean transformations in Eq. (1.1). These are the only terms that are noticeable for everyday speeds. This is why no one prior to Einstein realized that the correct transformations between two frames had anything to do with the speed of light.

> As he pondered the long futile fight
> To make Galileo's world right,
> In a new variation
> Of the old transformation,
> It was Einstein who first saw the light.

In short, the reasons why Maxwell's equations are in conflict with the Galilean transformations are: (1) The speed of light is what determines the scale at which the Galilean transformations break down, (2) Maxwell's equations inherently involve the speed of light, because light is an electromagnetic wave.

1.1.2 Michelson–Morley experiment

As mentioned above, it was known in the late 19th century, after Maxwell wrote down his equations, that light is an electromagnetic wave and that it moves with a speed of about $3 \cdot 10^8$ m/s.[1] Now, every other wave that people knew about at the time needed a medium to propagate in. Sound waves need air, ocean waves of course need water, waves on a string of course need the string, and so on. It was therefore natural to assume that light also needed a medium to propagate in. This proposed medium was called the *ether*.

However, if light propagates in a given medium, and if the speed in this medium is c, then the speed in a reference frame moving relative to the medium should be different from c. Consider, for example, sound waves in air. If the speed of sound in air is v_{sound}, and if you run toward a sound source with speed v_{you}, then the speed of the sound waves with respect to you (assuming it's a windless day) is $v_{\text{sound}} + v_{\text{you}}$. Equivalently, if you are standing at rest downwind and the speed of the wind is v_{wind}, then the speed of the sound waves with respect to you is $v_{\text{sound}} + v_{\text{wind}}$.

Assuming that the ether really exists (although we'll soon see that it doesn't), a reasonable thing to do is to try to measure one's speed with respect to it. This can be done as follows. We'll frame this discussion in terms of sound waves in air. Let v_s be the speed of sound in air. Imagine two people standing on the ends of a long platform of length L that moves at speed v_p with respect to the reference frame in which the air is at rest. One person claps, the other person claps immediately when he hears the first clap (assume that the reaction time is negligible), and then the first person records the

[1]The exact value of the speed is 299,792,458 m/s. A meter is actually *defined* to be 1/299,792,458 of the distance that light travels in one second in vacuum. So this speed of light is *exact*. There is no need for an error bar because there is no measurement uncertainty.

total time elapsed when she hears the second clap. What is this total time? Well, we can't actually give an answer without knowing which direction the platform is moving. Is it moving parallel to its length, or perpendicular to it (or somewhere in between)? Let's look at these two basic cases. For both of these, we'll view the setup and do the calculation in the frame in which the air is at rest.

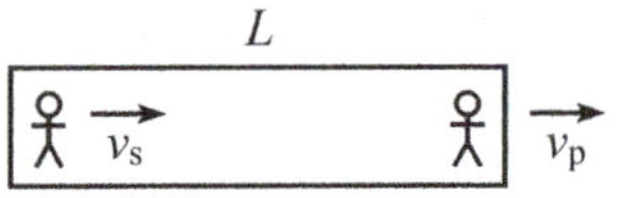

Figure 1.3

- Parallel motion: Consider first the case where the platform moves parallel to its length. In the reference frame of the air, assume that the person at the rear is the one who claps first; see Fig. 1.3. Then if v_s is the speed of sound and v_p is the speed of the platform, it takes a time of $L/(v_s - v_p)$ for the sound from the first clap to travel forward to the front person. This is true because the sound closes the initial gap of L at a relative speed of $v_s - v_p$, as viewed in the frame of the air. (Alternatively, relative to the initial position of the back of the platform, the position of the sound wave is $v_s t$, and the position of the front person is $L + v_p t$. Equating these gives $t = L/(v_s - v_p)$.) This time is longer than the naive answer of L/v_s because the front person is moving away from the rear person, which means that the sound has to travel farther than L.

  By similar reasoning, the time for the sound from the second clap to travel backward to the rear person is $L/(v_s + v_p)$. This time is shorter than the naive answer of L/v_s because the rear person is moving toward the front person, which means that the sound travels less than L.

  Adding the forward and backward times gives a total time of

$$t_1 = \frac{L}{v_s - v_p} + \frac{L}{v_s + v_p} = \frac{2Lv_s}{v_s^2 - v_p^2}. \tag{1.3}$$

  This correctly equals $2L/v_s$ when $v_p = 0$. In this case the platform is at rest, so the sound simply needs to travel forward and backward a total distance of $2L$ at speed v_p. And the result correctly equals infinity when $v_p \to v_s$. In this case the front person is receding as fast as the sound is traveling, so the sound from the first clap can never catch up.

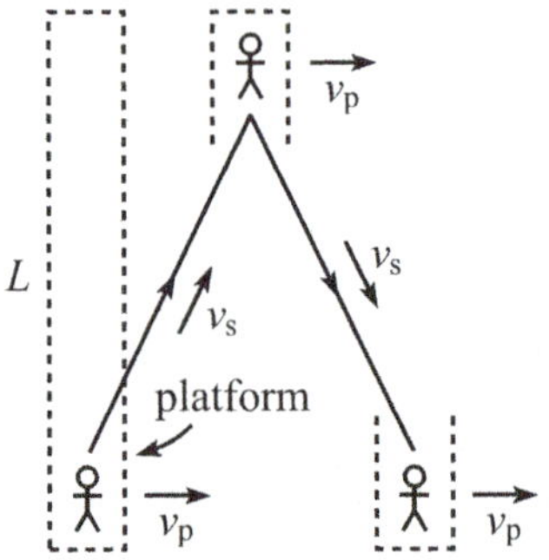

Figure 1.4

- Perpendicular motion: Now consider the case where the platform moves perpendicular to its length. In the reference frame of the air, we have the situation shown in Fig. 1.4. The sound moves diagonally with speed v_s. (The sound actually moves in all directions, of course, but it's only the part of the sound wave that moves in a particular diagonal direction that ends up hitting the other person.) Since the "horizontal" component of the diagonal velocity is the platform's speed v_p, the Pythagorean theorem gives the "vertical" component as $\sqrt{v_s^2 - v_p^2}$, as shown in Fig. 1.5. This is the speed at which the length L of the platform is traversed during both the out and back parts of the trip. So the total time is

$$t_2 = \frac{2L}{\sqrt{v_s^2 - v_p^2}}. \tag{1.4}$$

  Again, this correctly equals $2L/v_s$ when $v_p = 0$, and infinity when $v_p \to v_s$. The vertical component of the velocity is zero in the latter case, because the diagonal path is essentially horizontal.

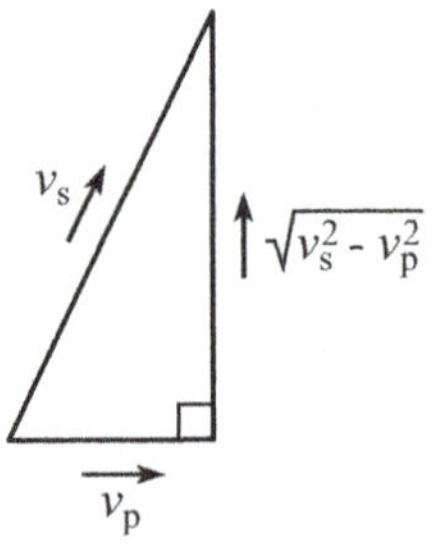

Figure 1.5

The times in Eqs. (1.3) and (1.4) are not equal; you can quickly show that $t_1 \geq t_2$. It turns out that (for given values of v_s and v_p) of all the possible orientations of the platform relative to the direction of motion (which we have been taking to be rightward), the t_1 in Eq. (1.3) is the largest possible time, and the t_2 in Eq. (1.4) is the smallest.

(The proof of this is somewhat tedious, but at least it is believable that if the platform is oriented between the above two special cases, the time lies between the associated times t_1 and t_2.) Therefore, if you are on a large surface that is moving with respect to the air, and if you know the value of v_s, then if you want to figure out what v_p is, all you have to do is repeat the above experiment with someone standing at various points along the circumference of a given circle of radius L around you. (Assume that it doesn't occur to you to toss a little piece of paper in the air, in order to at least determine the direction of the wind with respect to you.) If you take the largest total time observed and equate it with t_1, then Eq. (1.3) will give you v_p. Alternatively, you can equate the smallest total time with t_2, and Eq. (1.4) will yield the same v_p.

In the limiting case where $v_p \ll v_s$, we can make some approximations to the above expressions for t_1 and t_2. These approximations involve the Taylor-series expressions $1/(1-\epsilon) \approx 1+\epsilon$ and $1/\sqrt{1-\epsilon} \approx 1+\epsilon/2$. (See Appendix G for a discussion of Taylor series.) These expressions yield the following approximate result for the difference between t_1 and t_2 (after first rewriting t_1 and t_2 so that a "1" appears in the denominator):

$$\begin{aligned}
\Delta t = t_1 - t_2 &= \frac{2L}{v_s}\left(\frac{1}{1-v_p^2/v_s^2} - \frac{1}{\sqrt{1-v_p^2/v_s^2}}\right) \\
&\approx \frac{2L}{v_s}\left(\left(1+\frac{v_p^2}{v_s^2}\right) - \left(1+\frac{v_p^2}{2v_s^2}\right)\right) \\
&= \frac{Lv_p^2}{v_s^3}\,. \qquad (1.5)
\end{aligned}$$

The difference $t_1 - t_2$ is what we'll be concerned with in the Michelson–Morley experiment, which we will now discuss.

The strategy in the above sound-in-air setup is the basic idea behind Michelson's and Morley's attempt in 1887 to measure the speed of the earth through the supposed ether. (See Handschy (1982) for the data and analysis of the experiment.) There is, however, a major complication with light that doesn't arise with sound. The speed of light is so large that any time intervals that are individually measured will inevitably have measurement errors that are far larger than the difference between t_1 and t_2. Therefore, individual time measurements give essentially no information. Fortunately, there is a way out of this impasse. The trick is to measure t_1 and t_2 concurrently, as opposed to separately. More precisely, the trick is to measure only the difference $t_1 - t_2$, and not the individual values t_1 and t_2. This can be done as follows.

Consider two of the above "platform" scenarios arranged at right angles with respect to each other, with the same starting point. This can be arranged by having a (monochromatic) light beam encounter a beam splitter that sends two beams off at 90° angles. The beams then hit mirrors and bounce back to the beam splitter where they (partially) recombine before hitting a screen; see Fig. 1.6. The fact that light is a wave, which is what got us into this ether mess in the first place, is now what saves the day. The wave nature of light implies that the recombined light beam produces an interference pattern on the screen. At the center of the pattern, the beams will constructively or destructively interfere (or something in between), depending on whether the two light beams are in phase or out of phase when they recombine. This interference pattern is extremely delicate. The slightest change in travel times of the beams will cause the pattern to noticeably shift. This type of device, which measures the interference between two light beams, is known as an *interferometer*.

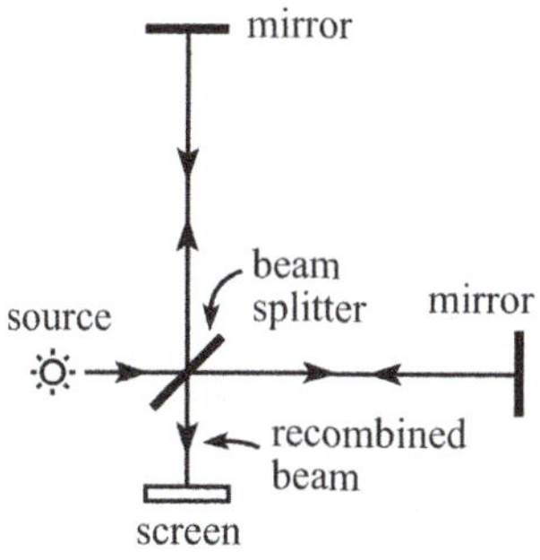

Figure 1.6

If the whole apparatus is rotated around, so that the experiment is performed at various angles, then the maximum amount that the interference pattern changes can be used to determine the speed of the earth through the ether (v_p in the platform setup above). In one extreme case, the time in a given arm is longer than the time in the other arm by Lv^2/c^3. (We have changed notation in Eq. (1.5) so that $v_p \to v$ is the speed of the earth through the supposed ether, and $v_s \to c$ is the speed of light.) But in the other extreme case, the time in the given arm is shorter by Lv^2/c^3. So the maximum shift in the interference pattern corresponds to a time difference of $2Lv^2/c^3$.

However, when Michelson and Morley performed their experiment, they observed no interference shift as the apparatus was rotated around. Their setup did in fact allow enough precision to measure a nontrivial earth speed through the ether, if such a speed existed. So if the ether did exist, their results implied that the speed of the earth through it was zero. This result, although improbable, was technically fine. It might simply have been the case that they happened to do their experiment when the relative speed was zero. However, when they performed their experiment a few months later, when the earth's motion around the sun caused it to be moving in a different direction, they still measured zero speed. It wasn't possible for both of these results to be zero (assuming that the ether exists), without some kind of modification to the physics known at the time.

Many people over the years tried to explain this null result, but none of the explanations were satisfactory. Some led to incorrect predictions in other setups, and some seemed to work fine but were a bit ad hoc.[2] The correct explanation, which followed from Einstein's 1905 theory of relativity, was that the ether simply doesn't exist.[3] In other words, light doesn't need a medium to propagate in. It doesn't move with respect to a certain special reference frame, but rather it moves with respect to whoever is looking at it.

> The findings of Michelson–Morley
> Allow us to say very surely,
> "If this ether is real,
> Then it has no appeal,
> And shows itself off rather poorly."

REMARKS:

1. We assumed above that the lengths of the two arms in the apparatus were equal. However, in practice there is no hope of constructing lengths that are equal, up to errors that are small compared with the wavelength of the light. But fortunately this doesn't matter. We're concerned not with the difference in the travel times associated with the two arms, but rather with the *difference in these differences* as the apparatus is rotated around. Using Eqs. (1.3) and (1.4) with different lengths L_1 and L_2, you can show (assuming $v \ll c$) that the maximum interference shift corresponds to a time of $(L_1 + L_2)v^2/c^3$. This is the generalization of the $2Lv^2/c^3$ result we derived in Eq. (1.5) (in different notation) when the lengths were equal. The measurement errors in L_1 and L_2 therefore need only be small compared with the (macroscopic) lengths L_1 and L_2, as opposed to small compared with the (microscopic) wavelength of light.

2. Assuming that the lengths of the arms are approximately equal, let's plug in some rough numbers to see how much the interference pattern shifts. The Michelson–Morley setup

[2] The most successful explanation was the Lorentz–FitzGerald contraction. These two physicists independently proposed that lengths are contracted in the direction of the motion by precisely the right factor, namely $\sqrt{1 - v^2/c^2}$, to make the travel times in the two arms of the Michelson–Morley setup equal, thus yielding the null result. This explanation was essentially correct, although the reason why it was correct wasn't known until Einstein came along.

[3] Although we've presented the Michelson–Morley experiment here for pedagogical purposes, the consensus among historians is that Einstein actually wasn't influenced much by the experiment, except indirectly through Lorentz's work on electrodynamics. See Holton (1988).

had arms with effective lengths of about 10 m. We'll take v to be on the order of the speed of the earth around the sun, which is about $3 \cdot 10^4$ m/s. We then obtain a maximal time difference of $t = 2Lv^2/c^3 \approx 7 \cdot 10^{-16}$ s. The large negative exponent here might make us want to throw in the towel, thinking that the effect is hopelessly small. However, the distance that light travels in the time t is $ct = (3 \cdot 10^8\ \text{m/s})(7 \cdot 10^{-16}\ \text{s}) \approx 2 \cdot 10^{-7}$ m, and this happens to be a perfectly reasonable fraction of the wavelength of visible light, which is around $\lambda = 6 \cdot 10^{-7}$ m, give or take. So we have $ct/\lambda \approx 1/3$. This maximal interference shift of about a third of a cycle was well within the precision of the Michelson–Morley setup. So if the ether had really existed, Michelson and Morley definitely would have been able to measure the speed of the earth through it.

3. One proposed explanation of the observed null effect was "frame dragging." What if the earth drags the ether along with it, thereby always yielding the observed zero relative speed? This frame dragging is quite plausible, because in the platform example above, the platform drags a thin layer of air along with it. And more mundanely, a car completely drags the air in its interior along with it. But it turns out that frame dragging is inconsistent with *stellar aberration*, which is the following effect.

   Depending on the direction of the earth's instantaneous velocity as it obits around the sun, it is an experimental fact that a given star might (depending on its location) appear at slightly different places in the sky when viewed at two times, say, six months apart. This is due to the fact that a telescope must be aimed at a slight angle relative to the actual direction to the star, because as the star's light travels down the telescope, the telescope moves slightly in the direction of the earth's motion. We're assuming (correctly) here that frame dragging does *not* exist.

   As a concrete analogy, imagine holding a tube while running through vertically falling rain. If you hold the tube vertically, then the raindrops that enter the tube won't fall cleanly through. Instead, they will hit the side of the tube, because the tube is moving sideways while the raindrops are falling vertically. However, if you tilt the tube at just the right angle, the raindrops will fall (vertically) cleanly through without hitting the side. At what angle θ should the tube be tilted? If the tube travels horizontally a distance d during the time it takes a raindrop to fall vertically a distance h, then the ratio of these distances must equal the ratio of the speeds: $d/h = v_{\text{tube}}/v_{\text{rain}}$ (see Fig. 1.7). The angle θ is then given by $\tan\theta = d/h \implies \tan\theta = v_{\text{tube}}/v_{\text{rain}}$. With respect to your frame as you run along, the raindrops come down at an angle θ; they don't come down vertically.

   Returning to the case of light, v_{tube} gets replaced with (roughly) the speed v of the earth around the sun, and v_{rain} gets replaced with the speed c of light. The ratio of these two speeds is about $v/c = 10^{-4}$, so the effect is small. But it is large enough to be noticeable, and it has indeed been measured; stellar aberration *does* exist. At two different times of the year, a telescope must be pointed at slightly different angles when viewing a given star. Now, if frame dragging *did* exist, then the light from the star would get dragged along with the earth and would therefore travel down a telescope that was pointed directly at the star, in disagreement with the observed fact that the telescope must point at the slight angle mentioned above. (Or even worse, the dragging might produce a boundary layer of turbulence that would blur the stars.) The existence of stellar aberration therefore implies that frame dragging doesn't occur.

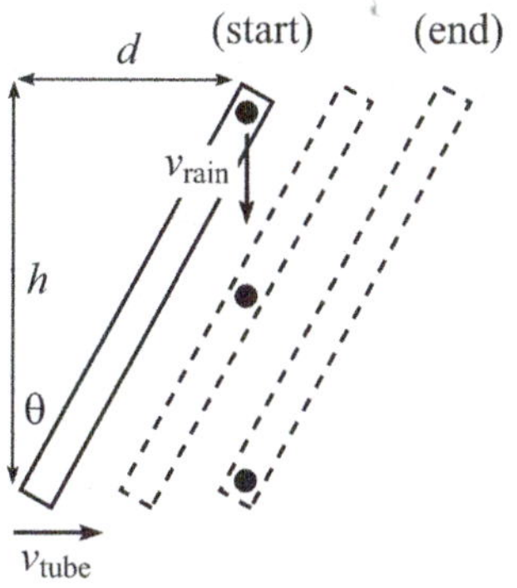

Figure 1.7

4. Note that it is the velocity of the telescope that matters in stellar aberration, and not its position. This aberration effect should not be confused with the *parallax* effect, where the direction of the actual position of an object changes, depending on the *position* of the observer. For example, people at different locations on the earth see the moon at slightly different angles (that is, they see the moon in line with different distant stars). As a more down-to-earth example, two students sitting at different locations in a classroom see the teacher at different angles. Although stellar parallax has been measured for nearby stars (as the earth goes around the sun), its angular effect is much smaller than the angular effect from stellar aberration. The former decreases with the distance to the star, whereas the latter doesn't. For further discussion of aberration, and of why it is only the earth's velocity (or rather, the change in its velocity) that matters, and not also the star's velocity (since

you might think, based on the fact that we're studying relativity here, that it is the relative velocity that matters), see Eisner (1966). ♣

1.2 The postulates

Let's now start from scratch and see what the theory of special relativity is all about. We'll take the route that Einstein took and use two postulates as the foundation of the theory. We'll start with the "relativity postulate" (also called the Principle of Relativity). This postulate is quite believable, so you might just take it for granted and forget to consider it. But like any other postulate, it is crucial. It can be stated in various ways, but we'll write it simply as:

- POSTULATE 1: *All inertial (non-accelerating) frames are "equivalent."*

This postulate says that a given inertial frame is no better than any other; there is no preferred reference frame. That is, it makes no sense to say that something is moving. It makes sense only to say that one thing is moving with respect to another. This is where the "relativity" in "special relativity" comes from. There is no absolute inertial frame; the motion of any frame is defined only relative to other frames.

This postulate also says that if the laws of physics hold in one inertial frame (and presumably they do hold in the frame in which I now sit), then they hold in all others. (Technically, the earth is spinning while revolving around the sun, and there are also little vibrations in the floor beneath my chair, etc., so I'm not *really* in an inertial frame. But it's close enough for me.) The postulate also says that if we have two frames S and S', then S should see things in S' in exactly the same way as S' sees things in S, because we can just switch the labels of S and S'. (We'll get our money's worth out of this statement in the next few sections.) It also says that empty space is homogeneous (that is, all points look the same), because we can pick any point to be, say, the origin of a coordinate system. It also says that empty space is isotropic (that is, all directions look the same), because we can pick any axis to be, say, the x axis of a coordinate system.

Unlike the second postulate below (the speed-of-light postulate), this first one is entirely reasonable. We've gotten used to having no special places in the universe. We gave up having the earth as the center, so let's not give any other point a chance, either.

Copernicus gave his reply
To those who had pledged to deny.
"All your addictions
To ancient convictions
Won't bring back your place in the sky."

The first postulate is nothing more than the familiar principle of Galilean invariance, assuming that this principle is written in the "The laws of physics hold in all inertial frames" form, and not in the form that explicitly mentions the Galilean transformations in Eq. (1.1), which are inconsistent with the second postulate below.

Everything we've said here about the first postulate refers to empty space. If we have a chunk of mass, then there is certainly a difference between the position of the mass and a point a meter away. To incorporate mass into the theory, we would have to delve into the subject of general relativity. But we won't have anything to say about that in this chapter. We will deal only with empty space, containing perhaps a few observant souls sailing along in rockets or floating aimlessly on little spheres. Although that might sound boring at first, it will turn out to be anything but.

The second postulate of special relativity is the "speed-of-light" postulate. This one is much less intuitive than the relativity postulate.

- Postulate 2: *The speed of light in vacuum has the same value c (approximately* $3 \cdot 10^8$ m/s*) in any inertial frame.*

This statement certainly isn't obvious, or even believable. But on the bright side, at least it's easy to understand what the postulate says, even if you think it's too silly to be true. It says the following. Consider a train moving along the ground with constant velocity. Someone on the train shines a light from one point on the train to another. The speed of the light with respect to the train is c. Then the above postulate says that a person on the ground also sees the light move at speed c.

This is a rather bizarre statement. It doesn't hold for everyday objects. If a baseball is thrown forward with a given speed on a train, then the speed of the baseball is different in the ground frame. An observer on the ground must add the velocity of the ball (with respect to the train) to the velocity of the train (with respect to the ground) in order to obtain the velocity of the ball with respect to the ground. Strictly speaking, this isn't quite true, as the velocity-addition formula in Section 1.5 shows. But it's true enough for the point we're making here.

The truth of the speed-of-light postulate cannot be demonstrated from first principles. No statement with any physical content in physics (that is, one that isn't purely mathematical, such as, "two apples plus two apples gives four apples") can be proven. In the end, we must rely on experiment. And indeed, all of the consequences of the speed-of-light postulate have been verified countless times during the past century. As discussed in the previous section, the most well-known of the early experiments on the speed of light was the one performed by Michelson and Morley. The zero interference shift they always observed implied that the v_p speed in Eq. (1.5) was always zero. This in turn implies that no matter what (inertial) frame you are in, you are always at rest with respect to a frame in which the speed of light is c. In other words, the speed of light is the same in any inertial frame.

In more recent years, the consequences of the second postulate have been verified continually in high-energy particle accelerators, where elementary particles reach speeds very close to c. The collection of all the data from numerous experiments over the years allows us to conclude with near certainty that our starting assumption of an invariant speed of light is correct (or is at least the limiting case of a more correct theory).

Remark: Given the first postulate, you might wonder if we even need the second. If all inertial frames are equivalent, shouldn't the speed of light be the same in any frame? Well, no. For all we know, light might behave like a baseball. A baseball certainly doesn't have the same speed in all inertial frames, and this doesn't ruin the equivalence of the frames.

It turns out (see Section 2.7) that nearly all of special relativity can be derived by invoking *only* the first postulate. The second postulate simply fills in the last bit of necessary information by stating that *something* has the same finite speed in every frame. It's actually not important that this thing is light. It could be mashed potatoes or something else, and the theory would still come out the same. (Well, the thing has to be massless, as we'll see in Chapter 3, so we'd need to have massless potatoes, but whatever.) The second postulate can therefore be stated more minimalistically as, "There is something that has the same speed in any inertial frame." It just so happens that in our universe this thing is what allows us to see.

To go a step further, it's not even necessary for there to exist something that has the same speed in any frame. The theory will still come out the same if we write the second postulate as, "There is a limiting speed of an object in any frame." (See Section 2.7 for a discussion of this.) There is no need to have something that actually travels at this speed. It's conceivable to have a theory that contains no massless objects, in which case everything travels slower than the limiting speed. ♣

Let's now see what we can deduce from the above two postulates. There are many different ways to arrive at the various kinematical consequences. Our road map for the initial part of the journey (through Section 2.2) is shown in Fig. 1.8. Additional kinematics topics are covered in Sections 2.3 through 2.7.

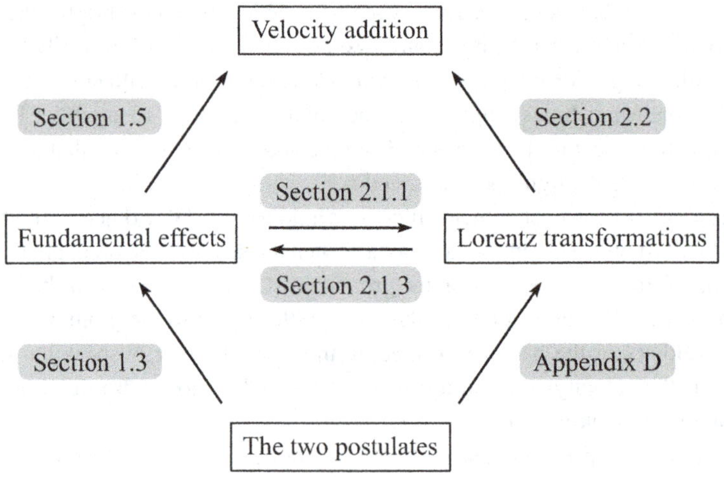

Figure 1.8

1.3 The fundamental effects

The most striking effects of the two postulates of relativity are (1) the *loss of simultaneity* (equivalently, the *rear-clock-ahead* effect), (2) *time dilation*, and (3) *length contraction*. In this section, we'll discuss these three effects by using some time-honored concrete setups. In Chapter 2, we'll use these three effects to derive the Lorentz transformations.

1.3.1 Loss of simultaneity

The basic effect

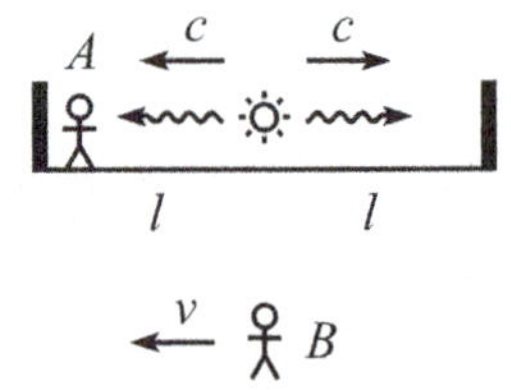

Figure 1.9

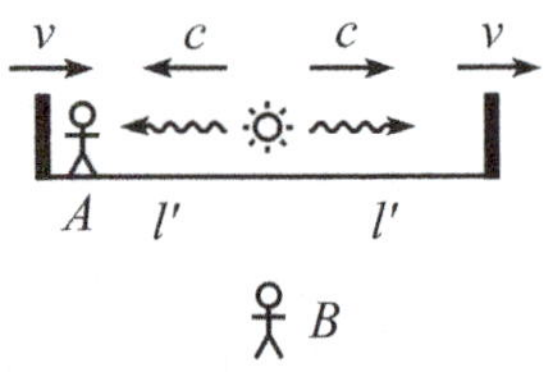

Figure 1.10

Consider the following setup. In person A's reference frame, a light source is placed midway between two receivers, a distance ℓ from each (see Fig. 1.9). The light source emits a flash. In A's reference frame, the light hits the two receivers at the same time, ℓ/c seconds after the flash. So if Event 1 is "light hitting the left receiver" and Event 2 is "light hitting the right receiver," then the two events are simultaneous in A's frame.

Now consider another observer, B, who travels to the left at speed v. In B's reference frame, does the light hit the receivers at the same time? That is, are Events 1 and 2 simultaneous in B's frame? We will show that surprisingly they are not.

In B's reference frame, the situation looks like that in Fig. 1.10. If you want, you can think of A as being on a train, and B as standing on the ground. With respect to B, the receivers (along with everything else in A's frame) move to the right with speed v. Additionally, with respect to B (and this is where the strangeness of relativity comes into play), the light travels in both directions at speed c, as indicated in the figure. Why is this the case? Because the speed-of-light postulate says so!

Note that everyday objects do *not* behave this way. Consider, for example, a train (A's frame) moving at 30 mph with respect to the ground (B's frame). If A stands in the middle of the train and throws two balls forward and backward, each with speed 50 mph *with respect to the train*, then the speeds of the two balls *with respect to the ground* are $50 - 30 = 20$ mph (backward) and $50 + 30 = 80$ mph (forward). (We're ignoring the minuscule corrections from the velocity-addition formula discussed in Section 1.5.)

These two speeds are *different*. In contrast with everyday objects like these balls, light has the bizarre property that its speed is always c (when viewed in an arbitrary inertial frame), independent of the speed of the source. Strange, but true.

Returning to our setup with the light beams and receivers, we can say that because B sees both light beams move with speed c, the *relative* speed (as viewed by B) of the light and the left receiver is $c + v$, and the *relative* speed (as viewed by B) of the light and the right receiver is $c - v$.

REMARK: Yes, it is legal to simply add or subtract these speeds to obtain the relative speeds *as viewed by B*. The reasoning here is the same as in the discussion of Fig. 1.3 in Section 1.1.2, where we obtained relative speeds of $v_s \pm v_p$. As a concrete example, if the v here equals $2 \cdot 10^8$ m/s, then in one second the left receiver moves $2 \cdot 10^8$ m to the right, while the left ray of light moves $3 \cdot 10^8$ m to the left. This means that they are now $5 \cdot 10^8$ m closer than they were a second ago. In other words, the relative speed (as measured by B) is $5 \cdot 10^8$ m/s, which is $c + v$ here. This is the rate at which the gap between the light and the left receiver closes. So in addition to calling it the "relative" speed, you can also call it the "gap-closing" speed. Note that the above reasoning implies that it is perfectly legal for the relative speed of two things, as measured by a third, to take any value up to $2c$.

Likewise, the relative speed between the light and the right receiver is $c - v$. The v and c in these results are measured with respect to the *same* person, namely B, so our intuition involving simple addition and subtraction works fine. Even though we're dealing with a relativistic speed v here, we don't need to use the velocity-addition formula from Section 1.5, which is relevant in a different setting. This remark is included just in case you've seen the velocity-addition formula and think it's relevant in this setup. But if it didn't occur to you, then never mind.

Note that the speed of the right photon[4] is *not* $c - v$. (And likewise the speed of the left photon is not $c + v$.) The photon moves at speed c, as always. It is the *relative* speed (as measured by B) of the photon and the front of the train that is $c - v$. No *thing* is actually moving with this speed in our setup. This speed is just the rate at which the gap closes. And a gap isn't an actual moving thing. ♣

Let ℓ' be the distance from the light source to each of the receivers, as measured by B.[5] Then in B's frame, the gap between the light beam and the left receiver starts with length ℓ' and subsequently decreases at a rate $c + v$. The time for the light to hit the left receiver is therefore $\ell'/(c + v)$. Similar reasoning holds for the right receiver along with the relative speed of $c - v$. The times t_L and t_R at which the light hits the left and right receivers are therefore given by

$$t_L = \frac{\ell'}{c + v} \quad \text{and} \quad t_R = \frac{\ell'}{c - v}. \tag{1.6}$$

These two times are not equal if $v \neq 0$. (The one exception is when $\ell' = 0$, in which case the two events happen at the same place and same time in all frames.) Since $t_L < t_R$, we see that in B's frame, the light hits the left receiver before it hits the right receiver. We have therefore arrived at the desired conclusion that the two events (light hitting back, and light hitting front) are *not* simultaneous in B's frame.

The moral of this is that it makes no sense to say that one event happens at the same time as another, unless you also state which frame you're dealing with. Simultaneity depends on the frame in which the observations are made.

---

[4]Photons are what light is made of. So "speed of the photon" means the same thing as "speed of the light beam." Sometimes it's easier to talk in terms of photons.

[5]We'll see in Section 1.3.3 that ℓ' is not equal to the ℓ in A's frame, due to length contraction. But this won't be important for what we're doing here. The only fact we need for now is that the light source is equidistant from the receivers, as measured by B. This is true because space is homogeneous, which implies that any length-contraction factor we eventually arrive at must be the same everywhere. More on this in Section 1.3.3.

Of the many effects, miscellaneous,
The loss of events, simultaneous,
Allows B to claim
There's no pause in A's frame,

REMARKS:

1. The strangeness of the $t_L < t_R$ loss-of-simultaneity result can be traced to the strangeness of the speed-of-light postulate. We entered the bizarre world of relativity when we wrote the c's above the photons in Fig. 1.10. The $t_L < t_R$ result is a direct consequence of the nonintuitive fact that light moves with the same speed c in *every* inertial frame.

2. The invariance of the speed of light led us to the fact that the relative speeds between the photons and the left and right receivers are $c + v$ and $c - v$. If we were talking about baseballs instead of light beams, then the relative speeds wouldn't take these general forms. If v_b is the speed at which the baseballs are thrown in A's frame, then as we noted above in the case where $v = 30$ mph and $v_b = 50$ mph, B sees the balls move with speeds of (essentially, ignoring the tiny corrections due to the velocity-addition formula) $v_b - v$ to the left (assuming $v_b > v$) and $v_b + v$ to the right. These are not equal (in contrast with what happens with light). By the same "gap-closing" reasoning we used above, the relative speeds (as viewed by B) between the balls and the left and right receivers are then $(v_b - v) + v = v_b$ and $(v_b + v) - v = v_b$. These are equal, so B sees the balls hit the receivers at the same time, as we know very well from everyday experience.

3. As explained in the remark prior to Eq. (1.6), it is indeed legal to obtain the times in Eq. (1.6) by simply dividing ℓ' by the relative speeds, $c + v$ and $c - v$. The gaps start with length ℓ' and then decrease at these rates. But if you don't trust this, you can use the following reasoning. In B's frame, the position of the right photon (relative to the initial position of the light source) simply equals ct, and the position of the right receiver (which has a head start of ℓ') equals $\ell' + vt$. The photon hits the receiver when these two positions are equal. Equating them gives

$$ct = \ell' + vt \quad \Longrightarrow \quad t_R = \frac{\ell'}{c - v}. \tag{1.7}$$

   Similar reasoning with the left photon gives $t_L = \ell'/(c + v)$.

4. There is always a difference between the time that an event *happens* and the time that someone *sees* the event happen, because light takes time to travel from the event to the observer. What we calculated above were the times t_L and t_R at which the events *actually happen* in B's frame. (These times are independent of *where* B is standing at rest in the frame.) If we wanted to, we could calculate the times at which B *sees* the events occur. (These times *do* depend on where B is standing at rest in the frame.) But such times are rarely important, so in general we won't be concerned with them. They can easily be calculated by adding on a (distance)$/c$ time for the photons to travel to B's eye. Of course, if B actually did the above experiment to find t_L and t_R, she would do it by writing down the times at which she sees the events occur, and then subtracting off the relevant (distance)$/c$ times, to find when the events actually happened.

   To sum up, the $t_L \neq t_R$ result in Eq. (1.6) is due to the fact that the events truly occur at different times in B's frame. The $t_L \neq t_R$ result *has nothing to do with the time it takes light to travel to B's eye.* ♣

Where this last line is not so extraneous.

The "rear clock ahead" effect

We showed in Eq. (1.6) that t_L is not equal to t_R, that is, the light hits the receivers at different times in B's frame. Let's now be quantitative and determine the degree to which two events that are simultaneous in one frame are not simultaneous in another frame.

Given the times t_L and t_R that we found in Eq. (1.6), the simplest quantitative number that we can produce, as a measure of the non-simultaneity, is the difference $t_R - t_L$. This tells us how *un*simultaneous the events are in B's frame (the ground frame), given that they are simultaneous in A's frame (the train frame). The interpretation of the resulting expression for $t_R - t_L$ is the task of Problem 1.1 (which relies on time dilation and length contraction, discussed below). But let's take a slightly different route here, which will end up being a little more useful. This route will lead us to the *rear-clock-ahead* effect, which is the standard quantitative statement of the loss of simultaneity.

Consider a setup where two clocks are positioned at the ends of a train of length L (as measured in its own frame). The clocks are synchronized in the train frame. That is, they have the same reading at any given instant, as observed in the train frame, as you would naturally expect. (Throughout this book, we will always assume that clocks are synchronized in the frame in which they are at rest.) The train travels past you at speed v. It turns out that if you observe the clocks at simultaneous times in *your* frame, the readings will *not* be the same. You will observe the rear clock showing a higher reading than the front clock, as indicated in Fig. 1.11.

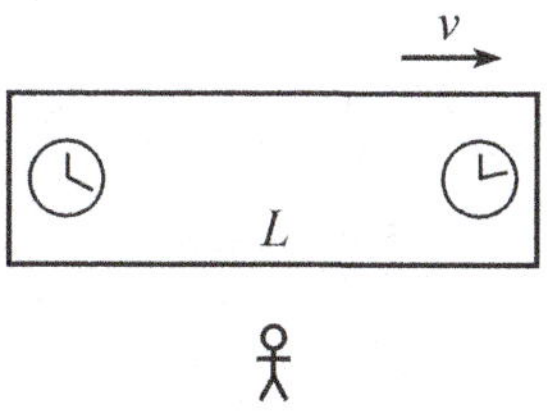

Figure 1.11

We'll explain why this is true momentarily, but first let us note that a nonzero difference in the readings is certainly a manifestation of the loss of simultaneity. To see why, consider a given instant in your (ground) frame when the rear and front clocks read, say, 12:01 and 12:00. (We'll find that the actual difference depends on L and v, but let's just assume it's one minute here, for concreteness.) Assume that you hit both clocks simultaneously (in your ground frame) with paintballs when they show these readings. Then in the train frame, the front clock gets hit when it reads 12:00, and then a minute later the rear clock gets hit when it reads 12:01. The simultaneous hits in your frame are therefore *not* simultaneous in the train frame. We have used the fact that the reading on a clock when a paintball hits it is frame independent. This is true because you can imagine that a clock breaks when a ball hits it, so that it remains stuck at a certain value. Everyone has to agree on what this value is.

Let's now find the exact difference in the readings on the two train clocks in Fig. 1.11. To do this, we will (as we did above in Fig. 1.9) put a light source on the train. But we'll now position it so that the light hits the clocks at the ends of the train at the same time in *your (ground) frame*. As in the discussion of Fig. 1.10, the relative speeds of the photons and the clocks are $c + v$ and $c - v$ (as viewed in your frame). We therefore need to divide the train into lengths in this ratio, in your frame, if we want the light to hit the ends at the same time. Now, because length contraction (discussed below in Section 1.3.3) is independent of position, the ratio in the train frame must also be $c + v$ to $c - v$. You can then quickly show that two numbers that are in this ratio, and that add up to L, are $L(c + v)/2c$ and $L(c - v)/2c$. (Mathematically, you're solving the system of equations, $x/y = (c + v)/(c - v)$ and $x + y = L$.) Dividing the train into these two lengths (in the train frame, as shown in Fig. 1.12) causes the light to hit the ends of the train simultaneously in the *ground* frame.

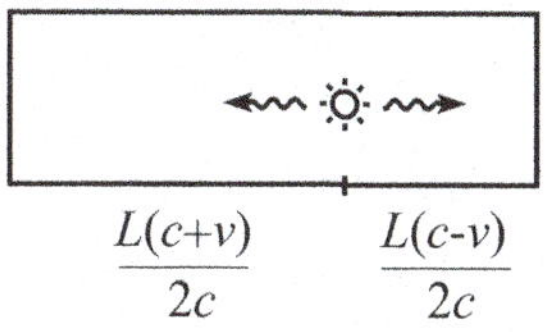

Figure 1.12

Let's now examine what happens during the process in the *train* frame. Compared with the forward-moving light, Fig. 1.12 tells us that the backward-moving light must travel an extra distance of $L(c+v)/2c - L(c-v)/2c = Lv/c$. The light travels at speed c (as always), so the extra time is Lv/c^2. The rear clock therefore reads Lv/c^2 more when it is hit by the backward photon, compared with what the front clock reads when it is hit by the forward photon. (Remember that the clocks are synchronized in the train frame.) This difference in readings has a frame-independent value, because the readings on the clocks when the photons hit them are frame independent, by the same reasoning as with the paintballs above.

Finally, let's switch back to your (ground) frame. Let the instant you look at the clocks be the instant the photons hit them. (That's why we constructed the setup with the hittings being simultaneous in *your* frame.) Then from the previous paragraph, we conclude that you observe the rear clock reading more than the front clock by an amount Lv/c^2:

$$\text{The rear clock is ahead by } Lv/c^2. \tag{1.8}$$

This result is important enough to spell out in full and put in a box:

> REAR-CLOCK-AHEAD: If a train with length L moves with speed v relative to you, then you observe the rear clock reading Lv/c^2 more than the front clock, at any given instant.

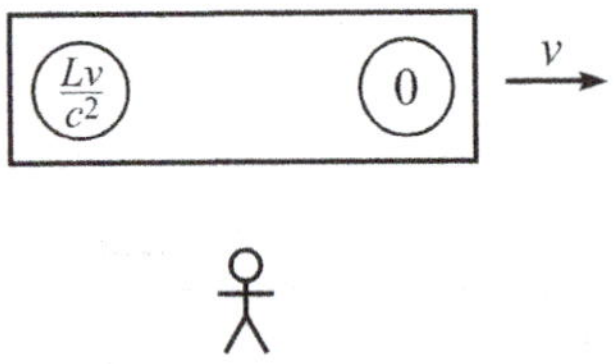

Figure 1.13

This statement corresponds to Fig. 1.13. For concreteness, we have chosen the front clock to read zero. But if the front clock reads, say, 9:47, then the rear clock reads 9:47 plus Lv/c^2. There is of course no need to have an actual train in the setup. In general, all we need are two clocks separated by some distance L and moving with the same speed v. But we'll often talk in terms of trains, since they're easy to visualize.

Note that the L in the Lv/c^2 result is the length of the train *in its own frame*, and not the shortened length that you observe in your frame (see Section 1.3.3). Appendix B gives a number of other derivations of Eq. (1.8), although they rely on material we haven't covered yet.

Example (Clapping first): Two people stand a distance L apart along an east-west road. They clap simultaneously in the ground frame. In the frame of a car driving eastward along the road, which person claps first?

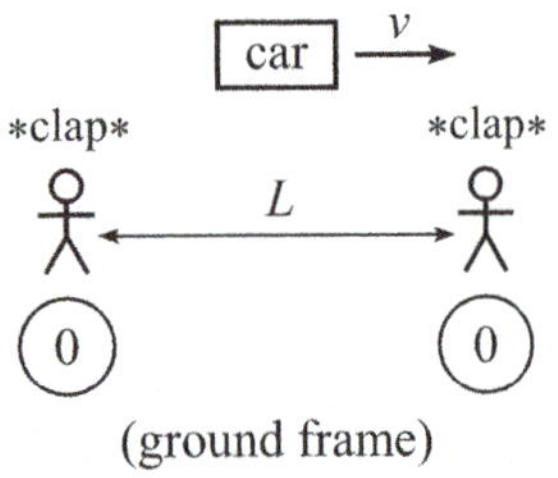

Figure 1.14

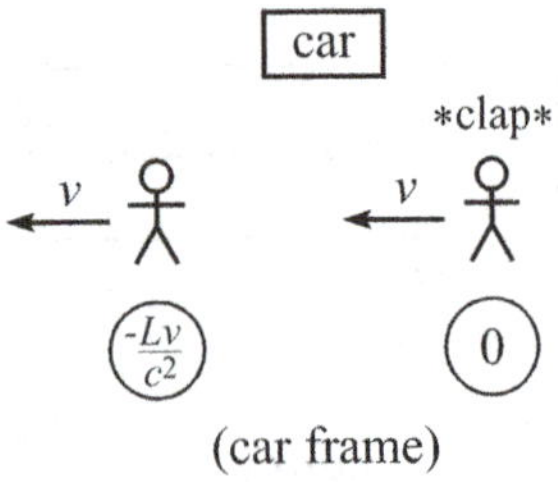

Figure 1.15

Solution: The eastern person claps first, for the following reason. Without loss of generality, let's assume that clocks on the two people read zero when the claps happen. Then a snapshot in the ground frame at the instant the claps happen is shown in Fig. 1.14. We've drawn the car in the middle as it travels past, but its exact location is irrelevant.

Now consider a snapshot in the car frame at the instant the eastern (right) person claps. This person's clock reads zero when he claps, because that is a frame-independent fact. Now, we can imagine that the two people are on a westward-traveling train, which means that the western person is the front person. By the rear-clock-ahead effect, the western (left) person's clock is *behind* by Lv/c^2. So it reads only $-Lv/c^2$, as shown in Fig. 1.15. (As we will see many times throughout this chapter, drawing pictures is extremely helpful when solving relativity problems!) Since this clock hasn't hit zero, the western person hasn't clapped yet. The eastern person therefore claps first, as we claimed.

REMARK: In the car frame, the distance between the two people is actually less than L (as we have indicated in Fig. 1.15), due to the length-contraction result we'll derive in Section 1.3.3. But this doesn't affect the result that the eastern person claps first. Similarly, the time-dilation result that we'll derive in Section 1.3.2 is relevant if we want to determine exactly how long the car observer needs to wait for the western person to clap. (It will turn out to be longer than Lv/c^2.) We'll talk about these matters shortly. ♣

REMARKS:

1. The Lv/c^2 result has nothing to do with the fact that the rear clock passes you at a later time than the front clock passes you. The train could already be past you, or it could even be moving directly toward or away from you. The rear clock will still be ahead by Lv/c^2, as observed in your frame.

2. The Lv/c^2 result does *not* say that you see the rear clock ticking at a faster *rate* than the front clock. They run at the same rate. (They both have the same time-dilation factor relative to you; see Section 1.3.2.) The rear clock is simply always a fixed time ahead of the front clock, as observed in your frame.
3. In the train setup (with the off-centered light source) that led to Eq. (1.8), the fact that the rear clock is *ahead* of the front clock in the ground frame means that in the train frame the light hits the rear clock *after* it hits the front clock.
4. The L in Eq. (1.8) is the separation between the clocks in the *longitudinal* direction, that is, the direction of the velocity of the train (or more generally, the velocity of the clocks, if we don't have a train). The height in the train doesn't matter; all clocks along a given line perpendicular to the train's velocity have the same reading at any given instant in the ground frame.
5. For everyday speeds v, the Lv/c^2 effect is extremely small. If $v = 30\,\text{m/s}$ (about 67 mph) and if $L = 100\,\text{m}$, then $Lv/c^2 \approx 3 \cdot 10^{-14}\,\text{s}$. This is completely negligible on an everyday scale.
6. What if we have a train that doesn't contain the above setup with a light source and two light beams? That is, what if the given events have nothing to do with light? The Lv/c^2 result still holds, because we *could* have built the light setup if we wanted to (arranging for the light-hitting-end events to coincide with the given events). It doesn't matter if the light setup actually exists.
7. It's easy to forget which of the clocks is the one that is ahead. But a helpful mnemonic for remembering "rear clock ahead" is that both the first and fourth letters in each word form the same acronym, "rca," which is an anagram for "car," which is sort of like a train. Sure. ♣

1.3.2 Time dilation

We showed above that if two clocks are separated by a distance L in the horizontal (that is, longitudinal) direction on a train, and if the train is moving with respect to you, then you observe different readings on the clocks, at any given instant in your frame. Note that this result relates the readings on two *different* clocks at a given *instant* in your frame. It says nothing about the *rate* at which a *single* clock runs in your frame. This is what we will now address.

We will demonstrate that if a given clock is moving with respect to you, then you will observe the clock running slowly. That is, if you use a stopwatch to measure how long it takes a given train clock to tick off 10 seconds, your stopwatch might read 20 seconds. The exact time on your watch depends on the speed v of the train, as we'll see shortly. But in any case, your clock will always read *more* than 10 seconds in this setup. (Or it will read exactly 10 seconds if $v = 0$ and the train is sitting at rest.) This effect is called *time dilation*. The name is appropriate, because the word "dilate" means to become larger. Since the moving clock runs slow (as viewed by you), a time T that ticks off on it takes *more* than a time T on your watch.

We will derive this time-dilation result by presenting a classic example of a light beam traveling in the vertical (that is, transverse) direction on a train. Let there be a light source on the floor of the train, and let there be a mirror on the ceiling, which is a height h above the floor. Let observer A be at rest on the train, and let observer B be at rest on the ground. The speed of the train with respect to the ground is v.[6] A flash of light is emitted upward. The light travels up to the mirror, bounces off it, and then

[6]Technically, the words "with respect to ..." should always be included when talking about speeds, because there is no absolute reference frame, and hence no absolute speed. But in the future, when it is clear what we mean (as in the case of a train moving with respect to the ground), we'll occasionally be sloppy and drop the "with respect to ...."

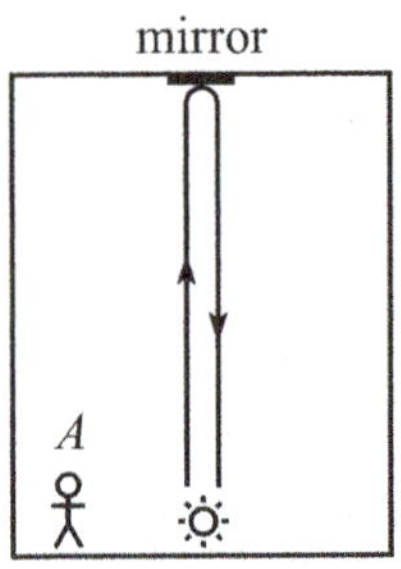

Figure 1.16

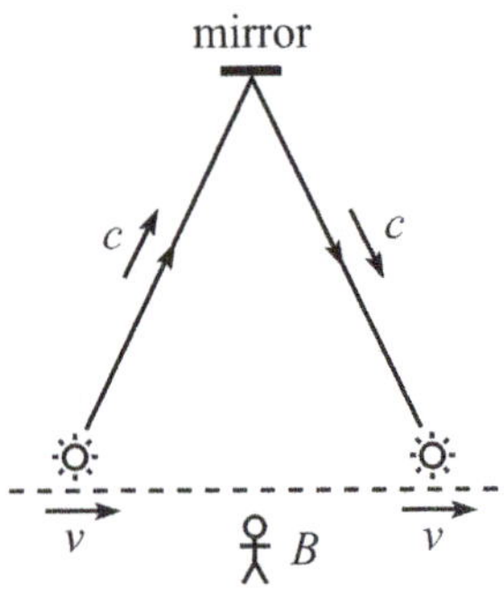

Figure 1.17

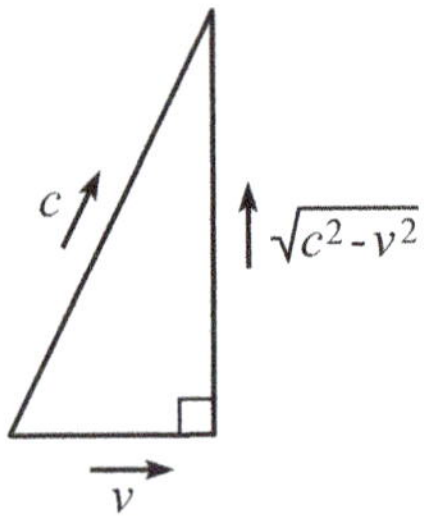

Figure 1.18

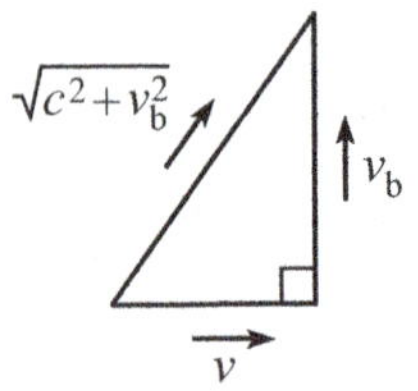

Figure 1.19

heads back down. Assume that right after the light is emitted, we replace the source with a mirror on the floor, so that the light keeps bouncing up and down indefinitely.

In A's frame, the train is at rest, so the path of the light is simple. It just goes straight up and straight down, as shown in Fig. 1.16. The light travels at speed c, so it takes a time of h/c to reach the ceiling and then a time of h/c to return to the floor. The roundtrip time in A's frame is therefore

$$t_A = \frac{2h}{c}. \tag{1.9}$$

There's nothing fancy going on here. All we have used is the fact that rate times time equals distance.

Now consider the setup in B's frame, where the train moves at speed v. In this frame, the path of the light is diagonal, as shown in Fig. 1.17. It is indeed diagonal, because in addition to moving upward, the light also gets carried rightward along with the train. You can imagine that the light travels up and down a vertical tube on the train. Since the light remains in the tube in the train frame (let's imagine that it's a laser beam that doesn't spread out), it also remains in the tube in the ground frame. ("The light remains in the tube" is a frame-independent statement. We could have a setup where a paint bomb explodes if the light touches the side of the tube. All observers must agree on whether the train is covered in paint.) Therefore, since in the ground frame the tube moves rightward along with the train, the light must also move rightward.

The crucial fact that we will now invoke is that the speed of light in B's frame is still c. Why? Because the speed-of-light postulate says so! The light therefore travels along its diagonally upward path in Fig. 1.17 at speed c. Since the horizontal component of the light's velocity is v, the vertical component must be $\sqrt{c^2 - v^2}$, as shown in Fig. 1.18. The horizontal component is in fact v, because the light always remains in the hypothetical vertical tube mentioned above, and this tube moves horizontally with speed v.

The Pythagorean theorem is indeed valid in Fig. 1.18, because it is valid for distances, and because speeds are just distances divided by time. Note that the vertical component of the light's velocity is *not* c, as would be the case if light behaved like a baseball. If you throw a baseball with speed v_b vertically on train, then its velocity with respect to the ground is shown in Fig. 1.19; the vertical speed remains v_b. The difference between this picture and the one in Fig. 1.18 is that for a baseball, the *vertical* speed is the thing that remains the same when shifting to the ground frame, and this leads to a larger *total* speed. But for light, it is the *total* speed that remains the same, and this leads to a smaller *vertical* speed. As with the c's in Fig. 1.10, we entered the bizarre world of relativity when we wrote the c's next to the diagonal paths in Fig. 1.17.

Having established that the vertical speed in B's frame is $\sqrt{c^2 - v^2}$, it follows that the time it takes the light to travel upward a height h to reach the mirror is $h/\sqrt{c^2 - v^2}$. Likewise for the downward trip. The roundtrip time is therefore

$$t_B = \frac{2h}{\sqrt{c^2 - v^2}}. \tag{1.10}$$

In this reasoning, we have assumed that the height of the train in B's frame is still h. Although we'll see in Section 1.3.3 that there is length contraction along the direction of motion, there is none in the direction perpendicular to the motion; we'll show this at the end of Section 1.3.3. So the height is indeed still h in B's frame.

Dividing Eq. (1.10) by Eq. (1.9) gives

$$\frac{t_B}{t_A} = \frac{c}{\sqrt{c^2 - v^2}} = \frac{1}{\sqrt{1 - v^2/c^2}}. \tag{1.11}$$

If we define γ by

$$\boxed{\gamma \equiv \frac{1}{\sqrt{1 - v^2/c^2}}} \tag{1.12}$$

then we arrive at

$$\boxed{t_B = \gamma t_A} \quad \text{(time dilation)} \tag{1.13}$$

The γ factor here is ubiquitous in special relativity. We'll occasionally add a subscript with the associated velocity v (like γ_v) if a setup involves more than one velocity (and hence γ factor), to avoid any confusion. In these cases we'll sometimes just use whatever fraction of c the speed is, as the subscript. For example, the γ factor associated with the speed $c/2$ is $\gamma_{1/2} = 2/\sqrt{3}$.

Note that γ is always greater than or equal to 1. This means that the roundtrip time in the above setup is *longer* in B's frame than in A's frame. The one exception occurs when $v = 0 \implies \gamma = 1$, in which case the two times are equal. But in this case A is at rest with respect to B, so they are both in the same frame, which isn't very interesting.

What are the implications of Eq. (1.13)? For concreteness, let $v/c = 3/5$, which yields $\gamma = 5/4$. (The numbers work out nicely here, because 3-4-5 is a Pythagorean triple.) We may then say the following. If A is standing at rest on the train next to the light source, and if B is standing on the ground, and if A claps his hands at $t_A = 4$ second intervals according to his watch, then B observes A's claps happening at $t_B = 5$ second intervals according to her watch. (As usual, it is understood that B subtracts off the time it takes the light to travel to her eye, to determine when the claps actually happen in her frame.) This is true because both A and B must agree on the number of roundtrips the light beam completes between claps. If we assume, for convenience, that a roundtrip takes one second in A's frame (yes, that would be a tall train), then Eq. (1.13) tells us that the four roundtrips between successive claps take five seconds in B's frame. B therefore sees A's clock running slow, by a factor 4/5.

We just made the claim that both A and B must agree on the number of roundtrips between successive claps. However, since A and B disagree on so many things (whether two events are simultaneous, the rate at which clocks tick, and the length of things, as we'll see below), you might be wondering if there's *anything* they agree on. Yes, there are still frame-independent statements we can hang on to. We noted on page 15 that the reading on a clock when a paintball hits it is frame independent. As another example, if a bucket of paint flies past you and dumps paint on your head, then everyone agrees that you are covered with paint. Likewise, if A is standing next to the light clock and claps when the light reaches the floor, then everyone agrees on this. If the light is actually a strong laser pulse, and if A's clapping motion happens to bring his hands over the mirror right when the pulse gets there, then everyone agrees that his hands get burned by the laser.

What if we have a train that doesn't contain one of our special light clocks? It doesn't matter. We *could* have built one if we wanted to, so the same results concerning the claps must still hold. Therefore, light clock or no light clock, B observes A moving strangely slowly. From B's point of view, A's heart beats slowly, his blinks are lethargic, and his sips of coffee are slow enough to suggest that he needs another cup.

> The effects of dilation of time
> Are magical, strange, and sublime.
> In your frame, this verse,
> Which you'll see is not terse,
> Can be read in the same amount of time it takes someone
>   else in another frame to read a similar sort of rhyme.

Our assumption that A is at rest on the train was critical in the above derivation. If A is moving with respect to the train, then Eq. (1.13) doesn't hold, because we *cannot* say that both A and B must agree on the number of roundtrips the light beam takes between claps, because there is now an issue with simultaneity. More precisely, if A is at rest on the train right next to the light source, then there is no issue with simultaneity, because the distance L in Eq. (1.8) is zero. And if A is at rest at a fixed distance from the source, then consider a person A' at rest on the train right next to the source. The distance L between A and A' is nonzero, so from the rear-clock-ahead effect, B sees their two clocks differ by Lv/c^2. But this difference is *constant*, so B sees A's clock tick at the same rate as A''s clock. Basically, since A and A' represent the same reference frame, there is again no issue with simultaneity. (More precisely, there *is* a loss of simultaneity, but it has no consequence here, because it is constant.) Equivalently, we can just build a second light clock next to A, and it will have the same speed v (and thus yield the same γ factor) as the original clock.

However, if A is moving with respect to the train, then we have a problem. If A' is again at rest on the train next to the source, then the distance L between A and A' is *changing*, so B can't use the reasoning in the previous paragraph to conclude that A's and A''s clocks tick at the same rate. And in fact they do not, because as above, we can build another light clock and have A hold it. In this case, A's speed is what goes into the γ factor in Eq. (1.12), and this speed is different from A''s speed (which is the speed of the train).

REMARKS:

1. The speed v needs to be fairly large in order for the γ factor in Eq. (1.12) to differ appreciably from 1. If $v = c/10$ (which is still quite fast), we only have $\gamma_{1/10} \approx 1.005$. A few other values are: $\gamma_{1/2} \approx 1.15$, $\gamma_{9/10} \approx 2.3$, and $\gamma_{99/100} \approx 7$.

2. The time-dilation result in Eq. (1.13) is a bit strange, no doubt, but there doesn't seem to be anything downright incorrect about it until we look at the situation from A's point of view. A sees B flying by at a speed v in the other direction. The ground frame is no more fundamental than the train frame, so the same reasoning we used above also applies to A's frame. Equivalently, we can just switch all the A and B labels in the above derivation. The time-dilation factor, γ, doesn't depend on the sign of v, so A sees the same time-dilation factor that B sees. That is, A sees B's clock running slow. But how can this be? Are we claiming that A's clock is slower than B's, and also that B's clock is slower than A's? Well ... yes and no.

   Remember that the above time-dilation reasoning applies only to a situation where something is motionless in the appropriate frame. In the second situation (where A sees B flying by), the statement $t_A = \gamma t_B$ holds only for two events (say, two ticks on B's clock) that happen at the same place in B's frame. But two such events are certainly not at the same place in A's frame, so the $t_B = \gamma t_A$ result in Eq. (1.13) does *not* hold. The conditions of being motionless in each frame can never both hold in a given setup (unless $v = 0$, in which case $\gamma = 1$ and $t_A = t_B$). So the answer to the question at the end of the previous paragraph is "yes" if you ask the questions in the appropriate frames, and "no" if you think the answer should be frame independent.

3. Concerning the fact that A sees B's clock run slow, *and* B sees A's clock run slow, consider the following statement. "This is a contradiction. It is essentially the same as saying, 'I have two apples on a table. The left one is bigger than the right one, and the right one is bigger than the left one.'" How would you respond to this statement?

   Well, it is not a contradiction. Observers A and B are using *different coordinates* to measure time. The times measured in each of their frames are quite different things. The seemingly contradictory time-dilation result is really no stranger than having two people run away from each other into the distance, and having them both say that the other person looks smaller. In short, we are not comparing apples and apples. We are comparing apples and oranges. A more correct analogy would be the following. An apple and an orange sit on a table. The apple says to the orange, "You are a much uglier apple than I am," and the orange says to the apple, "You are a much uglier orange than I am."

4. One might view the statement, "A sees B's clock running slow, and also B sees A's clock running slow," as somewhat unsettling. But in fact it would be a complete disaster for the theory if A and B viewed each other in different ways. A critical ingredient in the theory of relativity is that A sees B in exactly the same way that B sees A.
5. In everything we've done so far, we've assumed that A and B are in inertial frames, because these are the frames that the postulates of special relativity deal with. However, it turns out that the time-dilation result in Eq. (1.13) holds even if A is accelerating, as long as B isn't. In other words, if you are looking at a clock that is undergoing a complicated accelerated motion, then to figure out how fast it is ticking in your frame at a given instant, all you need to know is its speed at that instant; its acceleration is irrelevant. (This has plenty of experimental verification. Perhaps the quickest theoretical argument involves using a Minkowski diagram; see Section 2.4 and the third remark in the solution to Problem 2.12.) If, however, *you* are accelerating, then all bets are off, and it isn't valid for you to use the time-dilation result when looking at a clock. But it's still possible to get a handle on such situations, as we'll see in Chapter 5. Problem 2.12 also deals with this issue. ♣

In the second remark above, we noted that the time-dilation result in Eq. (1.13) holds in setups where two events happen at the same place in one of the frames. The γ factor in Eq. (1.13) appears on the side of the equation associated with the frame in which the two events happen at the same place. An equivalent (and simpler) way of stating how to properly use time dilation is: *If you look at a moving clock, you observe it running slowly.* This has the "happen at the same place" requirement built into it, because two ticks on a clock certainly happen at the same place in the clock's frame, since they both happen at the clock. (Well, the hands on an analog clock necessarily move a little bit, but let's ignore that.)

To summarize, we can state the time-dilation result with this equation:

$$\boxed{t_{\text{observed}} = \gamma t_{\text{proper}}} \qquad \text{(time dilation)} \tag{1.14}$$

The *proper time* is the time that elapses between two ticks on a given clock in the frame of the clock. (More generally, the proper time is the time between two events, as measured in the frame where the events happen at the same place.) The observed time in Eq. (1.14) is the time that elapses between the two ticks, as measured in the frame of an observer who is looking at the clock. We can also state the time-dilation result with these words:

> TIME DILATION: If you look at a clock moving with speed v relative to you, then you observe the clock running slowly by a factor $\gamma \equiv 1/\sqrt{1 - v^2/c^2}$.

When relating the times in different frames, it is easy to get confused about where to put the γ factor. The safest way to proceed is to (1) start with the fact that if you look at a moving clock, it runs slow, then (2) identify which time is larger/smaller, and then (3) put the γ factor (which is always larger than 1) where it needs to be so that the relative size of the times is correct. For example, let's say that during a time T that elapses on your watch, you want to determine how much time elapses on a clock that is flying past you. Since you see the clock running slow, the time on it must be less than T, which means that the answer is T/γ; we must divide by γ.

A common trap to fall into when applying time dilation is the following. (Every physics student is bound to make this error at least once.) Let's say that at a given moment, you look at a clock flying by (say, a clock at the back of a train). And then a little later you look at a *different* clock (say, a clock at the front of the train). You take the difference in the readings and then multiply this difference by γ (because you see the train clocks run slow), to find the time elapsed on your watch in the ground frame. This strategy is incorrect, because it uses the readings on two *different* clocks. (The Lv/c^2 rear-clock-ahead result is what messes things up. This is evident in the examples in Section 1.4.) To apply time dilation correctly, you must take the difference in the readings on a *single* clock. Remember, all that time dilation says is, "If you look at a moving clock, you observe it running slowly." The word "clock" here is singular.

Note well that it is *elapsed times* that get dilated, and not *readings* on clocks. If you look at a clock on a moving train and observe that it has a reading of t_1, then there isn't much you can do with that. But if you then look at the same clock later on and observe that it has a reading of t_2, then you can say something. You can say that the time that elapses on your own clock between your two observations (during which a time of $t_2 - t_1$ elapses on the train clock) equals $\gamma(t_2 - t_1)$, because you see the train clock run slow. Since time dilation deals only with elapsed times and not with actual readings, we should technically be writing Δt's instead of t's in Eq. (1.14), that is,

$$\Delta t_{\text{observed}} = \gamma \, \Delta t_{\text{proper}}. \tag{1.15}$$

But we'll usually drop the Δ's, for simplicity.

Let's now do two classic examples involving time dilation.

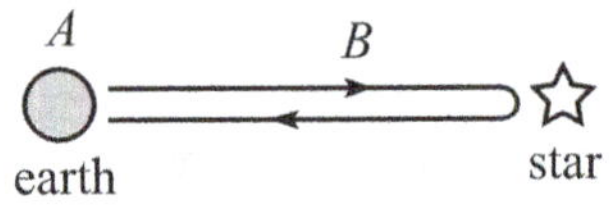

Figure 1.20

Example 1 (Twin paradox): Twin A stays on the earth, while twin B flies quickly to a distant star and back (see Fig. 1.20). After B returns, are the twins the same age? If not, who is younger?

Solution: From A's point of view, B's clock (and heartbeat, cell aging, and everything else) is running slow by a factor γ on both the outward and return parts of the trip. Therefore, B is younger than A when they meet up again. This is the answer, and that's that. So if getting the right answer is all we care about, then we can pack up and go home. But our reasoning leaves one large point unaddressed. The "paradox" part of this example's title comes from the following alternative reasoning. Someone might say that in B's frame, A's clock is running slow by a factor γ, so A should be younger than B when they meet up again.

It's definitely true that when the two twins are standing next to each other after B's journey concludes (that is, when they are eventually in the same frame), we can't have B younger than A, and also A younger than B. So what is wrong with the reasoning at the end of the preceding paragraph? The error lies in the fact that B doesn't remain in a single inertial frame. Her inertial frame for the outward trip is different from her inertial frame for the return trip. The derivation of our time-dilation result requires a single inertial frame.

Said in a different way, B accelerates when she turns around, and our time-dilation result holds only from the point of view of an *inertial* observer. The symmetry in the problem is broken by the acceleration. If both A and B are blindfolded, they can still tell who is doing the traveling, because B will feel the acceleration at the turnaround. Constant velocity cannot be felt, but acceleration can be. (However, see Chapter 5 on general relativity. Gravity complicates things.) For the entire outward and return parts of the trip, B *does* observe A's clock running slow, but enough strangeness occurs during the turning-around period (from B's point of view) to make A end up older.

The above paragraphs show what is wrong with the "A is younger" reasoning, but they don't show how to modify it quantitatively to obtain the correct answer. There are many different ways of doing this, and you can tackle some of them in the problems; see Exercises 1.30 and 2.32, Problems 1.21 and 2.11, and various problems in Chapter 5. Also, Appendix C gives a list of all the possible resolutions to the twin paradox that I can think of, although some rely on material we haven't covered yet.

Example 2 (Muon decay): Elementary particles called *muons* (which are identical to electrons, except that they are about 200 times as massive, and they decay) are created in the upper atmosphere when cosmic rays (energetic protons, mostly) collide with air molecules. The muons have an average lifetime of about $2 \cdot 10^{-6}$ seconds (this is the proper lifetime, that is, the lifetime as measured in the frame of the muon). They then decay into other particles (electrons and neutrinos). The muons move at nearly the speed of light. Assume for simplicity that a particular muon is created at a height of 20 km, moves straight downward, has speed $v = (0.9999)c$, decays in exactly $T = 2 \cdot 10^{-6}$ seconds, and doesn't collide with anything on the way down.[7] Will the muon reach the earth before it (the muon!) decays?

Solution: The naive thing to say is that the distance traveled by the muon is $d = vT \approx (3 \cdot 10^8\,\text{m/s})(2 \cdot 10^{-6}\,\text{s}) = 600\,\text{m}$, and that this is less than 20 km, so the muon doesn't reach the earth. This reasoning is incorrect, because of time dilation. We must remember that in the earth frame the muon lives longer by a factor of γ, which is $\gamma = 1/\sqrt{1 - v^2/c^2} \approx 70$ here. (You can imagine that the muon has a little clock, and when the clock hits $T = 2 \cdot 10^{-6}$ seconds, the muon decays.) So the actual lifetime in the earth frame is $\gamma T = (70)(2 \cdot 10^{-6}\,\text{s}) = 1.4 \cdot 10^{-4}\,\text{s}$. The correct distance traveled in the earth frame is therefore $v(\gamma T)$. This is $\gamma \approx 70$ times the $vT \approx 600\,\text{m}$ distance we found above, so we end up with 40 km. Hence, the muon travels the 20 km, with room to spare. The real-life fact that we actually do detect muons reaching the surface of the earth in the predicted abundances is one of the many experimental tests that support special relativity. The naive $d = vT$ reasoning would predict that we shouldn't see any (or at most a very small number, if we had based our calculation on more realistic assumptions).

1.3.3 Length contraction

Having discussed the loss-of-simultaneity (rear-clock-ahead) and time-dilation effects, we now come to the third of the fundamental effects of special relativity, namely *length contraction*. We will derive this effect by again looking at how a light beam travels in a train, except that now we will shine the light in the longitudinal (horizontal) direction instead of the transverse (vertical) direction. There is actually a much quicker derivation of length contraction than the one we will give here; see the third remark below. But we'll work through the present derivation because the calculation is instructive.

Consider the following setup. Person A is at rest on a train that he measures to have length L_A, and person B is at rest on the ground. The train moves at speed v with respect to the ground. A light source is located at the back of the train, and a mirror is located at the front. The source emits a flash of light that heads to the mirror, bounces off, then heads back to the source. By looking at how long this process takes in each of the two reference frames, we can determine the length of the train as measured by B. In A's frame (see Fig. 1.21), the light travels a total distance of $2L_A$ at speed c, so the roundtrip time is simply

$$t_A = \frac{2L_A}{c}. \tag{1.16}$$

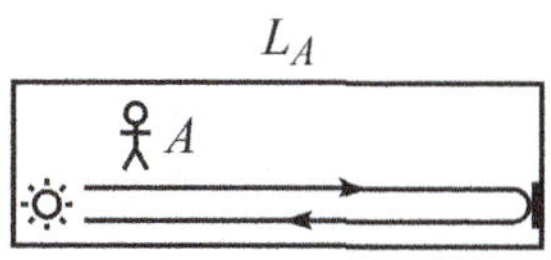

Figure 1.21

[7]In the real world, the muons are created at various heights, move in different directions, have different speeds, decay in lifetimes that vary according to a standard half-life formula, and may very well bump into air molecules. So technically we've got everything wrong here. But that's no matter. Our assumptions are good enough for the present purpose!

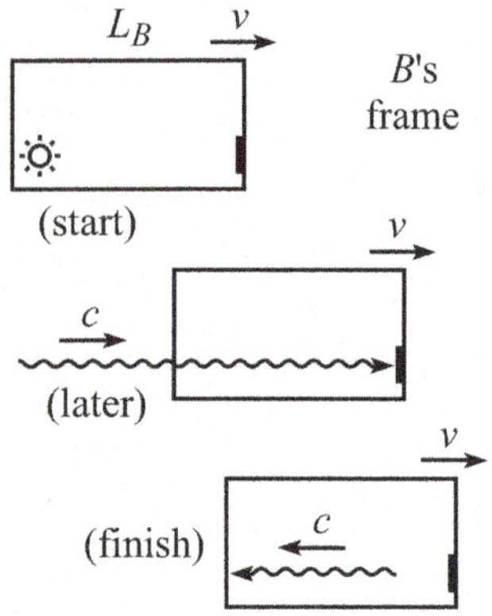

Figure 1.22

Things are a little more complicated in B's frame; see Fig. 1.22. Let the length of the train as measured by B be L_B. For all we know at this point, L_B might be equal to L_A, but we'll soon find that it is not. During the first part of the trip, the relative speed (as measured by B) of the light and the mirror at the front of the train is $c - v$. Since the initial gap between the light and the mirror is L_B, the time it takes to close this gap down to zero is $L_B/(c - v)$. This is the same type of reasoning we used on various occasions in Section 1.3.1.

Similarly, during the second part of the trip, the relative speed (as measured by B) of the light and the back of the train is $c + v$. Since the initial gap between the light and the back of the train is again L_B, the time it takes to close this gap down to zero is $L_B/(c + v)$. The total roundtrip time in B's frame is therefore

$$t_B = \frac{L_B}{c - v} + \frac{L_B}{c + v} = \frac{2cL_B}{c^2 - v^2} = \frac{2L_B/c}{1 - v^2/c^2} = \gamma^2 \frac{2L_B}{c}. \tag{1.17}$$

But we also know from Eq. (1.13) that

$$t_B = \gamma t_A. \tag{1.18}$$

This is a valid statement, because the two events we are concerned with (light leaving back, and light returning to back) happen at the same place in the train frame (A's frame), so it is legal to use the time-dilation result in Eq. (1.13). The γ factor goes on the side of the equation associated with the frame in which the two events happen at the same place, which is A's frame here. Equivalently, just imagine a clock ticking at the back of the train; B sees this clock run slow.

Substituting the results for t_A and t_B from Eqs. (1.16) and (1.17) into Eq. (1.18), we find

$$\gamma^2 \frac{2L_B}{c} = \gamma \cdot \frac{2L_A}{c} \quad \Longrightarrow \quad \boxed{L_B = \frac{L_A}{\gamma}} \qquad \text{(length contraction)} \tag{1.19}$$

Note that we could not have used this setup to derive length contraction if we had not already derived time dilation in Eq. (1.13).

Since $\gamma \ge 1$, we see that B measures the train to be shorter than A measures (or equal, if $v = 0$). The term *proper length* is used to describe the length of an object in its rest frame. So L_A is the proper length of the train, and the length L_B in any other frame is less than or equal to L_A. This length contraction is often called the *Lorentz–FitzGerald contraction,* for the reason given in Footnote 2.

> Relativistic limericks have the attraction
> Of being shrunk by a Lorentz contraction.
> But for readers, unwary,
> The results may be scary,
> When a fraction . . .

REMARKS:

1. The length-contraction result in Eq. (1.19) holds for lengths in the direction of the relative velocity between the two frames (the longitudinal direction). There is no length contraction in the perpendicular direction (the transverse direction), as we'll show at the end of this section.

2. As with time dilation, length contraction is a bit strange, but there doesn't seem to be anything actually paradoxical about it, until we look at things from A's point of view. To make a nice symmetrical situation, let's say B is standing on an identical train, which is at

rest with respect to the ground. Then A sees B flying by at speed v in the other direction. Neither train is any more fundamental than the other, so the same reasoning we used above also applies here. (Just switch all the A and B labels in the derivation.) We conclude that A sees the same length-contraction factor that B sees. That is, A measures B's train to be short. But how can this be? Are we claiming that A's train is shorter than B's, and also that B's train is shorter than A's? Is the actual setup the one shown in Fig. 1.23, or is it the one shown in Fig. 1.24? Well . . . it depends.

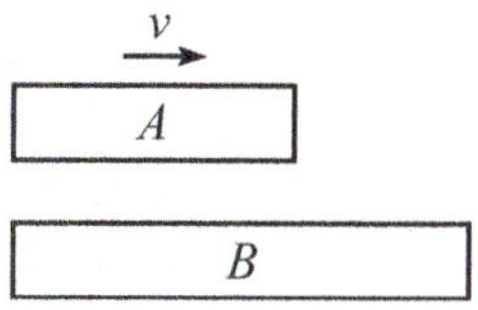

Figure 1.23

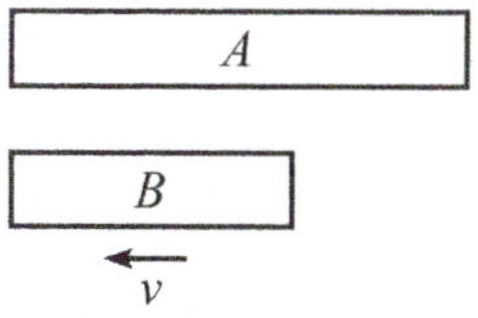

Figure 1.24

As with time dilation, it makes no sense to say what the length of a train really *is*. It makes sense only to say what the length is *in a given frame*. The situation doesn't really look like one thing in particular. The look depends on the frame in which the looking is being done.

Let's be a little more specific. How do you measure a length? You write down the position coordinates of the ends of something measured *simultaneously*, and then you take the difference between these coordinates. But the word "simultaneously" here should send up all sorts of red flags. Simultaneous events in one frame are not simultaneous in another. Stated more precisely, here is what we are claiming: Let B write down simultaneous coordinates of the ends of A's train, and also simultaneous coordinates of the ends of her own train. Then the difference between the former is smaller than the difference between the latter. Likewise, let A write down simultaneous coordinates of the ends of B's train, and also simultaneous coordinates of the ends of his own train. Then the difference between the former is smaller than the difference between the latter. There is no contradiction here, because the times at which A and B are writing down the coordinates don't have much to do with each other, due to the loss of simultaneity. We'll be quantitative about this in the second example in Section 1.4. As with time dilation, we are comparing apples and oranges.

3. As we mentioned at the beginning of this section, there is a quick argument that demonstrates why time dilation implies length contraction, and vice versa. Let A stand on the ground, next to a stick with (proper) length L. Let B fly past the stick at speed v; see Fig. 1.25. In A's frame, it simply takes B a time of L/v to traverse the length of the stick. Therefore (assuming that we have demonstrated the time-dilation result), since A sees B's clock run slow, a watch on B's wrist will advance by a time of only $L/\gamma v$ while he traverses the length of the stick.

   How does B view the situation? He sees A and the stick fly by at speed v. The time between the two ends passing him is $L/\gamma v$, because we found above that this is the time elapsed on his watch. (This is a frame-independent value. Imagine a switch that starts his watch when he coincides with one end of the stick, and stops it when he coincides with the other end.) To obtain the length of the stick in his frame, B simply multiplies the speed of the stick times the time. So he measures the length to be $(L/\gamma v)v = L/\gamma$. This is the desired length contraction. The same argument in reverse shows conversely that length contraction implies time dilation. In short, any theory that has one of these effects must have the other.

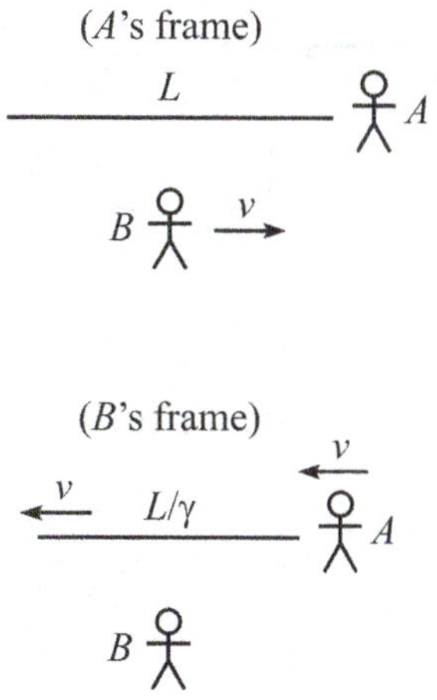

Figure 1.25

4. As mentioned in Footnote 5, the length-contraction factor γ is independent of the position on the train. That is, all parts of the train are contracted by the same factor. This follows from the fact that all points in space are equivalent. Equivalently, we could put a large number of small replicas of the above source-mirror system along the length of the train. They would all produce the same value of γ (because they all have the same v), independent of the position on the train.

5. If you still want to ask, "Is length contraction actually *real*?" then consider the following hypothetical undertaking. Imagine a sheet of paper moving sideways past the Mona Lisa, skimming the surface of the painting. A standard sheet of paper is plenty large enough to cover her face, so if the paper is moving slowly, and if you take a photograph at the appropriate time, then in the photo her entire face will be covered by the paper. However, if the sheet is flying by sufficiently fast, and if you take a photograph at the appropriate time, then in the photo you'll see a thin vertical strip of paper covering only a small fraction of her face. So you'll still see her smiling at you. ♣

To summarize, we can state the length-contraction result with this equation:

$$\boxed{L_{\text{observed}} = \frac{L_{\text{proper}}}{\gamma}} \qquad \text{(length contraction)} \tag{1.20}$$

As mentioned above, the proper length is the length measured in the frame of the stick (or whatever). The observed length is the length measured in any other frame. We can also state the length-contraction result with these words:

> LENGTH CONTRACTION: If you look at a stick moving longitudinally with speed v relative to you, then you observe the stick to be short by a factor $1/\gamma = \sqrt{1 - v^2/c^2}$.

As with time dilation, when relating distances in different frames, it is easy to get confused about where to put the γ factor. The safest way to proceed is to (1) start with the fact that if you look at a moving stick, it is short, then (2) identify which length is longer/shorter, and then (3) put the γ factor (which is always larger than 1) where it needs to be so that the relative size of the lengths is correct. For example, let's say that you want to determine the proper length of a moving train that has length L in your frame. Since you see the train as length contracted, its proper length must be longer than L (because that length is contracted down to the L that you observe). So the proper length is γL; we must multiply by γ.

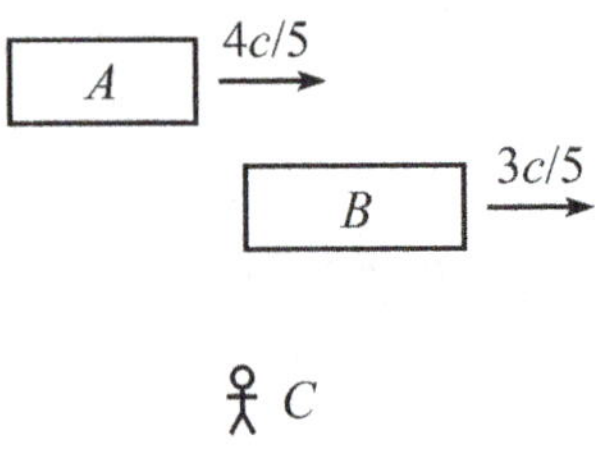

Figure 1.26

Example 1 (Passing trains): Two trains, A and B, each have proper length L and move in the same direction. A's speed is $4c/5$, and B's speed is $3c/5$. A starts behind B; see Fig. 1.26. How long, as measured by person C on the ground, does it take for A to overtake B? By this we mean the time between the front of A passing the back of B, and the back of A passing the front of B.

Solution: Relative to C on the ground, the γ factors associated with A and B are $5/3$ and $5/4$, respectively. Therefore, their lengths in the ground frame are $3L/5$ and $4L/5$. The overtaking begins when the back of A is a distance $7L/5$ (the sum of the lengths of the trains) behind the front B. The overtaking ends when the back of A reaches the front of B. So we need the initial gap of $7L/5$ to decrease to zero. The gap decreases at a rate of $c/5$ (the difference of the speeds in the ground frame). The overtaking therefore takes a time in the ground frame equal to

$$t_C = \frac{7L/5}{c/5} = \frac{7L}{c}. \tag{1.21}$$

Example 2 (Muon decay, again): Consider the "Muon decay" example in Section 1.3.2. From the muon's point of view, it lives for a time of $T = 2 \cdot 10^{-6}$ seconds, and the earth is speeding toward it at $v = (0.9999)c$. How, then, does the earth (which travels only $d = vT \approx 600$ m before the muon decays) reach the muon?

Solution: The important point here is that in the muon's frame, the distance to the earth is contracted by a factor $\gamma \approx 70$. Therefore, the earth starts only $(20\,\text{km})/70 \approx 300$ m away. (You can imagine that the muons are created next to the top of a hypothetical tower with height 20 km. This tower is at rest in the earth frame, so it is length contracted in the muon frame.) Since the earth can travel a distance of 600 m during the muon's lifetime, the earth collides with the muon, with room to spare.

As stated in the third remark above, time dilation and length contraction are intimately related. We can't have one without the other. In the earth's frame, we saw in the example in Section 1.3.2 that the muon's arrival at the earth is explained by time dilation. In the muon's frame, we just saw in the present example that the earth's arrival at the muon is explained by length contraction.

Observe that for muons created,
The dilation of time is related
To Einstein's insistence
Of shrunken-down distance
In the frame where decays aren't belated.

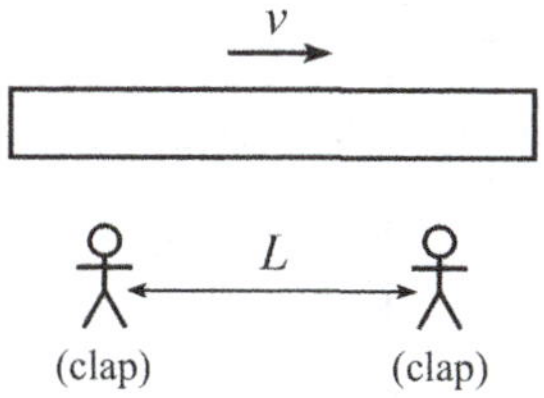

Figure 1.27

Example 3 (Two distances): In the ground frame, two people stand a distance L apart, and they clap simultaneously in the ground frame; see Fig. 1.27. A train moves to the right at speed v. In the train frame, what is the distance between the *people*, and what is the distance between the clapping *events*?

Solution: The train sees the proper distance L between the people as length contracted, so the people are only L/γ apart in the train frame.

Let the claps somehow make marks on the train. These marks are L apart in the ground frame, so they must be γL apart in the train frame, because from the ground's point of view, this distance is what is length contracted down to the distance L in the ground frame. The events are therefore γL apart in the train frame.

The non equality of the above two answers (L/γ and γL) is a consequence of the loss of simultaneity. In the train frame, as the people fly by to the left, the right person claps first. The people then travel leftward for some time before the left person claps. In the train frame, the distance between the events is therefore greater than the distance between the people. We'll be quantitative about this in the second example in Section 1.4, where we explain how the L/γ and γL distances are consistent with each other.

No transverse length contraction

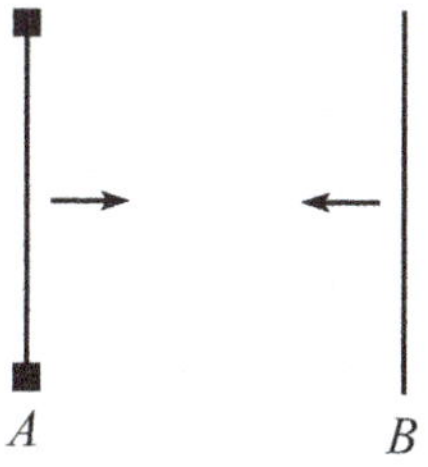

Figure 1.28

We have mentioned a few times that there is no length contraction in the direction perpendicular to the relative velocity of two frames (that is, the transverse direction). We'll now show why this is true, with the following setup. Consider two meter sticks, A and B, that move past each other as shown in Fig. 1.28. Stick A has paint brushes on its ends. If the paint brushes touch B, they leave marks on B. We can use this setup to show that in the frame of one stick, the other stick still has a length of one meter.

The key fact that we need to invoke is the first postulate of relativity, which says that all inertial frames are equivalent. In particular, the frames of the two sticks are equivalent. This implies that if A sees B shorter than (or longer than, or equal to) itself, then B *also* sees A shorter than (or longer than, or equal to) itself. The contraction factor must be the same when going each way between the frames. At first glance, this might seem backwards. After all, everyday life is full of statements such as, "If Alice is taller than Sue, then Sue is shorter than Alice," where the word "taller" is replaced with "smaller" (its opposite). But we're dealing with an entirely different thing here. We're talking about rules that hold in different reference frames. It would be a complete disaster for the theory if different frames had different rules. If the rule in one frame were "moving sticks are long" and the rule in another frame were "moving sticks are short," then this would be a violation of the fact that all inertial frames are equivalent.

Let's assume (in search of a contradiction) that A sees B shortened. Then B won't extend out to the ends of A, so there will be no paint marks on B; see the top picture

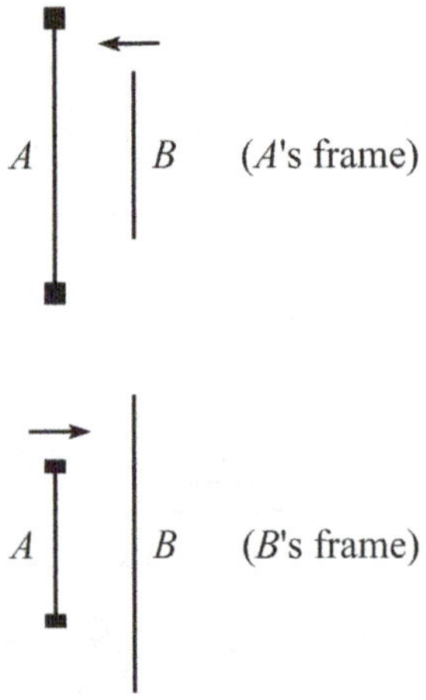

Figure 1.29

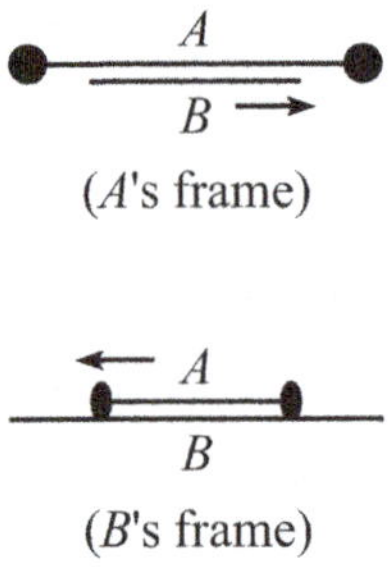

Figure 1.30

in Fig. 1.29. But in this case, B must *also* see A shortened, so there *will* be marks on B; see the bottom picture in Fig. 1.29. This is a contradiction, because the existence or nonexistence of marks on B is a frame-independent fact. Everyone must agree on whether or not there are marks on B. Likewise, if we assume that A sees B lengthened, we also reach a contradiction. We are therefore left with the only other possibility, namely that each stick sees the other stick as exactly one meter long. There is therefore no transverse length contraction.

Remark: Having used the above scenario to show that there is no *transverse* length contraction, you might wonder why we can't use the same kind of reasoning to show that there is no *longitudinal* length contraction. We had better *not* be able to use it, because we showed in Eq. (1.19) that longitudinal length contraction *does* exist. To see why we can't use the same kind of reasoning, first note that if the sticks are aligned longitudinally, then the paint brushes on A will each simply leave a streak along the entire length of B, independent of whether or not there is length contraction. This particular setup therefore can't be used to conclude anything about length contraction.

A possible improvement in the setup is to replace the paint brushes with paint bombs, so that they deposit paint only at a single instant. At first glance this seems to accomplish the task of producing a contradiction, because if the rule is that moving sticks are short, then the top picture in Fig. 1.30 shows no marks on B, whereas the bottom picture shows two marks. However, these two pictures don't describe the same scenario, because in the top picture the bombs explode simultaneously in A's frame, whereas in the bottom picture they explode simultaneously in B's frame. These two scenarios are inconsistent with each other, due to the loss of simultaneity. We can have one or the other picture, but not both. There is therefore no contradiction. Of course, this lack of a contradiction doesn't prove that longitudinal length contraction *does* in fact exist. But our goal here was only to show that the above type of reasoning can't be used to show that longitudinal length contraction *doesn't* exist. If you want to make some quantitative statements about how the loss of simultaneity relates to longitudinal length contraction, see the second example in Section 1.4 below.

In the above transverse case, there was no issue with simultaneity, because there was no extent in the longitudinal direction, which meant that the L in Lv/c^2 was zero. So we did in fact end up with a contradiction that ruled out transverse length contraction. ♣

1.3.4 A few other important topics

We have completed our treatment of the three fundamental effects, but let's discuss a few other important things before moving on.

Lattice of clocks and meter sticks

In everything we've done so far, we've taken the route of having observers sitting in various frames, making various measurements. But as mentioned earlier, this can cause some ambiguity, because you might think that the time when light reaches the observer is important, whereas what we are generally concerned with is the time when something actually happens.

A way to avoid this ambiguity is to remove the observers and define each frame by filling up space with a large rigid lattice of meter sticks and synchronized clocks, all at rest in the given frame. Different frames are defined by different lattices; assume that the lattices of different frames can somehow pass freely through each other. All of the meter sticks in a given frame are at rest with respect to all the others, so we don't have to worry about issues of length contraction within each frame. Likewise, we don't have to worry about time dilation within each frame. However, with respect to a given frame, the lattice of a different frame is squashed in the direction of its motion, because all the meter sticks in that direction are contracted. Likewise, all the clocks in the moving lattice run slow.

To measure the length of an object in a given frame, we just need to determine where the ends are (at simultaneous times, as measured in that frame) with respect to the lattice. As far as the synchronization of the clocks within each frame goes, this can be accomplished by putting a light source midway between any two clocks and sending out signals, and then setting the clocks to a certain value when the signals hit them. Alternatively, a more straightforward method of synchronization is to start with all the clocks synchronized right next to each other, and then move them very slowly to their final positions. Any time-dilation effects can be made arbitrarily small by moving the clocks sufficiently slowly. This is true because the time-dilation factor γ is second order in v, whereas the time it takes a clock to reach its final position is only first order in $1/v$; see Problem 1.3.

This lattice way of looking at things emphasizes that observers are not important, and that a frame is defined simply as a lattice of space and time coordinates. Anything that happens (an "event") is automatically assigned a space and time coordinate in every frame, independent of any observer. Just record the spatial coordinates of the lattice point where the event is located, along with the reading on the clock at that point. You can assume that the lattice spacing is arbitrarily small, constructed with, say, millimeter sticks instead of meter sticks.

Frame independence

The three fundamental effects that we derived above tell us that many "truths" from everyday life must be thrown out the window. We can no longer count on two people in different frames agreeing on matters of simultaneity, time, or length. In fact, so much has been thrown out the window, that you might be wondering if there's *anything* two people in different frames can agree on.

Fortunately, there are still some things you can hang on to, that is, things that are *frame independent*. We have encountered many examples of frame-independent statements in the above sections, and we have presented various arguments for why these statements were in fact frame independent. Let's revisit one example that came up – the reading on a clock when a ball hits it. All observers, no matter what frame they are in, must agree on what this reading is. If one person says noon, then everyone says noon. There are two ways to see why this is the case.

First, you can imagine that the ball breaks the clock, so that the clock is stuck on whatever value it had when the ball hit it. Everyone will agree on what this broken value is, because the clock can be arranged to eventually sit at rest next to any given person. Second, the two events, "ball hitting clock" and "clock reading noon (or whatever)," happen at the *same* location in any frame, namely, at the clock. (We'll assume that the clock has negligible spatial extent.) The two events are therefore separated by a distance of $L = 0$, which means that the Lv/c^2 rear-clock-ahead effect is zero. In other words, we don't have to worry about any issues involving the loss of simultaneity. In short, the two events are really just *one* event, the "ball hitting clock and clock reading noon" event, described by specific space and time coordinates. Note that we are talking here about the *reading* on the clock, and not the *time elapsed* on it. An elapsed time is the difference between two readings.

What other kinds of frame-independent statements can we make? Well, if a paint bomb explodes and leaves a mark on an object at the location of a dent that was already there, then everyone agrees that there is a mark, and that it is located at the dent. Another example is: If a person departs from one clock when it reads T_1 and arrives at another clock when it reads T_2, then the difference in these readings, $T_2 - T_1$, is frame independent. This is true because each reading is frame independent. Note that we are *not* saying that the time elapsed between these two events (person departing from one clock, person arriving at the other) is frame independent. The time elapsed *does* depend on the observer's frame (there will be a time-dilation factor involved), because the time elapsed in a given frame is the difference in readings on the clock of the *observer* (who is at rest in the given frame), and not on the two given clocks. Apples and oranges.

Draw pictures, and stick to a frame

When solving problems using the fundamental effects, a very important strategy is to draw a picture of the setup in whatever frame you have chosen to work in. You should draw a picture at every moment when something of importance happens, as we did above in Fig. 1.22, for example. Once we drew those pictures, it was reasonably clear what we needed to do. But without the pictures, we almost certainly would have gotten confused. This problem-solving strategy is so important that we'll display it in a box:

> DRAW PICTURES!

When drawing pictures at different times in a process, it is most informative to draw one picture above the other, with the x-axis origins of the two (or more) pictures vertically aligned. This way, objects that are at rest in the given frame are vertically aligned. See, for example, Figs. 1.32 and 1.34 below.

The importance of drawing pictures in relativity is analogous to the importance of drawing free-body diagrams when using $F = ma$ in mechanics problems. In both cases, the problem is usually easy once you draw the picture/diagram, but often hopeless if you don't (except in very simple cases).

A related strategy is to plant yourself in a frame and *stay there*. The only thoughts running through your head should be what *you* observe. That is, don't try to use reasoning like, "Well, the person I'm looking at in this other frame sees such-and-such." This will almost certainly cause an error somewhere along the way, because you will inevitably end up writing down an equation that combines quantities that are measured in different frames, which is a no-no. Or you might end up using time dilation backwards by putting the γ factor in the wrong place. The strategy of drawing pictures helps you avoid this kind of error, because when drawing a picture you necessarily have to pick a frame.

Of course, you might want to solve another part of the problem by working in another frame, or you might want to redo the whole problem in another frame. That's fine, but once you decide which frame you're going to use for a given line of reasoning, make sure you put yourself there and stay there. If you are drawing a picture in a train frame, then *be* the train. If you are drawing a picture in a ball frame, then *be* the ball. Sticking to a single frame is another problem-solving strategy that is so important that we'll display it in a box:

> CHOOSE A FRAME AND STICK WITH IT!

"Seeing" things

We mentioned above in the fourth remark on page 14 that there is always a difference between the time an event *happens* and the time someone *sees* the event happen, because light takes time to travel from the event to the observer. We will generally be concerned with the former and not the latter. You can avoid the issue of "seeing" if you use the lattice of clocks and meter sticks introduced above. The time associated with an event is the time on the clock at the location of the event. With a lattice setup, we never have to worry about the travel time of light.

In this chapter, we will often be a little sloppy and use language such as, "What time does B see event Q happen?" But we don't really mean, "When do B's eyes register that Q happened?" Instead, we mean, "What time does B *know* that event Q happened in her frame?" If we ever want to use "see" in the former sense, we will explicitly say so. Two such examples are the "Rotated square" setup in Problem 1.4 and the Doppler effect in Section 2.5.

1.4 Four instructive examples

We'll now present four examples that integrate everything we've done so far. Each of these examples involves all three of the fundamental effects we've discussed: rear clock ahead, time dilation, and length contraction. The first two examples address the paradoxical issues with time dilation and length contraction.

You should try to solve each of these problems on your own before looking at the solution. If you can't solve it right away, set it aside for a while and come back to it later. It will still be there when you come back. In the end, you'll find that these four problems are all quite similar. There are only three fundamental effects, so there are only so many ways to combine them!

Example 1 (Explaining time dilation): Two planets, A and B, are at rest with respect to each other, a distance L apart, with synchronized clocks. A spaceship flies with speed v past planet A toward planet B. Right when it passes A, it synchronizes its clock with A's; they both set their clocks to zero. The spaceship eventually flies past B and compares its clock with B's. We know, from working in the planets' frame, that when the spaceship reaches B, B's clock simply reads L/v. Additionally, the spaceship's clock reads $L/\gamma v$, because it runs slow by a factor of γ when viewed in the planets' frame. See Fig. 1.31.

How would someone on the spaceship quantitatively explain why B's clock reads L/v (which is *more* than its own $L/\gamma v$) when the spaceship and B coincide, considering that the spaceship sees B's clock running *slow*? Shouldn't a spaceship person conclude that B's clock reads only $(L/\gamma v)/\gamma = L/\gamma^2 v$?

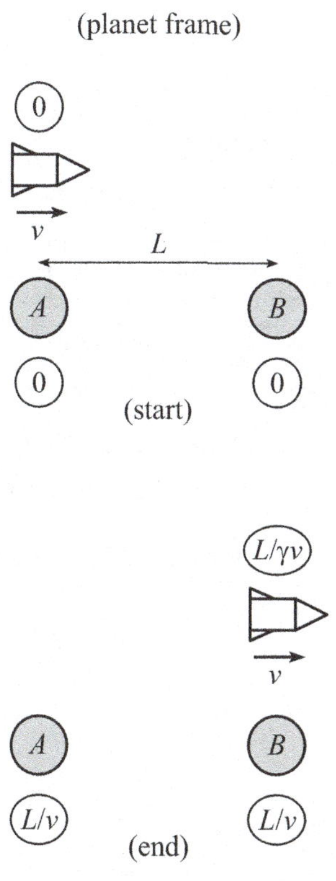

Figure 1.31

Solution: First note that if you want to work entirely in the spaceship's frame and not obtain the above $L/\gamma v$ result by using time dilation from the planets' point of view, you can use the fact that the spaceship says the distance between the planets is L/γ due to length contraction. Since the planets travel at speed v, the process therefore takes a time of $L/\gamma v$ on the spaceships's clock.

The resolution to the apparent paradox is the "head start" that B's clock has over A's clock, as seen in the spaceship frame. From Eq. (1.8), we know that in the spaceship frame, B's clock reads Lv/c^2 more than A's, because B is the rear person as they move leftward past the spaceship. See Fig. 1.32.

Therefore, what a person on the spaceship says is: "My clock advances by $L/\gamma v$ during the whole process. I see B's clock running slow by a factor γ, so I see B's clock advance by only $(L/\gamma v)/\gamma = L/\gamma^2 v$. However, B's clock started not at zero but at Lv/c^2. The final reading on B's clock when it reaches me is its initial reading of Lv/c^2 plus the elapsed time of $L/\gamma^2 v$, which gives

$$\frac{Lv}{c^2} + \frac{L}{\gamma^2 v} = \frac{L}{v}\left(\frac{v^2}{c^2} + \frac{1}{\gamma^2}\right) = \frac{L}{v}\left(\frac{v^2}{c^2} + \left(1 - \frac{v^2}{c^2}\right)\right) = \frac{L}{v}, \tag{1.22}$$

as we wanted to show." The final reading on A's clock is only $L/\gamma^2 v$, but that isn't relevant here.

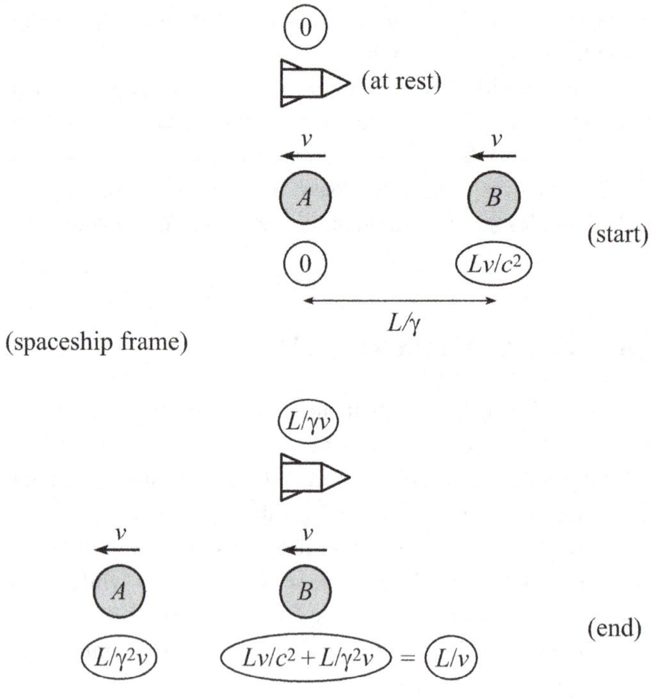

Figure 1.32

Note that in the sentence preceding Eq. (1.22), we used the phrase "on B's clock when it reaches me," as opposed to "on B's clock when I reach it." That latter would be incorrect, because we are working in the spaceship frame, where the spaceship is (of course) at rest; the two spaceships are vertically aligned in Fig. 1.32. Since the spaceship isn't moving, it therefore can't do any "reaching." B, however, is moving, so it can reach the spaceship.

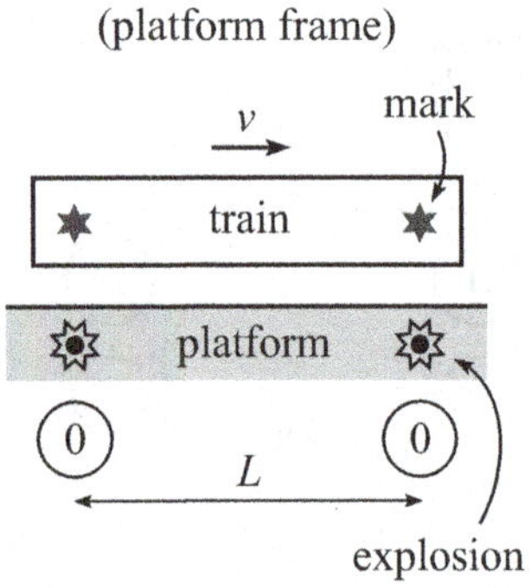

Figure 1.33

Example 2 (Explaining length contraction): Two paint bombs lie on a train platform, a distance L apart. As a train moves rightward at speed v, the paint bombs explode simultaneously (in the platform frame) and leave marks on the train; see Fig. 1.33. Due to the length contraction of the train, we know that the marks on the train are a distance γL apart when viewed in the train frame, because this distance is what is length-contracted down to the given distance L in the platform frame.

How would someone on the train quantitatively explain why the marks are a distance γL apart, considering that in the train frame the paint bombs are only a distance L/γ apart due to length contraction? (This example is the quantitative treatment of Example 3 on page 27.

Solution: The resolution to the apparent paradox is that the explosions do not occur simultaneously in the train frame. As the platform rushes past the train, the "rear" (right) paint bomb explodes before the "front" (left) one explodes.[8] The front one then gets to travel farther by the time it explodes and leaves its mark. The distance between the marks is therefore larger than the L/γ distance that you might naively expect. Let's be quantitative about this, to show that it all works out.

[8]Since we'll be working in the train frame here, we'll use the words "rear" and "front" in the way that someone on the train uses them as she watches the platform rush by. That is, if the train is heading east with respect to the platform, then from the point of view of the train, the platform is heading west. So the western paint bomb on the platform is the front one, and the eastern paint bomb is the rear one. They therefore have the opposite orientation compared with the way that someone on the platform labels the rear and front of the train. Using the same orientation would entail writing the phrase "front clock ahead" below, which would make me cringe.

For concreteness, let the two paint bombs contain clocks that read zero when they explode (they are synchronized in the platform frame), as shown above in Fig. 1.33. Then in the train frame, the front bomb's clock reads only $-Lv/c^2$ when the rear bomb explodes at the instant it reads zero; see Fig. 1.34. This is the rear-clock-ahead result from Eq. (1.8). The front bomb's clock must therefore advance by a time of Lv/c^2 before it explodes, because it is a frame-independent fact that it explodes when it reads zero. However, the train sees the bombs' clocks running slow by a factor γ, so in the train frame the front bomb explodes a time $\gamma Lv/c^2$ after the rear bomb explodes. (Note that it is the *elapsed* time of $0 - (-Lv/c^2)$ that gets dilated, and not the *reading* of Lv/c^2. Elapsed times get dilated, not readings.) During the time of $\gamma Lv/c^2$ in the train frame, the platform moves a distance $(\gamma Lv/c^2)v = \gamma Lv^2/c^2$ relative to the train, as shown in Fig. 1.34.

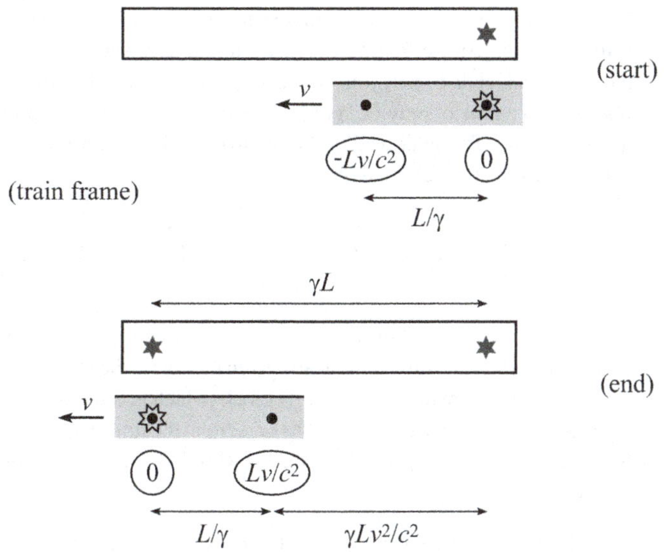

Figure 1.34

Therefore, what a person on the train says is: "Due to length contraction, the distance between the paint bombs is L/γ. The front (left) bomb is therefore a distance L/γ ahead of the rear (right) bomb when the latter explodes. The front bomb then travels an additional distance of $\gamma Lv^2/c^2$ by the time it explodes, at which point it is a distance of

$$\frac{L}{\gamma} + \frac{\gamma Lv^2}{c^2} = \gamma L\left(\frac{1}{\gamma^2} + \frac{v^2}{c^2}\right) = \gamma L\left(\left(1 - \frac{v^2}{c^2}\right) + \frac{v^2}{c^2}\right) = \gamma L \qquad (1.23)$$

ahead of the rear bomb's mark, as we wanted to show."

Example 3 (A passing stick): A stick with proper length L moves past you at speed v, as shown in Fig. 1.35. There is a time interval between the front end coinciding with you and the back end coinciding with you. What is this time interval in:

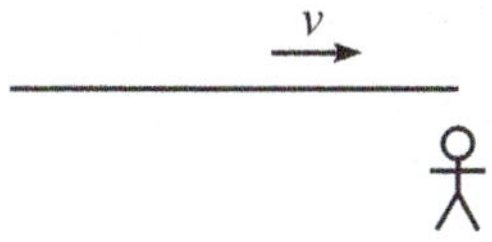

Figure 1.35

(a) your frame? (Calculate this by working in your frame.)

(b) your frame? (Work in the stick's frame.)

(c) the stick's frame? (Work in your frame. This is the tricky one.)

(d) the stick's frame? (Work in the stick's frame.)

Solution:

(a) The stick has length L/γ in your frame, and it moves with speed v. Therefore, the time taken in your frame to cover the distance L/γ is $(L/\gamma)/v = L/\gamma v$.

(b) The stick sees you fly by at speed v. The stick has length L in its own frame, so the time elapsed in the stick frame is L/v. During this time, the stick sees the watch on your wrist run slow by a factor γ. Therefore, a time of only $(L/v)/\gamma = L/\gamma v$ elapses on your watch, in agreement with part (a).

Logically, the two solutions in parts (a) and (b) differ in that one uses length contraction while the other uses time dilation. Mathematically, they differ simply in the order in which the divisions by γ and v occur.

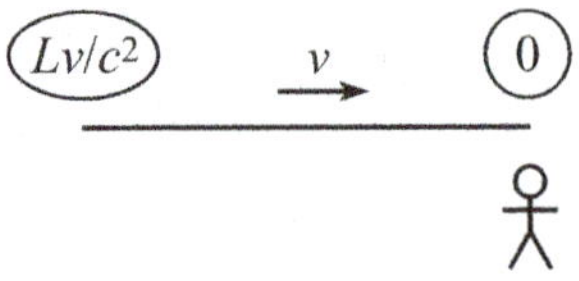

Figure 1.36

(c) Due to the rear-clock-ahead effect, you see the rear clock on the stick showing a time of Lv/c^2 more than the front clock. If we assume for concreteness that the front clock on the stick reads zero when it passes you, then at this same instant (in your frame), the rear clock on the stick reads Lv/c^2; see Fig. 1.36. In addition to this initial reading on the rear clock, more time will of course elapse on it by the time it reaches you. As we found in part (a), the time in your frame is $L/\gamma v$, because the stick has length L/γ in your frame and travels at speed v. But the stick's clocks run slow, so a time of only $(L/\gamma v)/\gamma = L/\gamma^2 v$ elapses on the rear clock by the time it reaches you. The final reading on the rear clock when it passes you is its initial reading plus the time elapsed on it, which gives

$$\frac{Lv}{c^2} + \frac{L}{\gamma^2 v} = \frac{L}{v}\left(\frac{v^2}{c^2} + \frac{1}{\gamma^2}\right) = \frac{L}{v}\left(\frac{v^2}{c^2} + \left(1 - \frac{v^2}{c^2}\right)\right) = \frac{L}{v}. \tag{1.24}$$

(This is the same calculation as in Eq. (1.22).) Since the two clocks are synchronized (that is, they show the same time at any given instant) in the stick frame, the difference between the initial reading on the front clock (which is zero) and the final reading on the rear clock (which we just found to be L/v) is the time elapsed in the stick frame. So the time elapsed is L/v, in agreement with the quick calculation that follows in part (d).

(d) The stick sees you fly by at speed v. The stick has length L in its own frame, so the time elapsed in the stick frame is simply L/v. (Of course, we already knew this from solving part (b).)

Example 4 (Photon on a train): A train with proper length L has clocks at the front and back. A photon is fired from the back to the front. Working in the train frame, we can easily say that if the photon leaves the back of the train when a clock there reads zero, then it arrives at the front when a clock there reads L/c.

Now consider this setup in the ground frame, where the train travels by at speed v. Rederive the above frame-independent result (namely, if the photon leaves the back of the train when a clock there reads zero, then it arrives at the front when a clock there reads L/c) by working *only* in the ground frame.

Solution: In the ground frame the train has length L/γ, so the photon starts the process a distance L/γ behind the front of the train. It must close this gap at a relative speed of $c - v$, because the front of the train is receding at speed v. (Simple subtraction of these speeds is valid because they are both measured with respect to the ground, and we are looking for the relative speed as viewed by the ground.) The time elapsed in the ground frame is therefore $(L/\gamma)/(c - v)$. But the ground frame sees the train clocks run slow, so only $(L/\gamma^2)/(c - v)$ elapses on any given train clock.

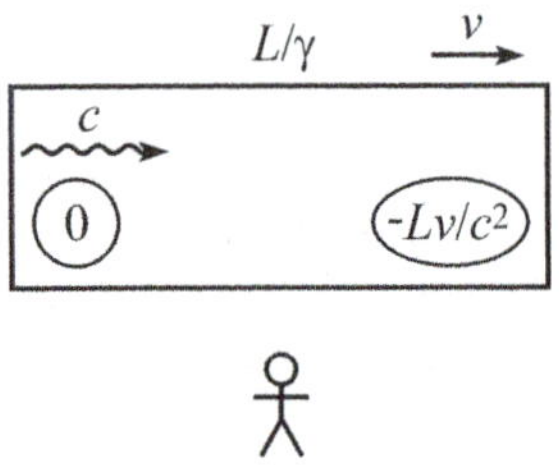

Figure 1.37

As viewed in the ground frame, when the photon is released next to the back clock when it reads zero, the front clock reads $-Lv/c^2$ due to the rear-clock-ahead effect. Fig. 1.37 shows the initial picture. The reading on the front clock when the photon hits it is the initial reading of $-Lv/c^2$ plus the time elapsed on it, which we found above to be $(L/\gamma^2)/(c - v)$. The final reading on the front clock is therefore

$$-\frac{Lv}{c^2} + \frac{L}{\gamma^2(c-v)} = -\frac{Lv}{c^2} + \frac{L\left(1-\frac{v^2}{c^2}\right)}{c\left(1-\frac{v}{c}\right)} = -\frac{Lv}{c^2} + \frac{L}{c}\left(1+\frac{v}{c}\right) = \frac{L}{c}, \tag{1.25}$$

as desired.

At this point, you might want to look at the "Qualitative relativity questions" in Appendix A, just to make sure there aren't any misconceptions lingering in your mind. The first half of the collection (through Question 25) deals with material we've covered so far.

1.5 Velocity addition

It's now time to derive the *velocity-addition formula*, which we have mentioned a few times in this chapter. If you want, you can consider the formula to be a fourth fundamental effect, in addition to rear clock ahead, time dilation, and length contraction.

Consider the following setup. A ball moves at speed u with respect to a train, and the train moves at speed v with respect to the ground (in the same direction as the motion of the ball; see Fig. 1.38). What is the speed V of the ball with respect to the ground? The result is the desired velocity-addition formula.

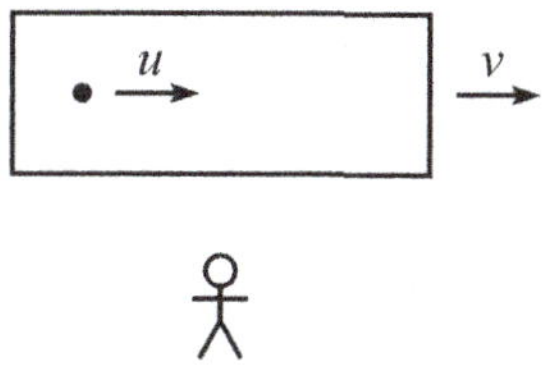

Figure 1.38

In the nonrelativistic limit (that is, when u and v are small compared with c), V is simply equal to $u + v$, as we know very well from everyday experience (although technically this result isn't *exactly* correct, as we'll see below). But the simple $u + v$ answer can't be correct for larger speeds, because if, for example, u and v are both equal to $(0.9)c$, then $u + v = (1.8)c$. This is certainly incorrect, because it is larger than c. The fact of the matter is that it is impossible for an object (or at least any object we can interact with) to move faster than c. There are various ways to demonstrate this. One is that it would require an infinite amount of energy to accelerate an object up to speed c. We'll see why in Chapter 3.

To find the correct general expression for V (that is, to derive the velocity-addition formula), let's consider a concrete setup where we look at the time it takes a ball to travel from the back of a train to the front. Let the train have proper length L. Our strategy for finding V will be to generate two different expressions for the time of this process in the ground frame. Equating these two expressions will allow us to solve for V. The setup is shown in Fig. 1.39. We've put the ball outside the train to emphasize that the speed V is with respect to the ground.

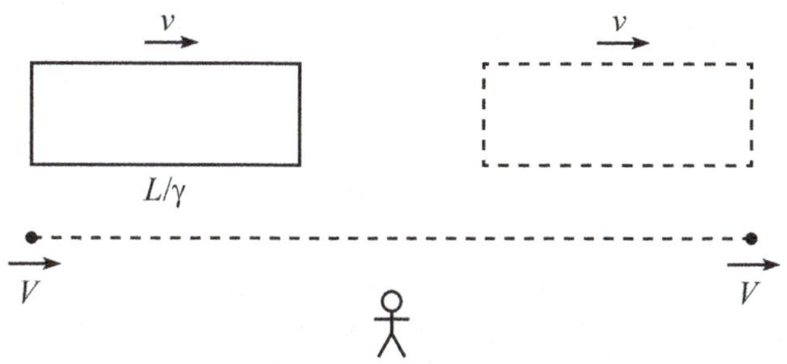

Figure 1.39

The first expression for the time (for the ball to go from the back of the train to the front) in the ground frame is found by noting that (a) the initial gap between the ball and the front of the train is L/γ due to length contraction, and (b) this gap is closed at a rate $V - v$, because this is the relative speed of the ball and the front of the train, as viewed in the ground frame. As we've seen a number of times in this chapter, it is legal to simply subtract these speeds, because they are both measured with respect to the same frame (the ground frame). The time in the ground frame for the ball to reach the front of the train is therefore

$$t_g = \frac{L/\gamma}{V - v}, \tag{1.26}$$

where V is not yet known. The γ factor here is associated with the speed v of the train (that is, not with V or u).

The second expression for the time in the ground frame is found by looking at a particular clock on the train and using time dilation. Assume that a clock at the back of the train reads zero when the ball is thrown. Then by working in the frame of the train, we quickly see that a clock at the front reads L/u when the ball gets there. (No relativity needed for this.) Now look at things in the ground frame. The readings we just mentioned are frame independent, so the starting and ending readings in the ground frame must be the ones shown in Fig. 1.40. At the start, the rear clock reads zero (frame independent), so the front clock reads $-Lv/c^2$, due to the rear-clock-ahead effect. And at the finish, the front clock reads L/u (frame independent), so the rear clock reads $L/u + Lv/c^2$, due to the rear-clock-ahead effect.

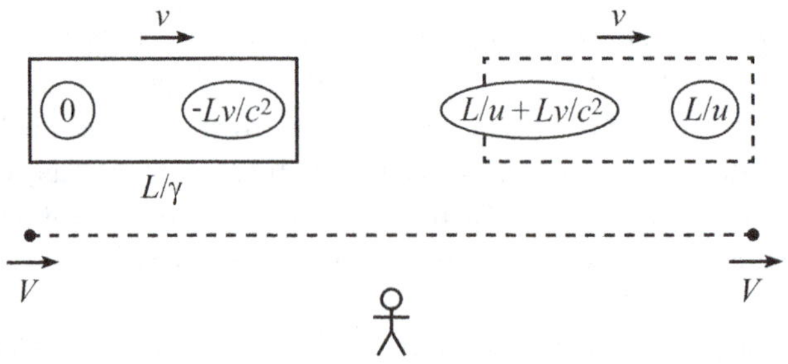

Figure 1.40

Looking at the rear clock (the front clock works just as well), we see that the time elapsed on this clock is $L/u + Lv/c^2$. Since we are looking at a single clock, it is legal to use time dilation, which tells us that the time elapsed on the ground is longer by a factor γ. (Again, this γ factor is associated with the speed v of the train.) So our second expression for the time in the ground frame is

$$t_g = \gamma\left(\frac{L}{u} + \frac{Lv}{c^2}\right). \tag{1.27}$$

Note that we *cannot* say that since the time elapsed on the train is L/u, the time elapsed on the ground should be $\gamma(L/u)$. This is incorrect because in the ground frame, L/u isn't the time elapsed on a *single* clock. Fig. 1.40 tells us that L/u is the final reading on the *front* clock minus the initial reading on the *rear* clock. It isn't legal to use time dilation when comparing the readings on two different clocks. In the above correct reasoning, we compared the readings on a *single* clock (either the front clock *or* the rear clock).

Now all we have to do is equate the two times in Eqs. (1.26) and (1.27) and then solve for V. There's a little algebra involved, but it isn't so bad. We have

$$\frac{L/\gamma}{V-v} = \gamma\left(\frac{L}{u} + \frac{Lv}{c^2}\right) \implies \frac{1}{\gamma^2}\frac{1}{\frac{1}{u}+\frac{v}{c^2}} = V - v$$

$$\implies V = \frac{1-\frac{v^2}{c^2}}{\frac{1}{u}+\frac{v}{c^2}} + v = \frac{1+\frac{v}{u}}{\frac{1}{u}+\frac{v}{c^2}}$$

$$\implies \boxed{V = \frac{u+v}{1+\frac{uv}{c^2}}} \qquad \text{(velocity-addition formula)} \qquad (1.28)$$

This is the *longitudinal* velocity-addition formula, relevant when adding velocities that are parallel to each other. There is also a *transverse* velocity-addition formula, relevant when the two given velocities are perpendicular. We'll derive that in Section 2.2.2. Note that we used all three of the fundamental effects (rear clock ahead, time dilation, and length contraction) in the above derivation.

Given a velocity v, the letter β is used to denote the ratio of v to c. That is, $\beta \equiv v/c$. Sometimes a subscript is added, as in β_v or β_1, if there is ambiguity about which speed is being referred to. In terms of β's, the velocity-addition formula for the speeds v_1 and v_2 takes the form,

$$\beta = \frac{\beta_1 + \beta_2}{1 + \beta_1\beta_2}\,. \qquad (1.29)$$

This expression is often simpler to work with than Eq. (1.28), because there are no c's to clutter things up. Alternatively, we will often simply drop the c's in calculations, to keep things from getting too messy. Eq. (1.28) then becomes $V = (u+v)/(1+uv)$. The c's can always be put back in at the end, by figuring out where they need to go in order to make the units correct (which is usually easy to see). Equivalently, you can just pretend that V, u, and v are actually β_V, β_u, and β_v.

Two other derivations of Eq. (1.28) are given in Exercises 1.47 and 1.48. They are similar to the above derivation, in the following way. There are (at least) three different methods for finding the time (for the ball to travel from the back of the train to the front) in the ground frame: (1) the front of the train has an initial head start over the ball, and we can find the time it takes the ball to close this gap down to zero, (2) we can apply time dilation to a given train clock, or (3) we can apply time dilation to the ball's clock. The above derivation used the first and second of these, Exercise 1.47 uses the first and third, and Exercise 1.48 uses the second and third.

Let's look at some of the properties of the V in Eq. (1.28).

- V is symmetric in u and v. We'll explain why when we discuss Fig. 1.41 below.
- If u and v are small compared with c, then $V \approx u + v$, because the uv/c^2 term in the denominator is negligible. This result makes sense, because we know very well that V equals $u + v$ for everyday speeds.
- If $u = c$ or $v = c$, then we obtain $V = c$. This is correct, because anything that moves with speed c in one frame moves with speed c in another.
- The maximum (or minimum) of V in the square region defined by the two inequalities, $-c \le u \le c$ and $-c \le v \le c$, equals c (or $-c$). This is due to the fact that the partial derivatives $\partial V/\partial u$ and $\partial V/\partial v$ are never zero in the interior

of the region (or anywhere, for that matter), as you can verify. The extrema must therefore occur on the boundary, and you can show that V always takes on the value of c or $-c$ there. (There are two exceptions at the two $u = -v = \pm c$ corners of the region, where V takes on the undefined value of $0/0$.)

The last of these bullet points tells us that if we take any two velocities that are less than c and add them according to Eq. (1.28), then we will obtain a velocity that is again less than c. If you don't want to rely on partial derivatives, you can alternatively demonstrate this fact with the following inequalities:

$$\frac{u+v}{1+uv/c^2} < c \iff u+v < c+uv/c \iff u(1-v/c) < c-v$$
$$\iff u < \frac{c-v}{(c-v)/c} \iff u < c. \tag{1.30}$$

Assuming that $c - v$ isn't zero (because otherwise we would have divided by zero), all of the above steps are reversible. So if we start with both $u < c$ and $v < c$, then we end up with $(u+v)/(1+uv/c^2) < c$. This means that no matter how much you keep accelerating an object (that is, no matter how many times you give the object a speed u with respect to the frame moving at speed v that it was just in), you can't bring the speed up to c. As mentioned earlier, this also follows from energy considerations, as we'll see in Chapter 3.

For a bullet, a train, and a gun,
Adding the speeds can be fun.
Take a trip down the path
Paved with Einstein's new math,
Where a half plus a half isn't one.

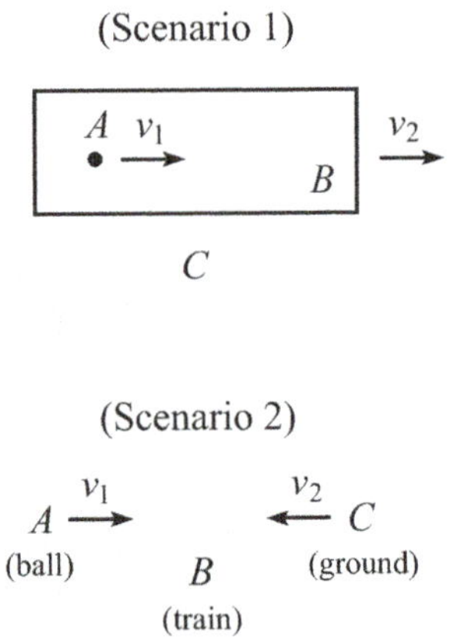

Figure 1.41

In addition to applying to the above "ball on train on ground" setup, there is another common scenario where the velocity-addition formula applies. Consider the two scenarios shown in Fig. 1.41. The first is the original "ball on train on ground" setup. These two scenarios are actually identical. The only difference is that the first one is shown in C's frame, while the second one is shown in B's frame. They are indeed identical, because in the first scenario B sees A approach him rightward at speed v_1 (in agreement with the second scenario). Similarly, in the first scenario, B sees C approach him leftward at speed v_2 (again in agreement with the second scenario). Basically, in the second scenario, A is the ball, B is the train, and C is the ground. Therefore, if the goal is to find the velocity of A with respect to C in the second scenario, then since we showed above that the velocity-addition formula applies in the first scenario, we conclude that it also applies to the second scenario.

We noted above that the V in Eq. (1.28) is symmetric in u and v, or equivalently in v_1 and v_2 in the present notation. This is clear in the second scenario in Fig. 1.41, because switching v_1 and v_2 is equivalent to switching A and C, and because the speed of A as viewed by C is the same as the speed of C as viewed by A (see the first remark below). Therefore, since the two scenarios in Fig. 1.41 are equivalent, V must also be symmetric in u and v in the original setup in Fig. 1.38.

In the second scenario in Fig. 1.41 (as in the first), the velocity-addition formula tells us the answer to the question, "What is the relative speed of A and C, *as viewed by C*?" The answer is $(v_1 + v_2)/(1 + v_1 v_2/c^2)$. However, the formula does *not* apply if we ask the more mundane question, "What is the relative speed of A and C, *as viewed by B*?" The answer to this is simply $v_1 + v_2$. In short, if the two velocities are given with respect to the *same* observer, B, and if you are asking for the relative velocity as measured by B, then you just have to add the velocities. But if you are asking for the relative velocity as measured by A or C, then you have to use the velocity-addition formula. Equivalently, in the first scenario in Fig. 1.41, it makes no sense to naively add velocities that are measured with respect to different observers. Doing so would involve adding things that are measured in different coordinate systems, which is meaningless. Taking the velocity of A with respect to B and adding it to the velocity of B with respect to C, hoping to obtain the velocity of A with respect to C, is invalid.

We see that relativistically the question, "What is the relative speed of A and C?" is ambiguous. We have to finish it with "... as measured by such and such a person." Nonrelativistically, though, there is no ambiguity. The answer is simply $v_1 + v_2$ in any frame.

Note that the $v_1 + v_2$ relative speed of A and C, as viewed by B, in Fig. 1.41 can certainly be greater than c. If I see a ball heading toward me at $(0.9)c$ from the right, and another one heading toward me at $(0.9)c$ from the left, then the relative speed of the balls in my frame is $(1.8)c$. In the frame of one of the balls, however, Eq. (1.28) gives the relative speed as $(1.8/1.81)c \approx (.9945)c$. This is correctly less than c, because you (or one of the balls, in this case) can never see another object (the other ball) move with a speed greater than c. This restriction doesn't rule out the above result of $(1.8)c$, because this is simply the rate at which the gap between the balls is closing. A gap isn't an actual object, so there is nothing wrong with the length of the gap decreasing at a rate faster than c. In the extreme, if one photon heads rightward and another one heads leftward, then as measured by you, their relative speed is $2c$. That is the rate at which the gap between them is closing in your frame.

REMARKS:

1. If two people, A and B, are moving with respect to each other in one dimension, why is the speed of B as viewed by A equal to the speed of A as viewed by B? This equality seems quite reasonable, of course, but how do we prove it rigorously from basic principles (that is, ignoring what we've derived about velocity addition)? The proof follows directly from the first postulate of relativity – that all inertial frames are equivalent (which implies that there is no preferred location or direction in space). Let's assume that the relative speed measured by the left person is larger than the relative speed measured by the right person. This implies that there is a preferred direction in space; apparently people on the left always measure a larger speed. This violates the first postulate. Likewise if the left person measures a smaller speed. The two speeds must therefore be equal.

2. Strictly speaking, the signs in the numerator and denominator in Eq. (1.28) are always plus signs, assuming that v_1 and v_2 are the signed *velocities* of the objects in the first scenario in Fig. 1.41. However, in practice it is often more convenient to let v_1 and v_2 be *speeds* (which are always positive). In this case, the same sign appears in the numerator and denominator in Eq. (1.28), and the correct choice of sign is determined by the sign you would use in the simple nonrelativistic case. For example, the nonrelativistic speed of A with respect to C in the first scenario in Fig. 1.41 is simply the sum of the speeds, $v_1 + v_2$, which means we must use positive signs in Eq. (1.28). If the ball is instead thrown backward on the train, then the nonrelativistic speed of A with respect to C is the difference of the speeds, $-v_1 + v_2$ (or $|-v_1 + v_2|$ if this quantity is negative), which means we must use negative signs in Eq. (1.28). In any case, the numerator in the relativistic case is always the naive nonrelativistic answer. If you get confused about the signs, it's best to just plug is some actual numbers for v_1 and v_2, to see what's going on.

3. For everyday speeds, the nonrelativistic and relativistic results for the speed of A as viewed by C in Fig. 1.41 are essentially the same. If $v_1 = 50\,\text{m/s}$ and $v_2 = 30\,\text{m/s}$, then the nonrelativistic result is simply $v_1 + v_2 = 80\,\text{m/s}$, while Eq. (1.28) gives the relativistic result as $(80\,\text{m/s})(1 - 1.67 \cdot 10^{-14})$. The nonrelativistic result is therefore incorrect by less than 2 parts in 10^{14}. That's plenty good for me.

4. The sum $v_1 + v_2$ doubles as the answer to two different questions concerning Fig. 1.41. It is the *approximate* (nonrelativistic) answer to the question, "What is the relative speed of A and C, as viewed by C?" It is also the *exact* (relativistic or nonrelativistic) answer to the question, "What is the relative speed of A and C, as viewed by B?" ♣

Let's now do two examples. The second one has a little bit of everything we've done so far – rear clock ahead, time dilation, length contraction, and velocity addition.

Example 1 (Passing trains, again): Consider again the scenario in the "Passing trains" example in Section 1.3.3.

(a) How long, as viewed by A and as viewed by B, does it take for A to overtake B?

(b) Let event E_1 be "the front of A passing the back of B", and let event E_2 be "the back of A passing the front of B." Person D walks at constant speed from the back of B to the front (see Fig. 1.42), such that he coincides with both events, E_1 and E_2. How long does the "overtaking" process take, as viewed by D?

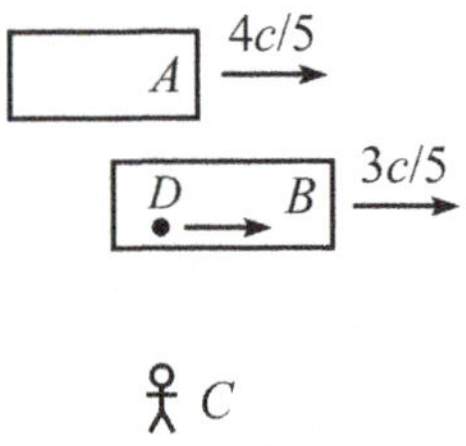

Figure 1.42

Solution:

(a) First consider B's point of view. From the velocity-addition formula, B sees A move with speed

$$u = \frac{\frac{4c}{5} - \frac{3c}{5}}{1 - \frac{4}{5} \cdot \frac{3}{5}} = \frac{5c}{13}. \tag{1.31}$$

This expression involves minus signs because the naive nonrelativistic relative speed involves subtracting the speeds. The γ factor associated with the speed u is $\gamma_{5/13} = 13/12$, as you can check. Therefore, B sees A's train contracted to a length $12L/13$. During the overtaking, A must travel a distance equal to the sum of the lengths of the trains in B's frame (see Fig. 1.43), which is $L + 12L/13 = 25L/13$. Since A moves with speed $5c/13$, the total time in B's frame is

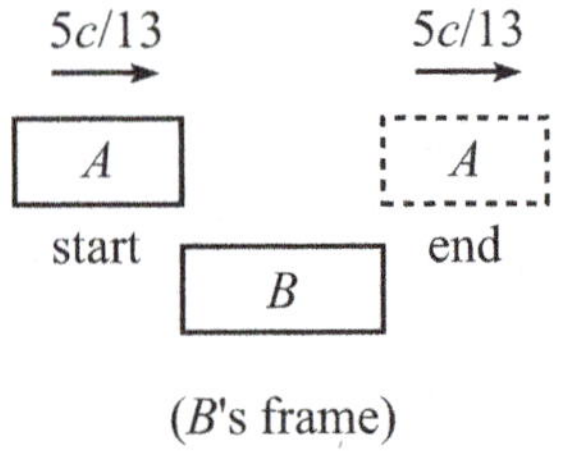

Figure 1.43

$$t_B = \frac{25L/13}{5c/13} = \frac{5L}{c}. \tag{1.32}$$

The exact same reasoning holds from A's point of view, so we have $t_A = t_B = 5L/c$.

(b) Look at things in D's frame. D is at rest, and the two trains move with equal and opposite speeds v (to be determined), because this causes the second event E_2 to be correctly be located at D; see Fig. 1.44. Our setup is equivalent to the second scenario in Fig. 1.41, so the speed of A as viewed by B is the relativistic addition of v with itself. But from part (a), we know that this relative speed equals $5c/13$. Therefore (with $\beta \equiv v/c$),

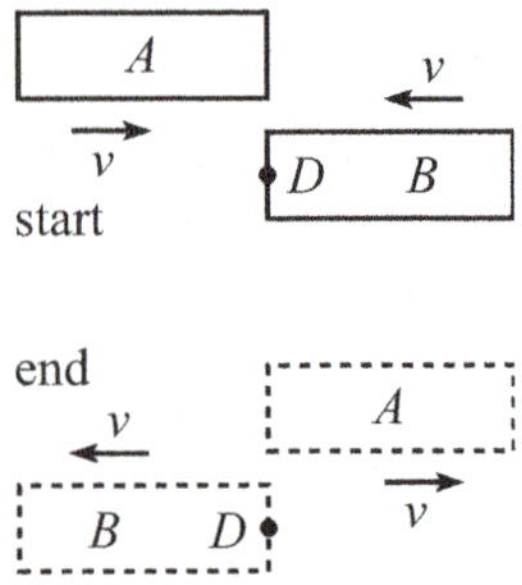

Figure 1.44

$$\frac{v + v}{1 + v^2/c^2} = \frac{5c}{13} \implies 5\beta^2 - 26\beta + 5 = 0$$
$$\implies (5\beta - 1)(\beta - 5) = 0 \implies \beta = \frac{1}{5}, \tag{1.33}$$

which gives $v = c/5$. We have ignored the unphysical solution, $\beta = 5 \implies v = 5c$, because v can't exceed c. The γ factor associated with $v = c/5$ is $\gamma_{1/5} = 5/2\sqrt{6}$. So D sees both trains contracted to a length $2\sqrt{6}L/5$. During the process, each train must travel a distance equal to its length (as shown in Fig. 1.44) because both events, E_1 and E_2, take place right at D. The time in D's frame is therefore

$$t_D = \frac{2\sqrt{6}L/5}{c/5} = \frac{2\sqrt{6}L}{c}. \tag{1.34}$$

Remarks: There are a few double checks we can perform. The speed of D with respect to the ground can be obtained either via B's frame by relativistically adding

$3c/5$ and $c/5$, or via A's frame by relativistically subtracting $c/5$ from $4c/5$. You can check that these methods give the same answer (as they must), namely $5c/7$. (The $v = c/5$ speed can actually be determined by demanding that these methods give the same answer, instead of by Eq. (1.33).) The γ factor between the ground and D is therefore $\gamma_{5/7} = 7/2\sqrt{6}$. We can then use time dilation to say that someone on the ground sees the overtaking take a time of $(7/2\sqrt{6})t_D$. (We can say this because both events happen right at D.) Using the t_D from Eq. (1.34), this gives a ground-frame time of $7L/c$, in agreement with our earlier result in Eq. (1.21). Likewise, since the γ factor between D and either train is $\gamma_{1/5} = 5/2\sqrt{6}$, the time of the overtaking as viewed by either A or B is $(5/2\sqrt{6})t_D = 5L/c$, in agreement with Eq. (1.32).

Note that we *cannot* use simple time dilation to relate the time on the ground to the time on either train, because the two events don't happen at the same place in either of the train frames (or in the ground frame). But since both events happen at the same place in D's frame, namely right at D, it is legal to use time dilation to go from D's frame to any other frame. And when doing so, the relevant γ factor always appears in front of t_D. ♣

Example 2 (Clock readings on trains):

(a) Two trains each have proper length L and travel on parallel tracks. They both move with speed v with respect to the ground, one rightward and one leftward. You notice that clocks at the fronts of the trains both read zero when the fronts coincide, as shown in Fig. 1.45. What do clocks at the backs of the trains read when the backs eventually coincide? Answer this by working in the ground frame.

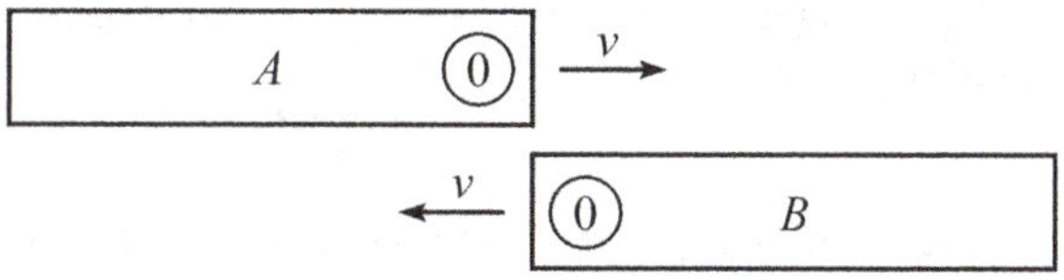

Figure 1.45

(b) Again find the readings on the two clocks at the backs of the trains when the backs coincide, but now by working in the frame of train B. To keep things from getting messy, you can let v take on the particular value of $c/2$ for this part.

Solution:

(a) At the given initial instant in the ground frame, a ground observer sees both of the rear clocks reading Lv/c^2, due to the rear-clock-ahead effect. The backs of the trains eventually coincide at the same location the fronts coincided, after each train has traveled a distance L/γ; this is the contracted length of each train in the ground frame. The time elapsed in the ground frame is therefore $(L/\gamma)/v$. But the ground observer sees the train clocks (in particular, the rear clocks) running slow by a factor γ. So the time elapsed on each rear clock is $(L/\gamma v)/\gamma$. The final reading on each rear clock is its initial reading of Lv/c^2 plus the elapsed time of $L/\gamma^2 v$. The final reading is therefore (as we've seen a few times in other examples)

$$\frac{Lv}{c^2} + \frac{L}{\gamma^2 v} = \frac{Lv}{c^2} + \left(1 - \frac{v^2}{c^2}\right)\frac{L}{v} = \frac{L}{v}. \tag{1.35}$$

REMARK: A quick way to see why this result is so simple is the following. Imagine a person standing at rest on the ground, at the initial location of the fronts of the trains. Since the person is at rest, the backs of the trains will be located at the person when they eventually coincide. Therefore, in the frame of one of the trains, the person simply travels the length L of the train (by the time the backs and the person all coincide), at speed v (because that is the relative speed of the ground and a train). So the time in the train frame is L/v. This elapsed time is the desired reading on the back clock, because the clock started at zero in the train frame, since clocks on a given train are synchronized in that train's frame. ♣

(b) The given setup is equivalent to the second scenario in Fig. 1.41, so the speed V of one train as viewed by the other is the relativistic addition of $v \equiv c/2$ with itself:

$$V = \frac{\frac{c}{2} + \frac{c}{2}}{1 + \frac{1}{2} \cdot \frac{1}{2}} = \frac{4c}{5}. \tag{1.36}$$

Since $\gamma_{4/5} = 5/3$, the contracted length of A in B's frame is $3L/5$. The setup in B's frame is therefore shown in Fig. 1.46 (ignore A's rear clock for now).

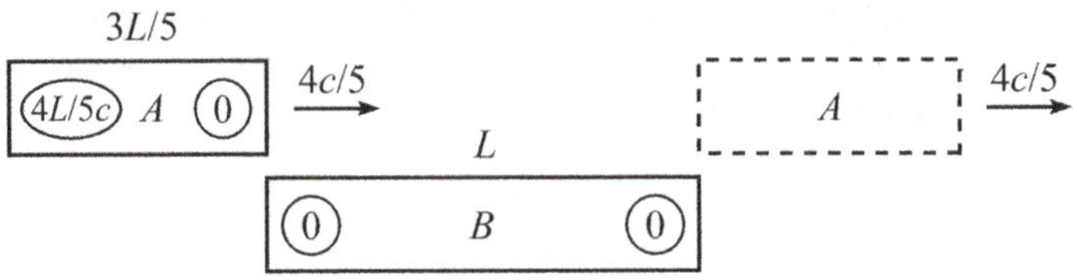

Figure 1.46

Let's first find the final reading on B's rear (right) clock when A's rear (left) clock reaches it. (B's right clock starts at zero, as shown, because B's clocks are synchronized in B's frame.) This is obtained by noting that A must travel a distance equal to the sum of the lengths of the trains. It does this at speed $4c/5$, so the final reading on B's right clock is

$$0 + \frac{3L/5 + L}{4c/5} = \frac{2L}{c}. \tag{1.37}$$

This agrees with the result in part (a) when $v = c/2$, as it must, because the reading is frame independent.

To find the final reading on A's left clock, we must remember that it starts the process reading $LV/c^2 = 4L/5c$, due to the rear-clock-ahead effect. It then advances by $(2L/c)/\gamma_{4/5}$, due to time dilation and due to the fact that $2L/c$ is the time that elapses in B's frame. (B sees A's clocks run slow, so we must divide by γ.) The final reading on A's left clock is therefore

$$\frac{4L}{5c} + \frac{3}{5} \cdot \frac{2L}{c} = \frac{2L}{c}. \tag{1.38}$$

Again this agrees with the result in part (a) when $v = c/2$. Note that while there is a symmetry between A's and B's rear clocks in the ground frame in part (a), there is no such symmetry between the clocks here in B's frame. The calculation for A's clock here is therefore different from the calculation for B's clock (although we know from part (a) that the readings must end up the same).

General problem-solving strategies

We'll end this chapter by collecting all of our problem-solving strategies in one place. We'll encounter additional kinematics strategies in Chapter 2, but for the types of problems we've solved in this chapter, things generally boil down to the following ingredients. If you look back at the examples we've done, you can verify that these strategies pretty much have everything covered. You should therefore keep this checklist on the tip of your mind.

1. LOSS OF SIMULTANEITY (REAR CLOCK AHEAD): As viewed in the ground frame, the rear clock on a train reads Lv/c^2 more than the front clock.

2. TIME DILATION: If you look at a moving clock, it runs slow by a factor γ.

3. LENGTH CONTRACTION: If you look at a stick moving longitudinally, it is short by a factor γ.

4. VELOCITY-ADDITION FORMULA: This gives the speed of A as viewed by C in the two scenarios in Fig. 1.41.

5. GAP-CLOSING SPEED: $v_1 + v_2$ (or $v_1 - v_2$, depending on directions and sign conventions) is the relative speed of A and C, as viewed by B in the second scenario (and in the first) in Fig. 1.41.

6. FRAME-INDEPENDENT STATEMENTS: If information is given in one frame, but you want to work out the problem in another frame, you need to identify which statements remain true in the new frame.

7. DRAW PICTURES AND STICK TO A FRAME! Drawing pictures in a given frame helps you properly implement the above strategies. You should draw a picture whenever anything of importance happens, labeling all speeds, lengths, clock readings, etc.

1.6 Summary

In this chapter we learned about the basics of special relativity. In particular, we learned:

- A few puzzles in 19th-century physics suggested that something was amiss. These puzzles included the inconsistency of the Galilean transformations with Maxwell's equations, and the null result of Michelson and Morley in their search for the ether. Einstein's theory of relativity solved these puzzles by showing that the Lorentz transformations replace the Galilean transformations, and that light waves propagate without the need for an ether medium.

- The theory of special relativity rests on two postulates:

  1. All inertial (non-accelerating) frames are equivalent; there is no preferred reference frame.
  2. The speed of light in vacuum has the same value in any inertial frame.

- The above postulates lead to many counterintuitive effects, the most fundamental of which are:

  1. Loss of simultaneity (rear-clock-ahead): As viewed in the ground frame, the rear clock on a train reads Lv/c^2 more than the front clock, where L is the proper length of the train.
  2. Time dilation: If you look at a moving clock, it runs slow by a factor γ:

  $$t_{\text{observed}} = \gamma t_{\text{proper}}. \tag{1.39}$$

  More generally, if you are considering the time between two events, the γ goes on the side of the equation associated with the frame in which the two events happen at the same place (which is automatically the case for two ticks on a clock). The time in this frame is by definition the proper time. Remember that it is *elapsed times* that get dilated, and not *readings* on clocks.

3. Length contraction: If you look at a stick moving longitudinally, it is short by a factor γ:

$$L_{\text{observed}} = \frac{L_{\text{proper}}}{\gamma}. \tag{1.40}$$

The proper length of an object is the length as measured in the frame in which the object is at rest. In any frame, the length of an object is defined to be the distance between the ends, measured simultaneously in that frame. There is no transverse length contraction.

- A reference frame can be defined by a lattice of meter sticks and clocks. The coordinates of an event are the spatial coordinates of the lattice point where the event is located, along with the reading on the clock at that point.

- The velocity-addition formula,

$$V = \frac{u+v}{1+\frac{uv}{c^2}}, \tag{1.41}$$

gives the speed of A as viewed by C in the two scenarios in Fig. 1.41 (with $u \leftrightarrow v_1$ and $v \leftrightarrow v_2$). When written in terms of β's, the formula takes the form in Eq. (1.29).

The velocity addition formula does *not* apply when finding the relative speed of A and C, as viewed by B, in the second scenario (and also in the first) in Fig. 1.41. This gap-closing speed is simply $v_1 + v_2$.

- The various problem-solving strategies we have used throughout this chapter are listed just before this summary.

1.7 Problems

Section 1.3: The fundamental effects

1.1. **Consistency with Lv/c^2** $*$

Show that the $t_R - t_L$ difference of the times in Eq. (1.6) (where ℓ' is half the length of the train in the ground frame) is consistent with the Lv/c^2 rear-clock-ahead result (where L is the proper length of the train).

1.2. **Here and there** $*$

A train with proper length L travels past you at speed v. A person on the train stands at the front, next to a clock that reads zero. At this moment (as measured by you), a clock at the back of the train reads Lv/c^2, due to the rear-clock-ahead effect. How would you respond to the following statement:

"In the train frame, the person at the front of the train can leave the front right after the clock there reads zero, and then run to the back and get there right before the clock there reads Lv/c^2. You (on the ground) will therefore see the person simultaneously at *both* the front and the back of the train when the clocks there read zero and Lv/c^2, respectively."

1.3. **Synchronizing clocks** $*$

Two synchronized clocks, A and B, are at rest in a given frame, a distance L apart. A third clock, C, is initially located right next to A. All three clocks have initial readings of zero, and then C is moved very slowly (with speed $v \ll c$) from A to B. Show that its final reading can be made arbitrarily close to B's, by making v be sufficiently small. (The Taylor series $\sqrt{1-\epsilon} \approx 1 - \epsilon/2$ will come in handy.)

1.4. Rotated square $**$

A square with sides of proper length L flies past you at speed v, in a direction parallel to two of its sides. You stand in the plane of the square. When you see the square at its nearest point to you (see Fig. 1.47), show that it *looks* to you like it is rotated instead of contracted, and find the apparent angle of rotation. Assume that L is small compared with the distance between you and the square. (This setup is one of the few cases where we are actually concerned with the time it takes light to travel to your eye.)

Figure 1.47

1.5. Deriving length contraction $**$

The derivation of length contraction in Section 1.3.3 relied on time dilation. This problem gives a derivation that is independent of time dilation.

Assume that the rule for (longitudinal) length contraction is: "If a stick with proper length L moves at speed v with respect to you, then its length in your frame is $a_v L$." (The subscript v signifies the possible dependence of a on v.) Your eventual goal is to show that $a_v = 1/\gamma_v$, but for all you know at the moment, a_v might be larger than, less than, or equal to 1. A critical point here is that the first postulate of relativity says that all inertial frames are equivalent. So the same $a_v L$ rule must apply to everyone.

Consider the following setup. A train with proper length L moves with speed v. When the back of the train passes a tree, a photon is fired from the back toward the front. It arrives at the front when the front passes a house. What is the distance between the tree and the house (in the ground frame)?

Now look at things in the train frame. Using the tree-house proper distance you just found, write down the relation that expresses the fact that the house meets the front of the train at the same time the photon does. This will allow you to solve for a_v.

1.6. Pole in barn $*$

A pole with proper length L moves rightward with speed v through a barn, also with proper length L. Assume that initially the door at the left end of the barn is open and the door at the right end is closed. Just after the left end of the pole enters the barn, the left door closes. And just before the right end of the pole leaves the barn, the right door opens. How would you respond to the following question: "Is the pole ever completely inside the barn (with both doors closed)?"

1.7. Train in a tunnel $**$

A train and a tunnel both have proper length L. The train moves toward the tunnel at speed v. A bomb is located at the front of the train. The bomb is designed to explode when the front of the train passes the far end of the tunnel. A deactivation sensor is located at the back of the train. When the back of the train passes the near end of the tunnel, the sensor sends a signal to the bomb, telling it to disarm itself. Does the bomb explode?

1.8. Bouncing stick $**$

A stick, oriented horizontally, falls and bounces off the ground. Qualitatively, what does this setup look like in the frame of someone running by at speed v?

1.9. Seeing behind the board $**$

A ruler is positioned perpendicular to a wall, and you stand at rest with respect to the ruler and the wall. A board with proper length L moves to the right with speed

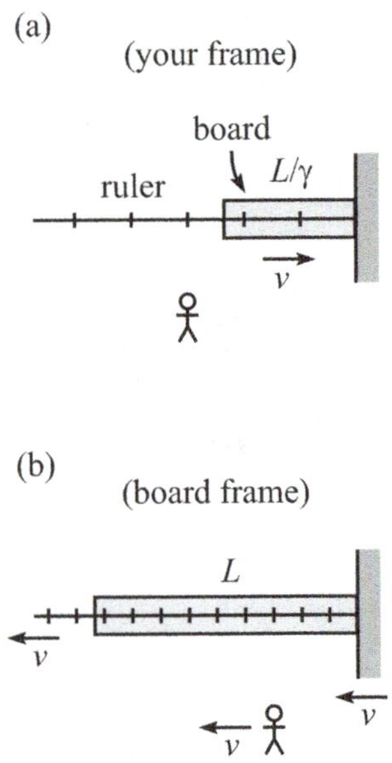

Figure 1.48

Figure 1.49

v. It travels in front of the ruler, so that it obscures part of the ruler from your view. The board eventually hits the wall. Which of the following two reasonings is correct (and what is wrong with the incorrect one)?

In your reference frame, the board is shorter than L, due to length contraction. Therefore, right before it hits the wall, you are able to see a mark on the ruler that is less than L units from the wall; see Fig. 1.48(a).

In the board's frame, the marks on the ruler are closer together, due to length contraction. Therefore, the closest mark to the wall that you will ever be able to see on the ruler is greater than L units; see Fig. 1.48(b).

1.10. **Cookie cutter** $**$

Cookie dough (chocolate chip, of course) lies on a conveyor belt that moves with speed v. A horizontal circular cookie cutter stamps out cookies as the dough rushes by beneath it. When you buy these cookies in a store, what shape are they? That is, are they squashed in the direction of the belt, stretched in that direction, or circular?

1.11. **Getting shorter** $**$

Two balls move with speed v along a line toward two people standing along the same line. The proper distance between the balls is γL, and the proper distance between the people is L. Due to length contraction, the people measure the distance between the balls to be L, so the balls pass the people simultaneously (as measured by the people), as shown in Fig. 1.49. Assume that the people's clocks both read zero at this time. If the people catch the balls, then the resulting proper distance between the balls becomes L, which is shorter than the initial proper distance of γL. Your task: By working in the frame in which the balls are initially at rest, explain how the proper distance between the balls decreases from γL to L. Do this in the following way.

(a) Draw the beginning and ending pictures for the process. Indicate the readings on both clocks in the two pictures, and label all relevant lengths.

(b) Using the *distances* labeled in your pictures, how far do the people travel? Using the *times* labeled in your pictures, how far do the people travel? Show that these two methods give the same result.

(c) Explain in words how the proper distance between the balls decreases.

1.12. **Transforming the length** $*$

A stick moves rightward with speed $3c/5$ with respect to the ground. The length of the stick *in the ground frame* is L. You move rightward with speed $c/2$ with respect to the ground. What is the length of the stick in your frame?

1.13. **Magnetic force** $***$

This problem demonstrates how the magnetic force arises from the combination of the electric force and length contraction. The interpretation of this problem (discussed in the solution) requires a familiarity with basic concepts of electricity and magnetism, although the problem itself does not.

Consider a current-carrying straight wire. The wire is neutral, that is, it has the same number of negatively charged electrons and positively charged protons in any given volume, on average. (If it weren't neutral, it would attract or repel electrons, thereby producing neutrality.) In any current-carrying wire, the current is caused by electrons moving along the wire; the protons are bolted down. So we have the situation shown in Fig. 1.50. (Ignore the charge q for a moment.) In reality, the electrons and protons are distributed throughout the wire, but we have drawn them separated for clarity. The equal and opposite charge densities $\pm\lambda_0$ (charge per meter) are indicated. The electrons' speed is v_0.

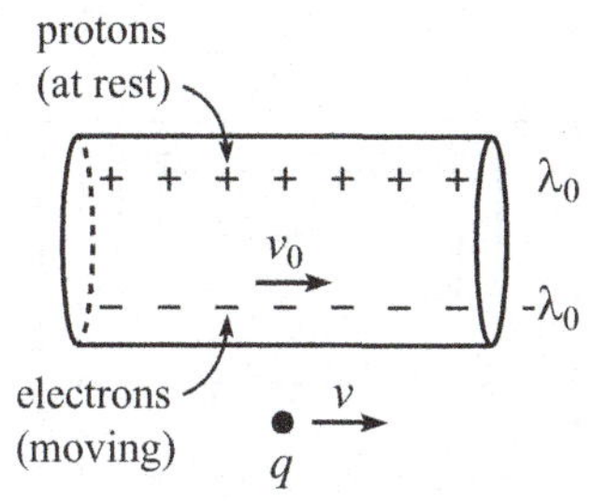

Figure 1.50

Now consider an electric charge q near the wire. If the charge q is at rest, then since the wire is neutral, q will feel no force. But let's assume that the charge q is moving to the right with speed v, as shown in Fig. 1.50. What are the charge densities of the protons and electrons in q's rest frame? What then is the net charge density of the wire in q's frame?

You will find that the charge density is nonzero, which means that the wire exerts a force on q in q's frame (and hence also in the original frame). Returning to the original frame, we conclude that an electric charge q that is *at rest* near a neutral current-carrying wire feels no force, whereas a charge q that is *moving* near a neutral current-carrying wire *does* feel a force. This force is known as the *magnetic force*. We will discuss it in the solution.

Section 1.5: Velocity addition

1.14. **Pythagorean triples** $*$

Let (a, b, h) be a Pythagorean triple. (We'll use h to denote the hypotenuse, instead of c, for obvious reasons.) Consider the relativistic addition or subtraction of two speeds with β values of $\beta_1 = a/h$ and $\beta_2 = b/h$. ($\beta \equiv v/c$ is a speed's fraction of the speed of light.) Show that the numerator and denominator of the result are the leg and hypotenuse of another Pythagorean triple, and find the other leg. What is the associated γ factor?

1.15. **Fizeau experiment** $**$

The second postulate of relativity says that the speed of light in vacuum is always c (in an inertial frame). However, the speed of light in a medium (such as water) is given by c/n, where n is the *index of refraction* of the medium. For water, n is about $4/3$.

Imagine aiming a beam of light rightward into a pipe of water moving rightward with speed v. Naively, the speed of the light with respect to the ground should be $c/n + v$. Find the correct speed by using the velocity-addition formula. Then in the case where $v \ll c$ (certainly a valid approximation in the case of moving water), show that to leading order in v, the speed takes the form of $c/n + Av$. What is the value of A?

1.16. **Equal speeds** $*$

A and B travel at $4c/5$ and $3c/5$ with respect to the ground, as shown in Fig. 1.51. How fast should C travel so that she sees A and B approaching her at the same speed? What is this speed?

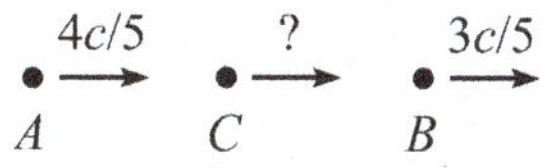

Figure 1.51

1.17. **More equal speeds** $**$

A travels at speed v with respect to the ground, and B is at rest, as shown in Fig. 1.52. How fast should C travel so that she sees A and B approaching her at the same speed? In the ground frame (B's frame), what is the ratio of the distances CB and AC? (Assume that A and C arrive at B at the same time.) The answer to this is nice and clean. Can you think of a simple intuitive explanation for the result?

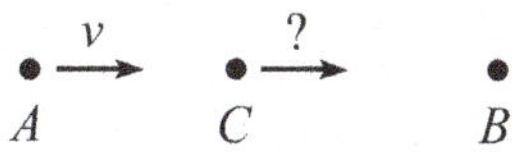

Figure 1.52

1.18. Many velocity additions **

An object moves at speed $v_1/c \equiv \beta_1$ with respect to S_1 (we'll call the β's "speeds" here), which moves at speed β_2 with respect to S_2, which moves at speed β_3 with respect to S_3, and so on, until finally S_{N-1} moves at speed β_N with respect to S_N (see Fig. 1.53). Show by mathematical induction that the speed $\beta_{(N)}$ of the object with respect to S_N can be written as

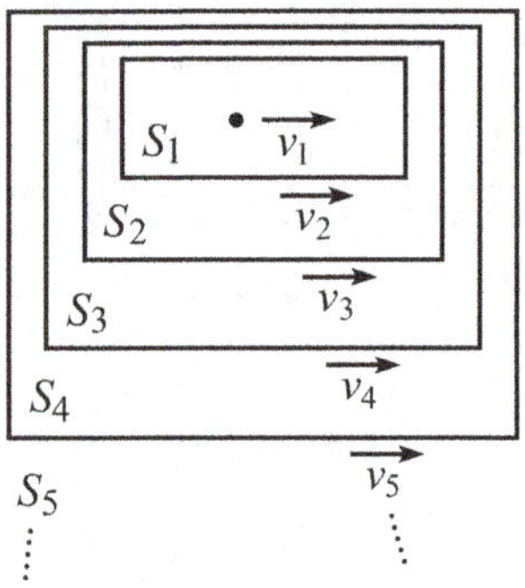

Figure 1.53

$$\beta_{(N)} = \frac{P_N^+ - P_N^-}{P_N^+ + P_N^-}, \quad \text{where } P_N^+ \equiv \prod_{i=1}^{N}(1+\beta_i) \text{ and } P_N^- \equiv \prod_{i=1}^{N}(1-\beta_i). \tag{1.42}$$

1.19. Velocity addition from scratch ***

A ball moves at speed v_1 with respect to a train. The train moves at speed v_2 with respect to the ground. What is the speed of the ball with respect to the ground? Solve this problem (that is, derive the velocity-addition formula, $V = (v_1 + v_2)/(1 + v_1 v_2/c^2)$) in the following way. (Don't use time dilation, length contraction, etc. Use only the two postulates of relativity.)

Let the ball be thrown from the back of the train. At the same instant, a photon is released next to it; see Fig. 1.54. The photon heads to the front of the train, bounces off a mirror, heads back, and eventually runs into the ball. In both the train frame and the ground frame, calculate the fraction of the way along the train where the meeting occurs, and then equate these fractions.

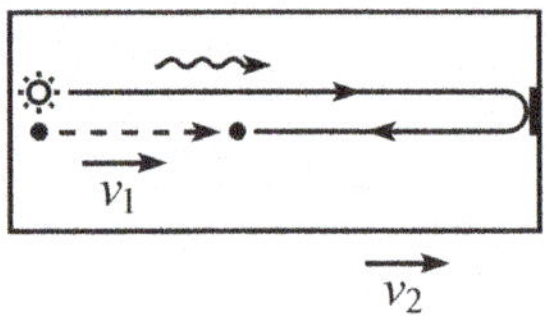

Figure 1.54

1.20. Time dilation and Lv/c^2 ***

A person walks very slowly at speed u from the back of a train of proper length L to the front. The time-dilation effect in the train frame can be made arbitrarily small by picking u to be sufficiently small, because the effect is second order in u, while the travel time is only first order in $1/u$. (See Problem 1.3.) Therefore, if the person's watch agrees with a clock at the back of the train when he starts, then it also (essentially) agrees with a clock at the front when he finishes.

Now consider this setup in the ground frame, where the train moves at speed v. The rear clock reads Lv/c^2 more than the front, so in view of the preceding paragraph, the time gained by the person's watch during the process must be Lv/c^2 less than the time gained by the front clock (because they agree in the end). By working in the ground frame, explain why this is the case.[9] Since we are assuming u is small, you may assume $u \ll c$.

1.21. Modified twin paradox ***

Figure 1.55

Consider the following variation of the twin paradox. A, B, and C each have a clock. In A's reference frame, B flies past A with speed v to the right, as shown in Fig. 1.55. When B passes A, they both set their clocks to zero. Also, in A's reference frame, C starts far to the right and moves to the left with speed v. When B and C pass each other, C sets his clock to read the same as B's. Finally, when

[9] If you line up a collection of these train systems around the circumference of a circular rotating platform, then the present result implies the following fact. Let person A be at rest on the platform at a point on the circumference, and let person B start at A and walk arbitrarily slowly around the circumference. Then when B returns to A, B's clock will read less than A's. This is true because the above reasoning shows (as you will figure out) that an inertial observer sees B's clock running slower than A's. This result, that you can walk arbitrarily slowly in a particular reference frame and have your clock lose synchronization with other clocks, is a consequence of the fact that in some accelerating reference frames it is impossible to produce a consistent method (that is, one without a discontinuity) of clock synchronization. See Cranor *et al.* (2000) for more details.

C passes A, they compare the readings on their clocks. At this moment, let A's clock read T_A, and let C's clock read T_C.

(a) Working in A's frame, show that $T_C = T_A/\gamma$, where $\gamma = 1/\sqrt{1 - v^2/c^2}$.

(b) Working in B's frame, show again that $T_C = T_A/\gamma$.

(c) Working in C's frame, show again that $T_C = T_A/\gamma$.

1.8 Exercises

Section 1.3: The fundamental effects

1.22. **Effectively speed c** $*$

A rocket flies between two planets that are one light-year apart. What should the rocket's speed be so that the time elapsed on the captain's watch is one year?

1.23. **A passing train** $*$

A train with length 15 cs moves at speed $3c/5$. (1 cs is one "light-second." It equals $(1)(3 \cdot 10^8\,\text{m/s})(1\,\text{s}) = 3 \cdot 10^8\,\text{m}$.) How much time does it take to pass a person standing on the ground, as measured by that person? Solve this by working in the frame of the person, and then again by working in the frame of the train.

1.24. **Simultaneous waves** $*$

Alice flies past Bob at speed v. Right when she passes, they both set their watches to zero. When Alice's watch shows a time T, she waves to Bob. Bob then waves to Alice simultaneously (as measured by him) with Alice's wave (so this is before he actually *sees* her wave). Alice then waves to Bob simultaneously (as measured by her) with Bob's wave. Bob then waves to Alice simultaneously (as measured by him) with Alice's second wave. And so on. What are the readings on Alice's watch for all the times she waves? And likewise for Bob?

1.25. **Overtaking a train** $**$

Train A has proper length L. Train B moves past A (on a parallel track, facing the same direction) with relative speed $4c/5$ (as measured by either train; so each one sees the other move at $4c/5$). The length of B is such that A says that the fronts of the trains coincide at exactly the same time as the backs coincide. What is the time difference between the fronts coinciding and the backs coinciding, as measured by B? Solve this in two ways: (a) by using length contraction, and (b) by using the rear-clock-ahead effect (among other things).

1.26. **Walking on a train** $**$

A train with proper length L and speed $3c/5$ approaches a tunnel with length L. At the moment the front of the train enters the tunnel, a person leaves from the front of the train and walks (briskly) toward the back. She arrives at the back of the train right when it (the back) leaves the tunnel.

(a) How much time does this take in the ground frame?

(b) What is the person's speed with respect to the ground?

(c) How much time elapses on the person's watch?

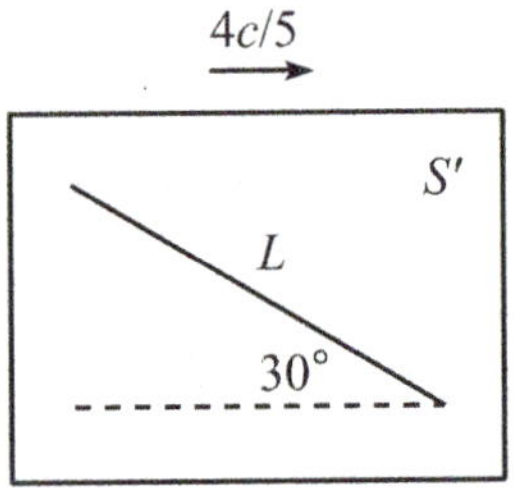

Figure 1.56

1.27. **Diagonal stick** ∗∗

In frame S' a stick with proper length L is at rest and is tilted at 30°. Frame S' moves to the right at speed $4c/5$ with respect to the ground. See Fig. 1.56.

(a) In the ground frame, what is the horizontal span of the stick?

(b) You stand far away from the stick (below it in the picture). When you see the stick at its closest point to you, what is the *apparent* horizontal span of the stick? That is, what would a photograph show, taken from a distant camera?

1.28. **Triplets** ∗∗

Triplet A stays on the earth. Triplet B travels at speed $4c/5$ to a planet (a distance L away) and back. Triplet C travels out to the planet at speed $3c/4$, and then returns at the necessary speed to arrive back exactly when B does. How much does each triplet age during this process? Who is youngest?

1.29. **Seeing the light** ∗∗

A and B leave from a common point (with their clocks both reading zero) and travel in opposite directions with relative speed v (that is, each sees the other move with speed v). When B's clock reads T, he sends out a light signal. When A receives the signal, what time does her clock read? Answer this by doing the calculation entirely in (a) A's frame, and then (b) B's frame. (This problem is basically a derivation of the longitudinal Doppler effect, discussed in Section 2.5.1. It is one of the few cases where we're actually concerned with the time is takes light to reach someone's eye.)

1.30. **Twin paradox and Lv/c^2** ∗∗

In the twin-paradox example near the end of Section 1.3.2, we noted that B (the traveler) is in different inertial frames for the outward and return trips. In B's frame during the outward trip (as the universe flies past B, from B's point of view), the star clock is the rear clock. But in B's (new) frame during the return trip, the earth clock is the rear clock. During B's turnaround, the earth clock therefore goes from being Lv/c^2 behind the star clock, to being Lv/c^2 ahead of it (where L is the earth-star proper distance). The earth clock must therefore quickly whip ahead by $2Lv/c^2$ during the turnaround, from B's (briefly noninertial) point of view. Explain quantitatively, by working in B's frame(s), how B puts everything together to conclude that at the end of the trip, the earth clock has advanced more than B's clock, by a factor γ.

1.31. **Backward photon** ∗∗

A train with proper length L has clocks at the front and back. A photon is fired from the front to the back. Working in the train frame, we can easily say that if the photon leaves the front of the train when a clock there reads zero, then it arrives at the back when a clock there reads L/c.

Now consider this setup in the ground frame, where the train travels by at speed v. Rederive the above frame-independent result (namely, if the photon leaves the front of the train when a clock there reads zero, then it arrives at the back when a clock there reads L/c) by working *only* in the ground frame.

1.32. **Pole's clocks in barn** ∗∗

Consider the pole-in-barn setup from Problem 1.6. In the solution to that problem, we stated that in the barn frame, the pole's right clock (when the right ends of the pole and barn coincide) reads *less* than the pole's left clock (when the left ends coincide), even though the right event happens *after* the left event. Demonstrate this by explicitly calculating the reading on the right clock when the right ends coincide. (Assume for simplicity that the left clock reads zero when the left ends coincide.) Do this by working in (a) the pole frame, and (b) the barn frame.

1.33. **Twice simultaneous** **

A train with proper length L moves at speed v with respect to the ground. When the front of the train passes a tree on the ground, a ball is simultaneously (as measured in the *ground frame*) thrown from the back of the train toward the front, with speed u with respect to the train. What should u be so that the ball hits the front simultaneously (as measured in the *train frame*) with the tree passing the back of the train? Show that in order for a solution for u to exist, we must have $v/c < (\sqrt{5} - 1)/2$, which happens to be the inverse of the golden ratio.

1.34. **People clapping** **

Two people stand a distance L apart along an east-west road. They clap their hands simultaneously in the ground frame. You are driving eastward along this road at speed $4c/5$. You notice that you are next to the western person at the same instant (as measured in your frame) that the eastern person claps. Later on, you notice that you are next to a tree at the same instant (as measured in your frame) that the western person claps. Where is the tree along the road? (Describe its location in the ground frame.)

1.35. **Photon, tree, and house** **

(a) A train with proper length L moves at speed v with respect to the ground. At the instant the back of the train passes a tree, someone at the back of the train shines a photon toward the front. The photon happens to hit the front of the train at the instant the front passes a house. As measured in the ground frame, how far apart are the tree and the house? Solve this by working in the ground frame.

(b) Now look at the setup from the point of view of the train frame. Using your result for the tree-house distance from part (a), verify that the house meets the front of the train at the same instant the photon meets it.

1.36. **Four clock readings** **

A train has proper length L. In the frame of the train, a photon is fired from the back of the train to the front. Assume that a clock at the back reads zero when the photon is fired. Then a clock at the front of course reads L/c when the photon arrives there.

Now consider the setup in the ground frame, where the train moves to the right at speed $3c/5$. In this frame, it is observed that the train enters a tunnel when the photon is fired, and the train leaves the tunnel when the photon arrives at the front, as shown in Fig. 1.57.

(a) What is the (proper) length of the tunnel?

(b) What are the four R_i readings on the train clocks at the two instants shown?

(c) As observed in the ground frame, verify that the time elapsed on a given train clock is related to the time elapsed on a ground clock by the appropriate γ factor. (So you will need to first determine the ground time via a method other than time dilation.)

(d) Now return to the train frame. Draw a reasonably accurate picture of what things look like at the instant the photon is fired. Label all distances necessary to describe the location of the tunnel.

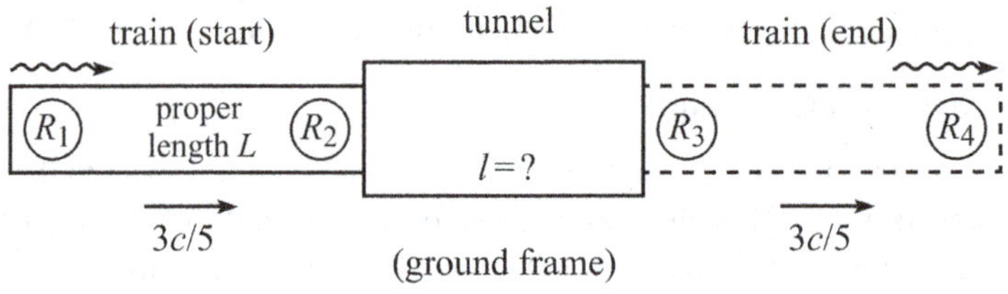

Figure 1.57

1.37. **Tunnel fraction** $**$

A person runs with speed v toward a tunnel of length L. A light source is located at the far end of the tunnel. At the instant the person enters the tunnel, the light source simultaneously (as measured in the tunnel frame) emits a photon that travels down the tunnel toward the person. When the person and the photon eventually meet, the person's location is a fraction f along the tunnel. What is f? Solve this by working in the tunnel frame, and then again by working in the person's frame.

1.38. **Through the hole?** $***$

A stick with proper length L moves at speed v in the direction of its length. It passes over a infinitesimally thin sheet that has a hole of diameter L cut in it. As the stick passes over the hole, the sheet is raised so that the stick passes through the hole and ends up underneath the sheet. Well, maybe . . .

In the lab frame, the stick's length is contracted to L/γ, so it appears to easily make it through the hole. But in the stick frame, the hole is contracted to L/γ, so it appears that the stick does *not* make it through the hole (or rather, the hole doesn't make it around the stick, since the hole is what is moving in the stick frame). So the question is: Does the stick end up on the other side of the sheet or not?

1.39. **Short train in a tunnel** $**$

Consider the scenario in Problem 1.7, with the only change being that the train now has length rL, where r is some numerical factor. What is the largest value of r, in terms of v, for which it is possible for the bomb not to explode? Verify that you obtain the same answer working in the train frame and in the tunnel frame.

1.40. **Charge density** $**$

Consider the setup in Problem 1.13. If the electrons' speed is $v_0 = 4c/5$ and the charge q's speed is $v = 3c/5$, what is the charge density of the wire in q's rest frame? Solve this from scratch; that is, don't just invoke the result in Eq. (1.59) (although you can of course check your answer with that result; likewise for any intermediate steps).

Section 1.5: Velocity addition

1.41. **γ's for relativistic addition** $*$

Show that the relativistic addition (or subtraction) of the velocities u and v has a γ factor given by $\gamma = \gamma_u \gamma_v (1 \pm uv)$, or $\gamma_u \gamma_v (1 \pm uv/c^2)$ with the c's.

1.42. Equal speeds *

A travels at speed $4c/5$ toward B, who is at rest. C is between A and B. How fast should C travel so that she sees both A and B approaching her at the same speed? (This problem is a special case of Problem 1.17. Solve this one from scratch, but feel free to check you answer with the one from Problem 1.17.)

1.43. Running away *

A and B both start at the origin and simultaneously head off in opposite directions, each with speed $3c/5$ with respect to the ground. A moves to the right, and B moves to the left. Consider a mark on the ground at $x = L$. As viewed in the ground frame, A and B are a distance $2L$ apart when A passes this mark. As viewed by A, how far away is B when A coincides with the mark?

1.44. Again simultaneous **

A train with proper length L moves at speed v with respect to the ground. When the front of the train passes a tree on the ground, a ball is simultaneously (as measured in the ground frame) thrown from the back of the train toward the front, with speed u with respect to the train. What should u be so that the ball hits the front simultaneously (as measured again in the ground frame) with the back of the train passing the tree? What is the maximum value of v for which a solution for u exists?

1.45. Overlapping trains **

An observer on the ground sees two trains, A and B, both with proper length L, move in opposite directions at speed v with respect to the ground. She notices that at the instant the trains overlap, clocks at the front of A and rear of B both read zero, as shown in Fig. 1.58. From the rear-clock-ahead effect, she therefore also notices that clocks at the rear of A and front of B read Lv/c^2 and $-Lv/c^2$, respectively. Now imagine riding along on A. When the rear of B passes the front of your train (A), clocks at both of these places read zero (a frame-independent statement). Explain, by working only in the frame of A, why clocks at the back of A and the front of B read Lv/c^2 and $-Lv/c^2$, respectively, when these points coincide.

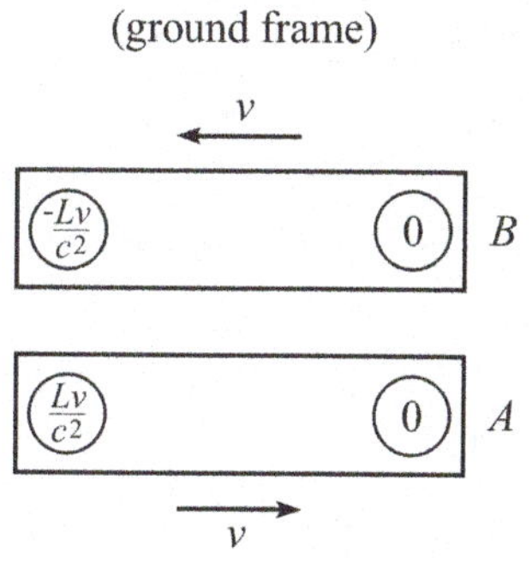

Figure 1.58

1.46. Running on a train **

A train with proper length L moves at speed v_1 with respect to the ground. A passenger runs from the back of the train to the front at speed v_2 with respect to the train. How much time does this take, as viewed by someone on the ground? Solve this in two different ways:

(a) Find the relative speed of the passenger and the train (as viewed by someone on the ground), and then find the time it takes for the passenger to erase the initial head start that the front of the train had.

(b) Find the time elapsed on the passenger's clock (by working in whatever frame you want), and then use time dilation to get the time elapsed on a ground clock.

1.47. Velocity addition **

The fact that the previous exercise can be solved in two different ways suggests a method of deriving the velocity-addition formula: A train with proper length L moves at speed v_1 with respect to the ground. A ball is thrown from the back of the train to the front at speed v_2 with respect to the train. Let the speed of the ball with respect to the ground be V. Calculate the time of the ball's journey, as measured by an observer on the ground, in the two different ways described in the previous exercise, and then equate the results to solve for V in terms of v_1 and v_2. (This gets rather messy. And yes, you have to solve a quadratic.)

1.48. **Velocity addition again** $**$

A train with proper length L moves at speed v with respect to the ground. A ball is thrown from the back of the train to the front at speed u with respect to the train.

(a) Find the time of the process in the ground frame by looking at how much a clock at rest in the train frame advances (for example, a clock at the front of the train), and then applying time dilation to this one clock.

(b) Find the time of the process in the ground frame by applying time dilation to the ball's clock. Your answer will contain the unknown speed V of the ball with respect to the ground.

(c) Equate your results from parts (a) and (b) to show that $\gamma_V = \gamma_u \gamma_v (1 + uv/c^2)$. Then solve for V to produce the velocity-addition formula.

1.49. **Bullets on a train** $**$

A train moves at speed v. Bullets are successively fired at speed u (relative to the train) from the back of the train to the front. A new bullet is fired at the instant (as measured in the train frame) the previous bullet hits the front. In the frame of the ground, what fraction of the way along the train is a given bullet, at the instant (as measured in the ground frame) the next bullet is fired? What is the maximum number of bullets that are in flight at a given instant, in the ground frame?

1.9 Solutions

1.1. **Consistency with Lv/c^2**

In the setup that led to Eq. (1.6), the two events (light hitting rear and light hitting front) were simultaneous in the train frame, because the light source was located at the center of the train. The difference $t_R - t_L$ is the time difference between these events, as measured in the ground frame. From Eq. (1.6) we have

$$t_R - t_L = \frac{\ell'}{c - v} - \frac{\ell'}{c + v} = \frac{2\ell' v}{c^2 - v^2} = \frac{2\ell' v}{c^2(1 - v^2/c^2)} = \frac{2\gamma^2 \ell' v}{c^2}. \tag{1.43}$$

Now, $2\ell'$ is the total length of the train, as measured in the ground frame. But due to length contraction, this length is shorter than the train's proper length L, by a factor γ. That is, $2\ell' = L/\gamma$. Substituting this into Eq. (1.43) gives

$$t_R - t_L = \frac{\gamma L v}{c^2}. \tag{1.44}$$

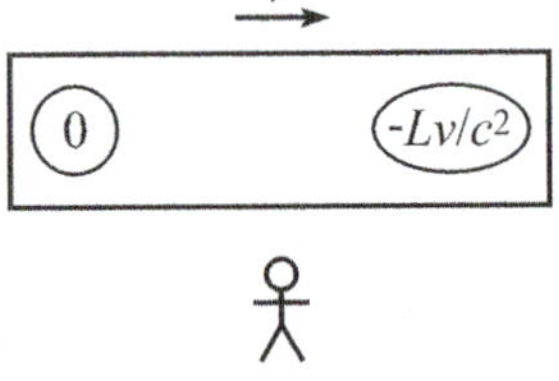

Figure 1.59

This is consistent with the rear-clock-ahead result of Lv/c^2, for the following reason. Since the events are simultaneous in the train frame, clocks at the front and rear of the train have the same reading when the photons hit. Assume for concreteness that this reading is zero. Then as viewed in the ground frame, at the instant the rear event occurs, the situation is shown in Fig. 1.59. The rear clock reads zero (a frame-independent statement), and the front clock reads $-Lv/c^2$ due to the rear-clock-ahead effect. The ground observer must then wait for the front clock to advance to zero, at which point the front event occurs (again a frame-independent statement). But the front clock (along with every other clock on the train) runs slow due to time dilation. So it takes a time (in the ground frame) of $\gamma Lv/c^2$ to advance its reading by Lv/c^2 to zero. This elapsed time of $\gamma Lv/c^2$ in the ground frame agrees with the $t_R - t_L$ result in Eq. (1.44).

1.2. **Here and there**

If the setup is to be possible, then in the train frame the person must run the length L of the train in a time Lv/c^2 (or slightly less). His speed with respect to the train must therefore be at least $L/(Lv/c^2) = c^2/v$. But $c^2/v = c(c/v) > c$, which is an impossible speed. So it is not possible for the person to perform the stated task. You will therefore *not* see him simultaneously at both the front and the back. This is good, because we could produce all sorts of paradoxes if someone were actually at two places at once in a given frame. Imagine a brick wall being constructed between the "two" people, and a bucket of paint being dropped on one of them.

1.3. **Synchronizing clocks**

In the frame of A and B, it takes a time of L/v for C to travel from A to B. During this time, C runs slow by a factor γ, so only $L/\gamma v$ elapses on C during the journey. Therefore, when C reaches B, the reading on C is $L/\gamma v$, and the reading on B is L/v. If v is small (more precisely, if $v \ll c$), we can use $\sqrt{1-\epsilon} \approx 1 - \epsilon/2$ to approximate C's reading as

$$\frac{L}{\gamma v} = \frac{L}{v}\sqrt{1-\frac{v^2}{c^2}} \approx \frac{L}{v}\left(1-\frac{v^2}{2c^2}\right) = \frac{L}{v} - \frac{Lv}{2c^2}. \tag{1.45}$$

The difference between C's and B's readings is therefore $Lv/2c^2$. This goes to zero as $v \to 0$, as desired. We see that even though the total time L/v in A's and B's frame goes to infinity as $v \to 0$, the $Lv/2c^2$ difference between C's and B's readings goes to zero. This is due to the fact that L/v has only one power of v in the denominator, whereas the γ factor depends quadratically on v. If the γ factor were instead equal to $1/\sqrt{1-v/c}$, then the difference in the readings would take on the fixed value of $L/2c$ in the $v \to 0$ limit.

1.4. **Rotated square**

Fig. 1.60 shows the square at the instant (in your frame) when it is closest to you. Its length is contracted along the direction of motion, so it takes the shape of a rectangle with sides L and L/γ. This is what the shape *is* in your frame (where *is*-ness is defined by where all the points of an object are at simultaneous times). But what does the square *look* like to you? That is, what is the nature of the photons hitting your eye at a given instant?

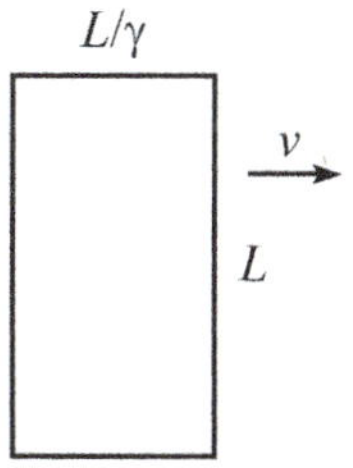

Figure 1.60

Photons from the far side of the square have to travel an extra distance L to get to your eye, compared with photons from the near side. This is true because we're assuming that you are far from the square, which means that the paths to you from the various points on the square are all essentially parallel. If you were instead close to the square, then we would have to use the Pythagorean theorem to obtain the distances, and things would be much more difficult and messy.

The photons from the far side need an extra time L/c of flight, compared with the photons from the near side. During this time L/c, the square moves a distance $v(L/c) \equiv L\beta$ sideways, where $\beta \equiv v/c$. Therefore, referring to Fig. 1.61, a photon emitted at point A reaches your eye at the same time as a photon emitted from point B, as do all the photons emitted from the near side, of course, and as do (as you can verify) all the photons emitted from the trailing (left) side, between A and B. This means that the trailing side of the square spans a distance $L\beta$ across your field of vision, while the near side spans a distance $L/\gamma = L\sqrt{1-\beta^2}$. But this is exactly what a rotated square of side L looks like, as shown in Fig. 1.62. From the figure, we see that the angle of rotation is given by $\sin\theta = \beta$, or equivalently $\cos\theta = \sqrt{1-\beta^2}$. So for $v \ll c$ the square is only slightly rotated, while for $v \to c$ the rotation angle approaches 90°. For the case of a circle instead of a square, see Hollenbach (1976).

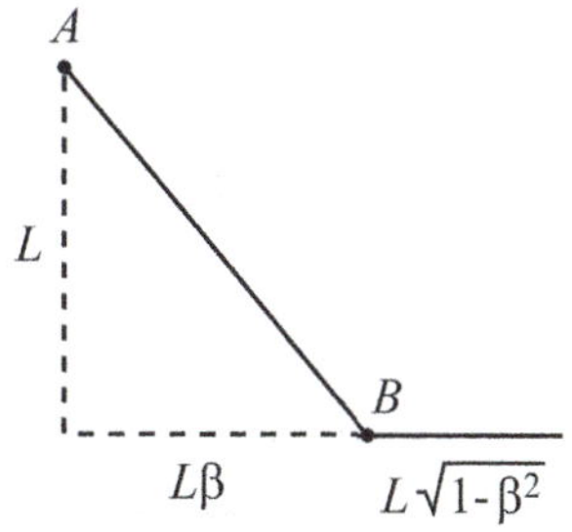

Figure 1.61

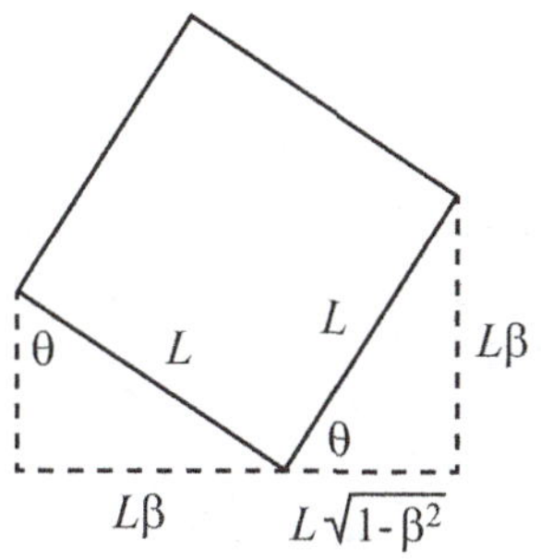

Figure 1.62

1.5. **Deriving length contraction**

In the ground frame, the given rule tells us that the length of the train is aL. (We'll drop the subscript v for convenience.) So the front of the train has a head start of aL over the photon. The photon closes this gap at a relative speed of $c - v$, as measured in the ground frame. The time of the process is therefore $t = aL/(c-v)$. During this time, the photon travels a distance $ct = caL/(c-v)$. This then is the tree-house distance in the ground frame.

Now consider the setup in the train frame. The starting and ending pictures are shown in Fig. 1.63.

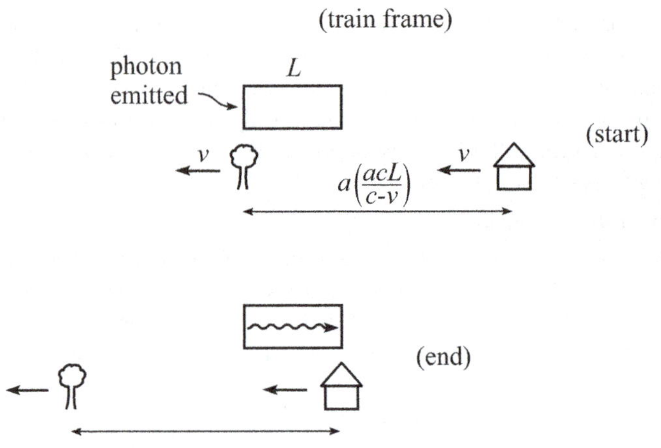

Figure 1.63

The photon is emitted when the tree coincides with the back of the train (a frame-independent statement). The train has length L (its proper length), and the tree-house distance is $a(acL/(c-v))$. This is true because our rule states that if an object is moving, the observed length equals a times the proper length, which we found above to be $acL/(c-v)$ for the tree-house separation. The initial distance between the house and the front of the train is therefore $a^2cL/(c-v) - L$ (subtracting off the train's length). The house covers this distance at speed v. The photon covers the length L of the train at speed c. Since the house and the photon arrive at the front of the train at the same time, we must therefore have

$$\frac{\frac{a^2cL}{c-v} - L}{v} = \frac{L}{c} \implies \frac{a^2}{1-\frac{v}{c}} - 1 = \frac{v}{c} \implies a^2 = 1 - \frac{v^2}{c^2}$$

$$\implies a = \sqrt{1-\frac{v^2}{c^2}} \equiv \frac{1}{\gamma}, \tag{1.46}$$

as desired.

1.6. **Pole in barn**

The proper answer to the question is: The question cannot be answered without more information. More precisely, the question is frame dependent; it must be finished with the qualifier, "... in the barn frame," or "... in the pole frame." The question is indeed frame dependent, because in the barn frame the pole is length contracted down to L/γ. So the pole certainly fits inside the barn; see Fig. 1.64. (For the purpose of drawing the figure, we have chosen γ to be a slightly larger than 3.) But in the pole frame the barn is length contracted down to L/γ. So the pole certainly *doesn't* fit in the barn.

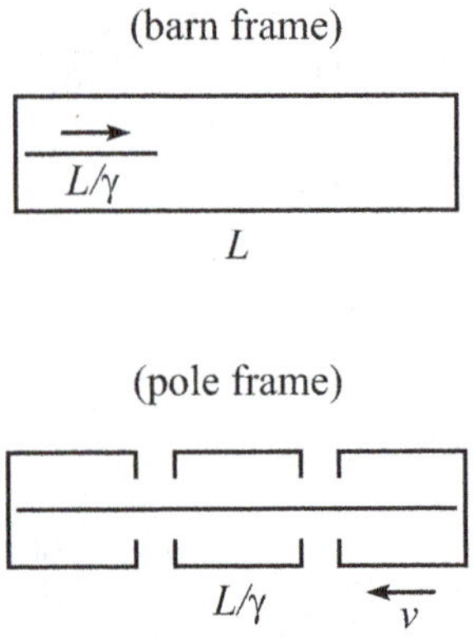

Figure 1.64

In retrospect, it is no surprise that the question is frame dependent, because the qualifier "with both doors closed" is shorthand for "with both doors closed *simultaneously*." And as soon as we start talking about simultaneity, we know that frame dependence will come into play, because events that are simultaneous in one frame are not simultaneous in another. We purposely didn't include the word "simultaneously" in the statement of the problem, in order to not give too much of a hint.

Remark: Fig. 1.64 makes it clear that the order of the two events, "left ends coinciding" and "right ends coinciding," is reversed in the two frames. In the barn frame the left ends coincide first, whereas in the pole frame the right ends coincide first. There is nothing wrong with the order of events in one frame being different from the order in another.

However, for this to be possible, the two events must be so-called *spacelike separated.* We'll discuss this term in Section 2.3.

Consider two clocks at the ends of the pole (synchronized in the pole frame). Assume that the left clock reads zero when it coincides with the left end of the barn. Then from the pole-frame view in Fig. 1.64, it is clear that the right clock has a *negative* reading when it coincides with the right end of the barn, because the right event happens *before* the left event (and because the pole clocks are synchronized in the pole frame; this fact is important.) These readings of zero (on the left clock when the left ends coincide) and a negative value (on the right clock when the right ends coincide) are frame independent, so they must be the same in the barn frame. In other words, in the barn frame the pole's right clock must read *less* than the left clock (at the two events at hand), even though the right event happens *after* the left event. The task of Exercise 1.32 is to explain explicitly why this is true, and to calculate the actual reading on the right clock. ♣

1.7. **Train in a tunnel**

Yes, the bomb explodes. This is clear in the frame of the train; see Fig. 1.65. In this frame, the train has length L, and the tunnel speeds past it. The tunnel is length contracted down to L/γ. Therefore, the far end of the tunnel passes the front of the train before the near end passes the back. So the bomb explodes.

(train frame)

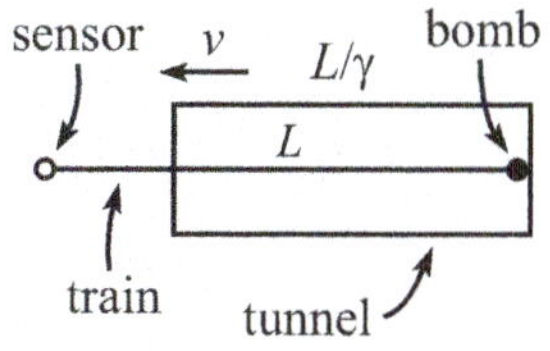

Figure 1.65

(tunnel frame)

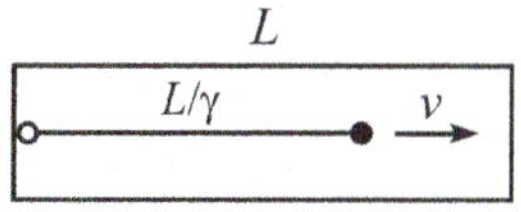

Figure 1.66

We can, however, also look at things in the frame of the tunnel; see Fig. 1.66. Here the tunnel has length L, and the train is length contracted down to L/γ. Therefore, the deactivation device gets triggered *before* the front of the train passes the far end of the tunnel. So you might think that the bomb does *not* explode. However, all observers must agree on whether or not the bomb explodes; the explosion (or lack thereof) is frame independent. So we appear to have a paradox.

The resolution to this paradox is that the deactivation device cannot instantaneously tell the bomb to deactivate itself; the signal can't travel faster than the speed of light. (If signals could travel faster than c, we would be able to generate setups that violate causality. We'll talk about this in Section 2.3.) It therefore takes a nonzero time for the signal to travel the length of the train (or actually a longer distance, since the train is moving) from the sensor to the bomb. And it turns out that this transmission time makes it impossible for the deactivation signal to get to the bomb before the bomb gets to the far end of the tunnel, no matter how fast the train is moving. The bomb therefore explodes. Let's quantitatively demonstrate this.

The signal has the best chance of winning the "race" if it has speed c, so let's assume this is the case. The time it takes the signal to reach the bomb is $(L/\gamma)/(c-v)$, because the train has length L/γ in the ground frame, and because the relative speed of the light and the bomb is $c-v$ in the ground frame. The time it takes the bomb to get to the far end of the tunnel is $(L-L/\gamma)/v$, because the bomb is already a distance L/γ through the tunnel, and because it is moving at speed v. So if the bomb is *not* to explode, the former of these times must be less than the latter. With $\beta \equiv v/c$, this gives

$$\begin{aligned} \frac{L/\gamma}{c-v} < \frac{(L-L/\gamma)}{v} \quad &\Longleftrightarrow \quad \frac{1}{\gamma}\left(\frac{1}{1-\beta}+\frac{1}{\beta}\right) < \frac{1}{\beta} \\ \Longleftrightarrow \quad \sqrt{1-\beta^2}\cdot\frac{1}{(1-\beta)\beta} < \frac{1}{\beta} \quad &\Longleftrightarrow \quad \sqrt{1-\beta^2} < 1-\beta \\ &\Longleftrightarrow \quad \sqrt{1+\beta} < \sqrt{1-\beta}. \end{aligned} \tag{1.47}$$

This is never true. Therefore, the signal always arrives too late, and the bomb always explodes, consistent with the conclusion in the train frame.

1.8. **Bouncing stick**

Assume that a series of clocks are lined up along the stick, and assume that in the ground frame they all read zero when the stick bounces. In the frame of someone running leftward at speed v, the stick is moving rightward (and vertically). The rear (left) clock on the stick is ahead of all the other clocks, so it will reach zero and bounce off the ground first. (It is a frame-independent fact that a clock reads zero when it bounces off the ground.) Clocks along the stick will successively reach zero and the stick will bounce at those points, until finally the clock at the front (right) end reads zero and that end bounces. Snapshots of the stick therefore look like the ones shown in Fig. 1.67.

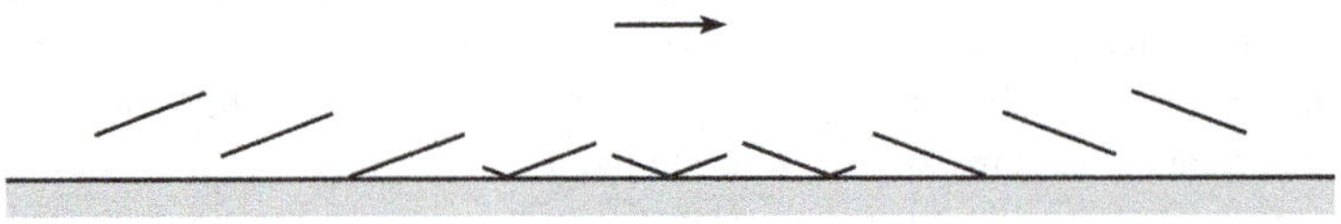

Figure 1.67

There is nothing wrong with the stick having a sharp bend in it. The stick doesn't break in the ground frame, so it doesn't break in the person's frame, either. The sharp bend doesn't imply any severe forces in the stick. The molecules in the stick think everything is perfectly normal; they have no clue that someone in running by to the left and that the stick is bent in this person's frame.

REMARK: If we want to get quantitative, we can calculate the angle the stick makes with respect to the horizontal, in the person's frame. Although the above reasoning involved clocks that were at rest in the *stick* frame, it will be easier in the following reasoning to work with a set of clocks that are at rest in the *ground* frame.

Let's work in the ground frame for a moment. Assume that the two ends of the stick slide down two vertical rails. If the proper length of the stick is L, then the rails are a distance L apart, because there is no transverse length contraction. Imagine that a large number of clocks are attached to the two rails, to make two towers of clocks (at rest in the ground frame). Let all of the clocks in the towers read zero when the two ends of the stick (along with the rest of the stick) bounce off the ground simultaneously in the ground frame. For future reference, note that at a time Lv/c^2 (as measured in the ground frame) before the stick bounces (that is, when all of the clocks in the towers read $-Lv/c^2$), the stick is at a height $u(Lv/c^2)$ above the ground, where u is the stick's (vertical) speed in the ground frame. (Assume that this speed is essentially constant; ignore the vertical acceleration near the ground.)

Now go back to the person's frame. When the back end of the stick hits the ground, all of the clocks in the back vertical tower read zero (a frame-independent statement), and all of the clocks in the front vertical tower read $-Lv/c^2$, due to the rear-clock-ahead effect. But from the preceding paragraph, we know that the front of the stick is at a height $u(Lv/c^2)$ above the ground (in the ground frame and hence also in the person's frame, since there is no transverse length contraction) when it is next to a tower clock that reads $-Lv/c^2$. The horizontal distance between the ends is L/γ_v, because this is the length-contracted distance between the two rails. The angle that the stick makes with the horizontal in the person's frame is therefore given by

$$\tan\theta = \frac{Luv/c^2}{L/\gamma_v} = \frac{\gamma_v uv}{c^2}. \quad ♣ \tag{1.48}$$

1.9. **Seeing behind the board**

First note that the reasonings can't both be correct, because the closest mark you can see has a frame-independent value. It can't depend on which frame we arbitrarily choose to do the calculation in.

The first reasoning is the correct one. You will be able to see a mark on the ruler that is less than L units from the wall. You will actually be able to see a mark even closer to the wall than L/γ, as we'll show below.

The error in the second reasoning (in the board's frame) is that the second picture in Fig. 1.48 is *not* what *you* see. This second picture shows where things are at simultaneous times in the *board's* frame, which are not simultaneous times in *your* frame. Alternatively, the error is the implicit assumption that signals travel instantaneously; but in fact the back (left) end of the board cannot know that the front (right) end has been hit by the wall until a nonzero time has passed. During this time, the ruler (and the wall and you) travels farther to the left, allowing you to see more of the ruler. Let's be quantitative about this and calculate (in both frames) the closest mark to the wall that you can see.

YOUR FRAME: In your reference frame, the board has length L/γ. Therefore, when the board hits the wall, you can see a mark a distance L/γ from the wall. You will, however, be able to see a mark even closer to the wall, because the back end of the board will keep moving forward, since it doesn't know yet that the front end has hit the wall. The stopping signal (shock wave, etc.) takes time to travel.

Let's assume that the stopping signal travels along the board at speed c. (We could instead work with a general speed u, but the speed c is simpler, and it yields an upper bound on the closest mark you can see.) Where will the signal reach the back end of the board? Starting from the time the board hits the wall, the signal travels backward from the wall at speed c, while the back end of the board travels forward at speed v (from a point L/γ away from the wall). The relative speed (as viewed by you) of the signal and the back end is $c + v$. This is the rate at which the initial gap of L/γ is closed. Therefore, the signal hits the back end after a time $(L/\gamma)/(c + v)$. During this time, the signal has traveled a distance $c \cdot (L/\gamma)/(c + v)$ from the wall. This is where the back end stops. The closest point to the wall that you can see on the ruler is therefore the mark with the value (with $\beta \equiv v/c$)

$$\frac{c(L/\gamma)}{c+v} = \frac{L}{\gamma(1+\beta)} = \frac{L\sqrt{1-\beta^2}}{1+\beta} = L\sqrt{\frac{1-\beta}{1+\beta}}. \tag{1.49}$$

BOARD FRAME: In the board's reference frame, the wall is initially moving leftward with speed v. After the wall hits the right end of the board, the signal moves to the left with speed c, while the wall keeps moving to the left with speed v (because the wall/earth is much more massive than the board). Where is the wall when the signal reaches the left end of the board (at which point the left end starts moving leftward along with the ruler)? The wall travels v/c as fast as the signal, so it travels a distance Lv/c in the time that the signal travels the distance L. This means that the wall is $L(1 - v/c)$ away from the left end of the board when the signal reaches the left end. This distance in the board's frame (or rather, in the board's original frame) corresponds to a distance $\gamma L(1 - v/c)$ on the ruler, because the (moving) ruler is length contracted. So the left end of the board is at the mark with the value

$$\gamma L(1 - v/c) = L\gamma(1 - \beta) = \frac{L(1-\beta)}{\sqrt{1-\beta^2}} = L\sqrt{\frac{1-\beta}{1+\beta}}, \tag{1.50}$$

in agreement with Eq. (1.49).

1.10. **Cookie cutter**

Let the diameter of the cookie cutter be L, and consider the two following reasonings.

- In the lab frame, the dough is length contracted, so the cutter's diameter L corresponds to a distance larger than L (namely γL) in the dough frame. Therefore, when you buy a cookie, it is stretched by a factor γ in the direction of the belt.[10]
- In the dough frame, the cookie cutter is length contracted down to L/γ in the direction of motion. So in the frame of the dough, the cookies have a length of only L/γ. Therefore, when you buy a cookie, it is squashed by a factor γ in the direction of the belt.

Which reasoning is correct? The first one is. The cookies are stretched out. The fallacy in the second reasoning is that the various parts of the cookie cutter do *not* strike the dough

[10]The shape is an ellipse, since that's what a stretched circle is. The eccentricity of an ellipse is the focal length divided by the semi-major axis length. As an exercise, you can show that this equals $\beta \equiv v/c$ here.

simultaneously in the dough frame. What the dough sees is this: Assuming that the cutter moves to the left, the rightmost point on the cutter stamps the dough, then nearby points on the cutter stamp it, and so on, until finally the leftmost point stamps it. But by this time the front (that is, the left) of the cutter has moved farther to the left. So the cookie turns out to be longer than L. Let's now show (by working in the dough frame) that the length of the cookie is in fact γL, as the first of the above reasonings correctly states.

Assume that all points on the cutter have little clocks associated with them. Since all points on the cutter strike the dough simultaneously in the lab frame (since the cutter is horizontal), all of the clocks have the same reading when they strike the dough. Let's assume that this value is zero.

Now let's look at what happens in the dough frame. Consider the moment when the rightmost (rear) point on the cutter strikes the dough. The clock there reads zero (a frame-independent reading). The clock at the leftmost (front) point on the cutter therefore reads $-Lv/c^2$, due to the rear-clock-ahead effect. This leftmost clock must then advance by Lv/c^2 by the time it strikes the dough when it reads zero (again a frame-independent reading). However, due to time dilation, this takes a time $\gamma(Lv/c^2)$ in the dough frame. During this time, the cutter travels a distance $v(\gamma Lv/c^2)$. But the front of the cutter was initially a distance L/γ (due to length contraction) ahead of the back. The total length of the cookie in the dough frame is this initial L/γ distance plus the extra $v(\gamma Lv/c^2)$ distance traveled by the front. The total length is therefore

$$\ell = \frac{L}{\gamma} + \frac{\gamma L v^2}{c^2} = \gamma L\left(\frac{1}{\gamma^2} + \frac{v^2}{c^2}\right) = \gamma L\left(\left(1 - \frac{v^2}{c^2}\right) + \frac{v^2}{c^2}\right) = \gamma L, \tag{1.51}$$

as we wanted to show. (This is the same calculation as in Eq. (1.23).) If the dough is then slowly decelerated, the shape of the cookies won't change. So this is the shape you see in the store.

1.11. **Getting shorter**

(a) In the frame in which the balls are initially at rest, the people move rightward with speed v. The beginning picture is shown in Fig. 1.68(a). The left person catches the left ball when his clock reads zero (a frame-independent fact). The right person's clock therefore reads $-Lv/c^2$, due to the rear-clock-ahead effect. The balls are a distance γL apart, and the people's separation is length contracted down to L/γ.

The ending picture is shown in Fig. 1.68(b). The right person catches the right ball when his clock reads zero (again a frame-independent fact). The left person's clock is ahead, so it reads Lv/c^2. By the time the right person catches the ball, the left person has moved to the right while holding the left ball. Both distances are now L/γ.

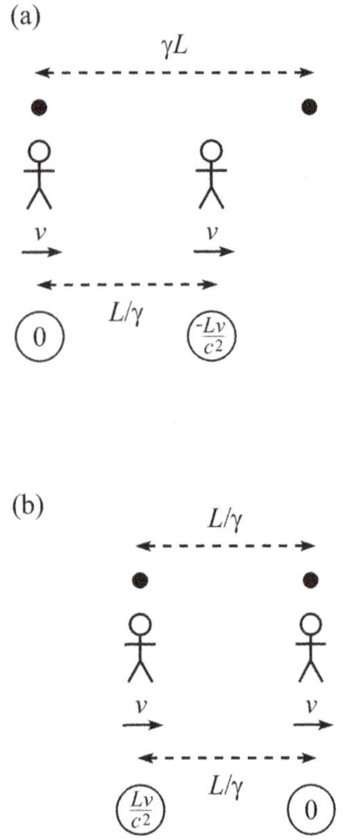

Figure 1.68

(b) By looking at the two distances in Fig. 1.68(a), we see that the people travel a distance $\gamma L - L/\gamma$ between the beginning and ending moments.

Let's now use the clock readings to obtain the distance the people travel. The total time of the process is $\gamma(Lv/c^2)$ because each person's clock advances by Lv/c^2, but these clocks run slow in the frame we're working in. Since the speed of the people is v, the distance they travel is $v(\gamma Lv/c^2)$. This had better be equal to $\gamma L - L/\gamma$. And it is, because

$$\gamma L - \frac{L}{\gamma} = \gamma L\left(1 - \frac{1}{\gamma^2}\right) = \gamma L\left(1 - \left(1 - \frac{v^2}{c^2}\right)\right) = \frac{\gamma L v^2}{c^2}. \tag{1.52}$$

If we then shift to the people's frame where which everything is at rest, we see that the proper distance between the balls is L, because this is what gets length contracted down to the L/γ in Fig. 1.68(b).

(c) To sum up, the proper distance between the balls decreases because in the frame in which the balls are initially at rest, the left person catches the left ball first and then drags it closer to the right ball by the time the right person catches that ball. So it all comes down to the loss of simultaneity.

1.12. **Transforming the length**

The proper length of the stick is $\gamma_{3/5}L = 5L/4$, because this is the length that is contracted down to L in the ground frame. Using the velocity-addition (or rather, subtraction) formula, the speed at which you see the stick move rightward is

$$\frac{\frac{3c}{5} - \frac{c}{2}}{1 - \frac{3}{5}\cdot\frac{1}{2}} = \frac{\frac{c}{10}}{\frac{7}{10}} = \frac{c}{7}. \tag{1.53}$$

The length that you observe is obtained by contracting the proper length by the γ factor associated with this $c/7$ speed, which is $\gamma_{1/7} = 7/\sqrt{48}$. So the length you observe is

$$\frac{L_{\text{proper}}}{\gamma_{1/7}} = \frac{\gamma_{3/5}L}{\gamma_{1/7}} = \frac{5L}{4}\cdot\frac{\sqrt{48}}{7} = \frac{5\sqrt{3}}{7}L \approx (1.24)L. \tag{1.54}$$

This is larger than the length L in the ground frame, because the stick is moving slower in your frame (at $c/7$) than in the ground frame (at $3c/5$). So it is contracted less in your frame, from its proper length $5L/4$.

Remark: Note that it is *not* correct to say, "Since the length of the stick in the ground frame is L, and since you are moving with speed $c/2$ with respect to the ground, you observe the length of the stick to be $L/\gamma_{1/2} = \sqrt{3}L/2$." This is incorrect because the standard length-contraction result applies only to the *proper* length. Length contraction says, "If a stick is moving at speed v with respect to you, then in your frame it is short (relative to its proper length) by a factor γ_v." This is why we had to first find the proper length in the above solution. You can't contract a non-proper length. (Equivalently, if you are using length contraction to go from one frame to another, the stick must be at rest in one of the frames.) This is clear in a special case: If in the above problem you are moving rightward also with speed $3c/5$, then the stick is at rest with respect to you, so you must observe a length that is *longer* than L (namely the proper length, $\gamma_{3/5}L = 5L/4$). The incorrect naive application of length contraction (contracting the ground frame's length L) would yield a length that is *shorter* than L (namely $L/\gamma_{3/5} = 4L/5$). ♣

1.13. **Magnetic force**

In q's rest frame, the situation is shown in Fig. 1.69. The protons are moving leftward with speed v (because they were at rest in the original frame). The distance between them is contracted by the factor $\gamma_v \equiv \gamma$, so the density is increased by the factor γ. The protons' charge density in q's frame is therefore

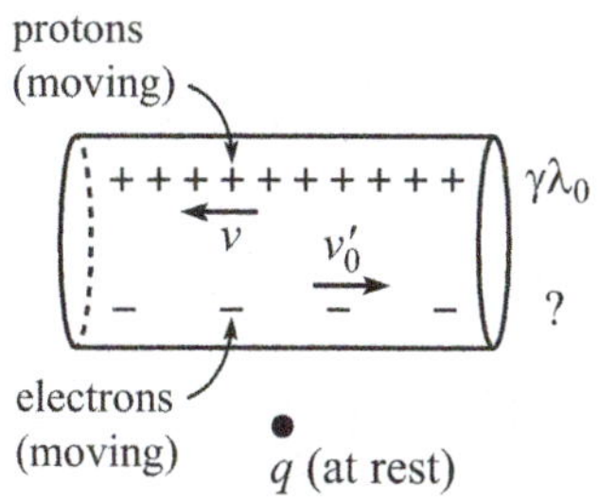

Figure 1.69

$$\lambda_{\text{protons}} = \gamma\lambda_0. \tag{1.55}$$

To determine the electrons' charge density in q's frame, we need to find the electrons' velocity (call it v_0') in q's frame. This is obtained via the velocity-addition (or subtraction) formula, which gives

$$v_0' = \frac{v_0 - v}{1 - v_0 v/c^2}. \tag{1.56}$$

If this is negative, then the electrons are actually moving leftward. Note that there are three different velocities that appear in this solution:

v: velocity of charge q in lab frame,

v_0: velocity of electrons in lab frame,

v_0': velocity of electrons in q's frame.

v is also the leftward speed of the protons in q's frame. The γ factors associated with each of the above three velocities will appear at various places in this solution.

Before getting quantitative, let's give a qualitative argument for why the net charge density of the wire is nonzero in q's frame. For the sake of drawing Fig. 1.69, we have assumed that $v < v_0$. The electrons are therefore still moving rightward, but with a speed (the v_0' in

Eq. (1.56)) that is smaller than v_0. Since this speed is smaller than the electrons' speed v_0 in the lab frame, the electrons are farther apart in q's frame than in the lab frame (because the proper distance between them isn't contracted by as large a factor). The electrons' charge density in q's frame is therefore *smaller* (in magnitude) than the density in the lab frame, which was λ_0. Since we found in Eq. (1.55) that the protons' charge density in q's frame is *larger* than λ_0, we see that the wire has a net positive charge. That is, it is *not* neutral in q's frame. Let's now be quantitative about what the charge density actually is.

To determine the separation between the electrons (and thereby their density) in q's frame, we can't simply contract the separation in the original frame. (See Problem 1.12 for a discussion of this erroneous method.) Instead, we must first find the proper separation (that is, in the rest frame of the electrons) and then length contract this distance by the appropriate γ factor. The proper separation between the electrons is $\gamma_{v_0} \equiv \gamma_0$ times the separation in the lab frame (because it is then contracted down to the distance in the lab frame shown in Fig. 1.50). The proper density is therefore smaller than λ_0 by a factor γ_0. So it equals $-\lambda_0/\gamma_0$, with the negative sign due to the fact that electrons are negatively charged.

The electrons' separation in q's frame is then obtained by dividing the proper separation by the γ factor associated with the speed v_0' in Eq. (1.56); let's call this γ_0'. Equivalently, the electrons' density in q's frame is obtained by multiplying the proper density (which is $-\lambda_0/\gamma_0$) by γ_0'. So the electrons' density in q's frame equals $\gamma_0'(-\lambda_0/\gamma_0)$. We must therefore determine γ_0'. A straightforward but slightly tedious calculation gives (switching to the β notation to that we can avoid writing all the c's)

$$\begin{aligned}
\gamma_0' = \frac{1}{\sqrt{1-\beta_0'^2}} &= \frac{1}{\sqrt{1-\left(\frac{\beta_0-\beta}{1-\beta_0\beta}\right)^2}} = \frac{1-\beta_0\beta}{\sqrt{(1-\beta_0\beta)^2-(\beta_0-\beta)^2}} \\
&= \frac{1-\beta_0\beta}{\sqrt{1+\beta_0^2\beta^2-\beta_0^2-\beta^2}} = \frac{1-\beta_0\beta}{\sqrt{1-\beta_0^2}\sqrt{1-\beta^2}} \\
&= \gamma_0\gamma(1-\beta_0\beta).
\end{aligned} \tag{1.57}$$

The electrons' density in q's frame is therefore

$$\lambda_{\text{electrons}} = \gamma_0' \cdot \frac{-\lambda_0}{\gamma_0} = \gamma_0\gamma(1-\beta_0\beta)\cdot\frac{-\lambda_0}{\gamma_0} = -\gamma\lambda_0(1-\beta_0\beta). \tag{1.58}$$

Recalling Eq. (1.55), the *total* charge density of the positive protons and the negative electrons is then

$$\begin{aligned}
\lambda_{\text{total}} &= \lambda_{\text{protons}} + \lambda_{\text{electrons}} \\
&= \gamma\lambda_0 - \gamma\lambda_0(1-\beta_0\beta) \\
&= \gamma\beta\beta_0\lambda_0.
\end{aligned} \tag{1.59}$$

This is nonzero, as we noted above. Assuming that β_0 is positive, λ_{total} is positive if β is positive (that is, q is moving rightward), but negative if β is negative (that is, q is moving leftward).

Remark: We have solved the stated problem, but let's now see how the result in Eq. (1.59) leads to the magnetic force. We'll mostly just work with proportionalities instead of equalities here, lest we get bogged down with various constants and definitions that belong more in a book on electromagnetism. We'll invoke a few facts from electromagnetism in the following discussion.

In q's frame, the force on q equals the product of q and the electric field E from the charged wire. We'll just accept here (quite reasonably) that the electric field from a wire is proportional to the charge density. So the force on q in q's frame is (using Eq. (1.59))

$$F_{\text{in } q \text{ frame}} = qE \propto q\lambda_{\text{total}} = q\gamma\beta\beta_0\lambda_0. \tag{1.60}$$

If this is positive (that is, if q and λ_{total} have the same sign), then the force is directed away from the wire. If it is negative, the force is directed toward the wire. In Chapter 3 we'll show that the transverse force on a particle is larger in the particle's frame than in any other frame (see Eq. (3.71)). The force in the original lab frame is then

$$F_{\text{in lab}} = \frac{F_{\text{in } q \text{ frame}}}{\gamma} = q\beta\beta_0\lambda_0 \propto qvv_0\lambda_0. \tag{1.61}$$

The current I in the wire in the lab frame is proportional (actually equal) to $v_0\lambda_0$. This makes sense; the larger that v_0 or λ_0 is, the more charge that passes by a given point on the wire. Using $v_0\lambda_0 \propto I$ in Eq. (1.61) yields

$$F_{\text{in lab}} \propto qvI. \tag{1.62}$$

Finally, the magnitude of the magnetic field B due to the current in the wire in the lab frame is proportional to the current I (we'll just accept this). So we finally have

$$F_{\text{in lab}} \propto qvB. \tag{1.63}$$

Although we dealt only with proportionalities in the above reasoning, we got a bit lucky; it turns out that qvB is either exactly the correct force, or off by only a factor of c, depending on which of the two most common systems of units you use.

In the present case where q is moving parallel to the wire, the force is directed either toward or away from the wire. For other directions of q's motion, the direction of the force in the lab frame is given by the general cross-product expression: $\mathbf{F} \propto q\mathbf{v} \times \mathbf{B}$. It can be shown that the magnetic field $\mathbf{B}$ (which is a vector) points in the tangential direction around the wire (that is, the field lines form circles around the wire). In the present setup, the velocity $\mathbf{v}$ and the magnetic field $\mathbf{B}$ are perpendicular, so the magnitude of $q\mathbf{v} \times \mathbf{B}$ is simply qvB, as we found above. Note that if $\mathbf{v}$ is parallel to $\mathbf{B}$, the magnetic force is zero. ♣

1.14. **Pythagorean triples**

The relativistic addition or subtraction of the two given β's has a β value of

$$\frac{\frac{a}{h} \pm \frac{b}{h}}{1 \pm \frac{ab}{h^2}} = \frac{(a \pm b)h}{h^2 \pm ab}. \tag{1.64}$$

The numerator and denominator here are two lengths in a Pythagorean triple, because

$$(h^2 \pm ab)^2 - ((a \pm b)h)^2 = h^4 + a^2b^2 - (a^2 + b^2)h^2 = a^2b^2, \tag{1.65}$$

where we have used the given information that $a^2 + b^2 = h^2$. The other leg is therefore ab, for both the addition and subtraction cases. So the full triple is

$$\big((a \pm b)h,\ ab,\ h^2 \pm ab\big). \tag{1.66}$$

The γ factor associated with the speed in Eq. (1.64) is

$$\gamma = \frac{1}{\sqrt{1 - \left(\frac{(a \pm b)h}{h^2 \pm ab}\right)^2}} = \frac{h^2 \pm ab}{\sqrt{(h^2 \pm ab)^2 - ((a \pm b)h)^2}} = \frac{h^2 \pm ab}{ab}, \tag{1.67}$$

which is the hypotenuse divided by the second leg. (You can show that this γ is consistent with the result from Exercise 1.41.) As an example, the initial triple $(3, 4, 5)$ (or $(4, 3, 5)$ if we take a to be the longer leg) gives the addition triple $(35, 12, 37)$ with $\gamma = 37/12$, and the subtraction triple $(5, 12, 13)$ with $\gamma = 13/12$.

1.15. **Fizeau experiment**

Since the light moves at speed c/n with respect to the water, and the water moves at speed v with respect to the ground, the velocity-addition formula gives the speed of the light with respect to the ground as

$$V = \frac{c/n + v}{1 + \frac{(c/n)v}{c^2}} = \frac{c/n + v}{1 + v/nc}. \tag{1.68}$$

To produce an approximate form of this answer when $v \ll c$, we can multiply the numerator and denominator each by $1 - v/nc$ and keep terms only up to order v/c. With $O(v^2/c^2)$ denoting terms of order v^2/c^2, we obtain

$$\begin{aligned} V &= \frac{c(1/n + v/c)}{1 + v/nc} \cdot \frac{1 - v/nc}{1 - v/nc} = c \cdot \frac{1/n + (v/c)(1 - 1/n^2) - O(v^2/c^2)}{1 - O(v^2/c^2)} \\ &\approx c \cdot \left(\frac{1}{n} + \frac{v}{c}\left(1 - \frac{1}{n^2}\right)\right) = \frac{c}{n} + v\left(1 - \frac{1}{n^2}\right). \end{aligned} \tag{1.69}$$

The desired value of A is therefore $1 - 1/n^2$. We see that the speed of light in moving water increases with the velocity v of the water, but not as fast as the naive answer of $c/n + v$ would imply. Instead of adding v to c/n, we add only $(1 - 1/n^2)v$.

We can check the result in Eq. (1.69) in a few special cases. If $n = 1$, which means that we have vacuum instead of water, we obtain a speed of $c/1 + v(1 - 1) = c$. This is correct because we know that light always moves with speed c in vacuum. If n is very large, we obtain a speed of $c/n + v(1 - 0) = c/n + v$. This is correct because it is the naive addition of the speeds, which we know works perfectly fine when both speeds are much less than c.

In 1851, which was well before Einstein's velocity-addition formula was known, Fizeau performed an experiment to measure the speed (with respect to the ground) of light in moving water. His setup involved an interferometer similar to the one Michelson and Morley used in their experiment. He obtained a result consistent with our approximate formula in Eq. (1.69), so he conjectured that the formula held (exactly) in general. Many people then made unsuccessful attempts (involving frame dragging of the "ether," for example) to explain why the parameter A took on the value of $1 - 1/n^2$ instead of the naive value of 1. In retrospect, of course, failure was the likely result of their (commendable) efforts to generate an exact theory from an approximate result. It was more than half a century until Einstein produced the theory of special relativity in 1905, from which the correct explanation of A's value followed via the velocity-addition formula (along with the approximations we made in Eq. (1.69)). Conversely, the result of Fizeau's experiment was highly influential in Einstein's formulation of special relativity.

1.16. **Equal speeds**

First solution: Let C move at speed v with respect to the ground, and let the relative speed of C and both A and B be u (as viewed by C). Then two different expressions for u are the relativistic subtraction of v from $4c/5$, and the relativistic subtraction of $3c/5$ from v. Therefore,

$$\frac{\frac{4}{5} - v}{1 - \frac{4}{5}v} = \frac{v - \frac{3}{5}}{1 - \frac{3}{5}v}, \tag{1.70}$$

where we have temporarily ignored the c's, or equivalently used v to stand for $\beta \equiv v/c$, or equivalently pretended that c equals 1. (We'll do this in all three solutions here, since it keeps things from getting too messy.) After some algebra, you can show that Eq. (1.70) reduces to $0 = 35v^2 - 74v + 35 = (5v - 7)(7v - 5)$. Since the $v = 7/5$ root represents a speed larger than c, we want the other root:

$$v = \frac{5}{7}c, \tag{1.71}$$

where we have brought the c back in. This is the speed of C with respect to the ground. Plugging this back into either expression for u in Eq. (1.70) gives $u = c/5$. This is how fast C sees both A and B approaching her. Note that C's speed with respect to the ground *cannot* be obtained by simply taking the average of A's and B's speeds, which would give $7c/10$. Taking the average works for nonrelativistic speeds, but not for relativistic ones.

SECOND SOLUTION: With u and v defined as above, two different expressions for v are the relativistic subtraction of u from $4c/5$, and the relativistic addition of u to $3c/5$. Therefore,

$$\frac{\frac{4}{5} - u}{1 - \frac{4}{5}u} = \frac{\frac{3}{5} + u}{1 + \frac{3}{5}u}. \tag{1.72}$$

After some algebra, you can show that this reduces to $0 = 5u^2 - 26u + 5 = (5u - 1)(u - 5)$. Since the $u = 5$ root represents a speed larger than c, we want the other root:

$$u = \frac{c}{5}. \tag{1.73}$$

Plugging this back into either expression for v in Eq. (1.72) gives $v = 5c/7$.

THIRD SOLUTION: The relative speed of A and B (as viewed by either A or B) is

$$\frac{\frac{4}{5} - \frac{3}{5}}{1 - \frac{4}{5}\cdot\frac{3}{5}} = \frac{5}{13}. \tag{1.74}$$

In C's frame, A approaches with speed u from one side, and B approaches with speed u from the other. The relative speed of A and B (as viewed by either A or B) is therefore obtained by relativistically adding u with another u. But we just found that this relative speed is $5/13$. Therefore,

$$\frac{u + u}{1 + u^2} = \frac{5}{13} \quad \Longrightarrow \quad 5u^2 - 26u + 5 = 0, \tag{1.75}$$

as in the second solution.

1.17. **More equal speeds**

Let u be the speed at which C sees A and B approaching her. Then u is the desired speed of C with respect to B (that is, the ground). From C's point of view, the given speed v is the result of relativistically adding u with another u. Therefore,

$$v = \frac{2u}{1 + u^2/c^2} \quad \Longrightarrow \quad \left(\frac{v}{c^2}\right)u^2 - 2u + v = 0. \tag{1.76}$$

Solving this quadratic equation for u gives

$$u = \frac{c^2(1 - \sqrt{1 - v^2/c^2})}{v} = \frac{c^2(1 - 1/\gamma)}{v}. \tag{1.77}$$

The quadratic equation also has a solution with a plus sign in front of the square root, but this solution cannot be correct, because it is greater than c, as you can verify (and in fact goes to infinity as v goes to zero). The above solution for u has the correct limit as v goes to zero, namely $u \to v/2$ (the expected nonrelativistic result); this can be obtained by using the Taylor approximation, $\sqrt{1 - \epsilon} \approx 1 - \epsilon/2$.

The ratio of the distances CB and AC in the ground frame is the same as the ratio of the differences in velocities as measured in the ground frame (because both A and C arrive at B at the same time, so you could imagine running the scenario backward in time). Therefore,

$$\begin{aligned}\frac{CB}{AC} = \frac{V_C - V_B}{V_A - V_C} &= \frac{\frac{c^2(1 - 1/\gamma)}{v} - 0}{v - \frac{c^2(1 - 1/\gamma)}{v}} = \frac{1 - 1/\gamma}{v^2/c^2 - 1 + 1/\gamma} \\ &= \frac{1 - 1/\gamma}{1/\gamma - (1 - v^2/c^2)} = \frac{1 - 1/\gamma}{1/\gamma - 1/\gamma^2} = \gamma. \end{aligned} \tag{1.78}$$

We see that C is γ times as far from B as she is from A, as measured in the ground frame. Note that for nonrelativistic speeds, we have $\gamma \approx 1$, so C is midway between A and B, as expected.

An intuitive reason for the simple factor of γ is the following. Imagine that A and B are carrying identical jousting sticks as they run toward C (in C's frame). In C's frame, the tips of the sticks reach C simultaneously, because in C's frame A and B are always the same distance from C. This is true because we are told that all three people eventually coincide at some instant. But since the sticks reach C simultaneously in C's frame, they do also in B's frame (the ground frame). This is true because since we're talking about the ends of sticks reaching C, everything happens right at C. The L in the Lv/c^2 rear-clock-ahead result is therefore zero, so we don't have to worry about any loss of simultaneity. Consider then the instant in B's frame (the ground frame) when both sticks reach C. B's stick is at rest, so it is uncontracted. But A's stick is moving with speed v, so it is length contracted by a factor γ. Therefore, in the ground frame, A is closer to C than B is, by a factor γ.

1.18. **Many velocity additions**

Let's first check the formula for $N = 1$ and $N = 2$. When $N = 1$, it gives

$$\beta_{(1)} = \frac{P_1^+ - P_1^-}{P_1^+ + P_1^-} = \frac{(1+\beta_1) - (1-\beta_1)}{(1+\beta_1) + (1-\beta_1)} = \beta_1, \tag{1.79}$$

as it should. And when $N = 2$, it gives

$$\beta_{(2)} = \frac{P_2^+ - P_2^-}{P_2^+ + P_2^-} = \frac{(1+\beta_1)(1+\beta_2) - (1-\beta_1)(1-\beta_2)}{(1+\beta_1)(1+\beta_2) + (1-\beta_1)(1-\beta_2)} = \frac{\beta_1 + \beta_2}{1 + \beta_1\beta_2}, \tag{1.80}$$

in agreement with the velocity-addition formula. You can check that the factors of c work out correctly when the β's are swapped for v's.

Let's now prove the formula for a general N. We will use induction. That is, we will assume that the result holds for a given N and then show that it also holds for $N + 1$. To find the speed, $\beta_{(N+1)}$, of the object with respect to S_{N+1}, we can relativistically add the speed of the object with respect to S_N (which is $\beta_{(N)}$) with the speed of S_N with respect to S_{N+1} (which is β_{N+1}). This gives

$$\beta_{(N+1)} = \frac{\beta_{N+1} + \beta_{(N)}}{1 + \beta_{N+1}\beta_{(N)}}. \tag{1.81}$$

Under the assumption that our formula holds for N, this becomes

$$\begin{aligned}
\beta_{(N+1)} = \frac{\beta_{N+1} + \dfrac{P_N^+ - P_N^-}{P_N^+ + P_N^-}}{1 + \beta_{N+1}\dfrac{P_N^+ - P_N^-}{P_N^+ + P_N^-}} &= \frac{\beta_{N+1}(P_N^+ + P_N^-) + (P_N^+ - P_N^-)}{(P_N^+ + P_N^-) + \beta_{N+1}(P_N^+ - P_N^-)} \\
&= \frac{P_N^+(1+\beta_{N+1}) - P_N^-(1-\beta_{N+1})}{P_N^+(1+\beta_{N+1}) + P_N^-(1-\beta_{N+1})} \\
&\equiv \frac{P_{N+1}^+ - P_{N+1}^-}{P_{N+1}^+ + P_{N+1}^-},
\end{aligned} \tag{1.82}$$

as we wanted to show. We have therefore shown that if the result holds for N, then it also holds for $N + 1$. Since we know that the result does indeed hold for $N = 1$, it therefore holds for all N.

The expression for $\beta_{(N)}$ has some expected properties. It is symmetric in the β_i. And if the given object is a photon with $\beta_1 = 1$, then $P_N^- = 0$, which yields $\beta_{(N)} = 1$ as it should. And if the given object is a photon with $\beta_1 = -1$, then $P_N^+ = 0$, which yields $\beta_{(N)} = -1$ as it should. Likewise, if any one of the β_i's equals 1 (or -1), then $P_N^- = 0$ (or $P_N^+ = 0$), which correctly yields $\beta_{(N)} = 1$ (or $\beta_{(N)} = -1$).

1.19. **Velocity addition from scratch**

As stated in the problem, we will use the fact that the meeting of the photon and the ball occurs at the same fraction of the way along the train, independent of the frame. This is true because, although distances may change depending on the frame, fractions remain the same, since length contraction doesn't depend on position. We'll compute the desired fraction in the train frame S', and then in the ground frame S.

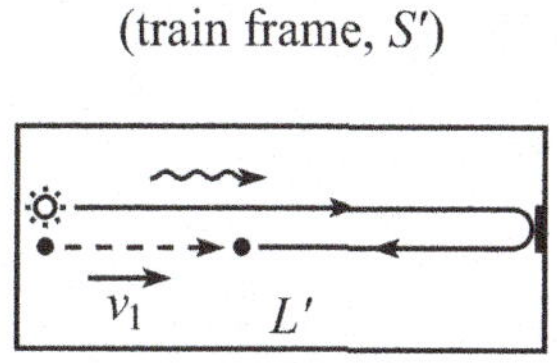

Figure 1.70

TRAIN FRAME: Let the train have length L' in the train frame, S'. Let's first find the time at which the photon meets the ball. From Fig. 1.70, we see that the sum of the distances traveled by the ball and the photon, which is $v_1t' + ct'$, must equal twice the length of the train, which is $2L'$. The time of the meeting is therefore

$$t' = \frac{2L'}{c + v_1}. \tag{1.83}$$

The distance the ball has traveled is then $d' = v_1t' = 2v_1L'/(c+v_1)$, so the desired fraction F' is

$$F' = \frac{d'}{L'} = \frac{2v_1}{c + v_1}. \tag{1.84}$$

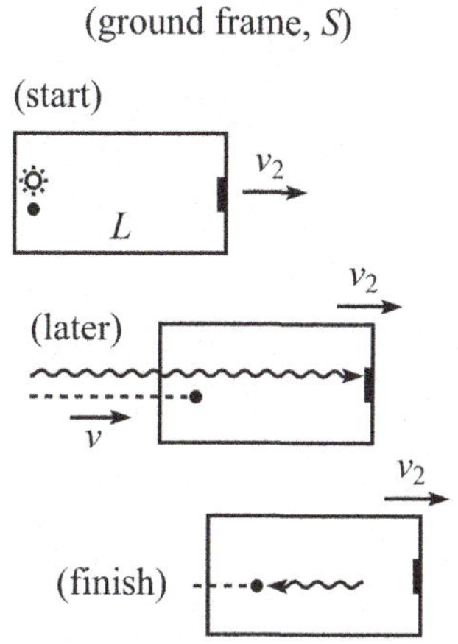

Figure 1.71

GROUND FRAME: Let the speed of the ball with respect to the ground be v, and let the train have length L in the ground frame. (L equals L'/γ, but we're not going to use this.) Again, let's first find the time at which the photon meets the ball. From Fig. 1.71, we see that the photon takes a time $L/(c - v_2)$ to reach the mirror, because the initial gap of L is closed at a rate $c - v_2$ in the ground frame. At this time, the photon has traveled a distance $cL/(c - v_2)$. From the figure, we see that we can use the same reasoning we used in the train frame, but with the sum of the distances traveled by the ball and the photon, which is $vt + ct$, now equal to $2[cL/(c - v_2)]$. The time of the meeting in the ground frame is therefore

$$t = \frac{2cL/(c - v_2)}{(c + v)}. \tag{1.85}$$

The relative speed of the ball and the back of the train (as viewed in the ground frame) is $v - v_2$. This is the rate at which the gap between them is increasing. So the distance between the ball and the back of the train at the time of the meeting is $d = (v - v_2)t = (v - v_2) \cdot 2cL/[(c - v_2)(c + v)]$. The desired fraction F is therefore

$$F = \frac{d}{L} = \frac{2(v - v_2)c}{(c - v_2)(c + v)}. \tag{1.86}$$

We can now equate the above expressions for F' and F. For convenience, define $\beta \equiv v/c$, $\beta_1 \equiv v_1/c$, and $\beta_2 \equiv v_2/c$. Then $F' = F$ yields

$$\frac{\beta_1}{1 + \beta_1} = \frac{\beta - \beta_2}{(1 - \beta_2)(1 + \beta)}. \tag{1.87}$$

Solving for β in terms of β_1 and β_2 gives, after some algebra,

$$\beta = \frac{\beta_1 + \beta_2}{1 + \beta_1\beta_2}, \tag{1.88}$$

as desired. This problem is solved in Mermin (1983).

1.20. **Time dilation and Lv/c^2**

The velocity-addition formula gives the person's speed in the ground frame as $(u + v)/(1 + uv)$, where we have dropped the c's. So in the ground frame, the person must close the initial gap of L/γ_v that the front of the train had, at a relative speed of $(u + v)/(1 + uv) - v$. The time in the ground frame is therefore

$$t_g = \frac{L/\gamma_v}{\dfrac{u + v}{1 + uv} - v} = \frac{L(1 + uv)}{u\sqrt{1 - v^2}}. \tag{1.89}$$

Compared with this ground-frame time, the front clock on the train runs slow by the factor γ_v, and the person's watch runs slow by the γ factor associated with the speed

$(u+v)/(1+uv)$, which you can show equals $\gamma_u\gamma_v(1+uv)$; see Exercise 1.41. The difference in the elapsed times on the front clock and the person's watch is therefore

$$\begin{aligned}\Delta T_{\text{front}} - \Delta T_{\text{person}} &= \frac{L(1+uv)}{u\sqrt{1-v^2}}\left(\frac{1}{\gamma_v} - \frac{1}{\gamma_u\gamma_v(1+uv)}\right)\\ &= \frac{L}{u}\left(1+uv-\frac{1}{\gamma_u}\right)\\ &= \frac{Lv}{c^2} + \frac{L}{u}\left(1-\frac{1}{\gamma_u}\right),\end{aligned} \tag{1.90}$$

where we have put the c's back in to make the units correct. The second term here is negligible for the following reason. For small u, we can use the Taylor series $\sqrt{1-\epsilon} \approx 1-\epsilon/2$ to write $1/\gamma_u = \sqrt{1-u^2/c^2} \approx 1-u^2/2c^2$. The $(L/u)(1-1/\gamma_u)$ term then becomes $(L/u)(u^2/2c^2) = Lu/2c^2$. Since u is assumed to be small (more precisely, $u \ll c$), this term is negligible. So Eq. (1.90) becomes $\Delta T_{\text{front}} - \Delta T_{\text{person}} \approx Lv/c^2$. The front clock therefore gains essentially Lv/c^2 more time than the person's watch, as we wanted to show. Since the front clock started Lv/c^2 behind the person's watch, we conclude that they end up showing the same time when the watch reaches the front, as we already knew from working in the train frame. The point here is that no matter how small u is, the result for $\Delta T_{\text{front}} - \Delta T_{\text{person}}$ is nonzero (namely Lv/c^2) because u appears at *first* order in the γ factor, $\gamma_u\gamma_v(1+uv)$, associated with $(u+v)/(1+uv)$, while it appears only at *second* order in γ_u. The difference between the γ factors is therefore first order in u, and this difference combines with the $1/u$ factor in the time to yield a nonzero result.

The result in Eq. (1.90) holds perfectly well for non-small u too, so it implies that the final readings on the front clock and the person's watch differ by $(L/u)(1-1/\gamma_u)$, for any u. In retrospect, this is clear from the train-frame calculation which gives the difference as $(L/u) - (L/u)/\gamma_u$, due to the time dilation of the watch.

1.21. **Modified twin paradox**

(a) To help visualize the setup in each frame, we'll draw the positions of the three people as functions of time. The resulting lines (or more generally, curves) are known as *worldlines*. In relativity, it is customary to put time on the vertical axis and space on the horizontal axis (the opposite of what is normally done). It is also customary to plot the value of ct, instead of t. This leads to the nice fact that light is represented by a lines with slope $\pm 45°$. The worldline of any (massive) object will always have a slope that is larger than 45°, because $v < c$.

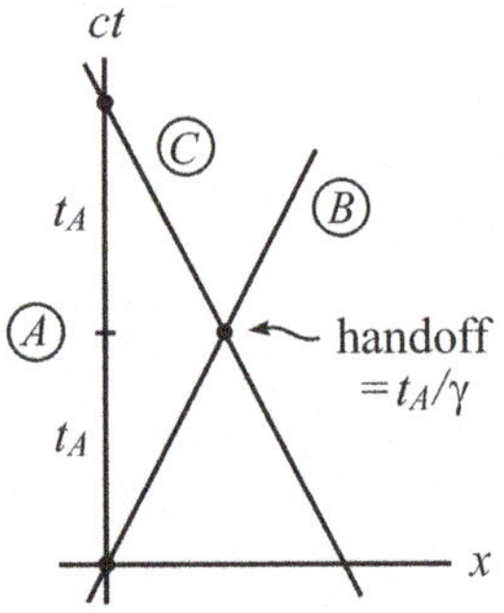

Figure 1.72

In A's frame, the worldlines of A, B, and C are shown in Fig. 1.72. A is at rest, so his worldline is vertical. B moves to the right at speed v, and C moves to the left at speed v. In A's frame, B's clock runs slow by a factor $1/\gamma$. Therefore, if A's clock reads t_A when B meets C, then B's clock reads only t_A/γ when he meets C. So the time he hands off to C is t_A/γ.

In A's frame, the time between the B-meets-C event and the C-meets-A event is again t_A, because B and C travel at the same speed. And A sees C's clock run slow by a factor $1/\gamma$, so A sees C's clock increase by only t_A/γ. Therefore, when A and C meet, A's clock reads $2t_A$, and C's clock reads $2t_A/\gamma$. In other words, $T_C = T_A/\gamma$.

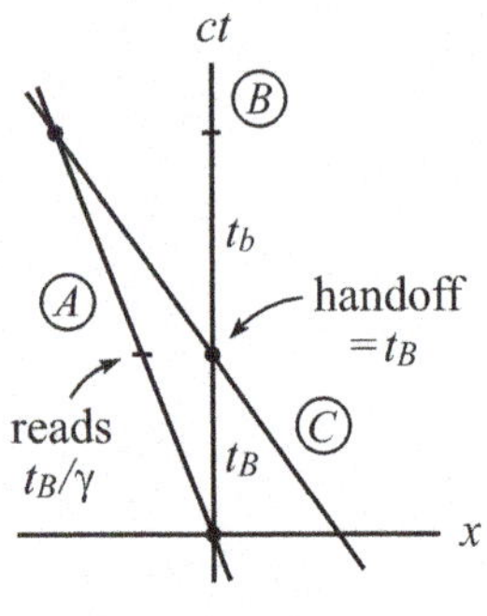

Figure 1.73

(b) Let's now look at things in B's frame. The worldlines of A, B, and C are shown in Fig. 1.73. A moves to the left at speed v, and C moves to the left at speed $2v/(1+v^2)$. This is the velocity addition of v with itself; we have ignored the c's, to keep things from getting cluttered.

From B's point of view, there are two competing effects that lead to the relation $T_C = T_A/\gamma$. The first is that B sees A's clock run slow, so the time that B hands off to C is *larger* than the time on A's clock at that moment. So C's clock reads more than A's at the handoff moment. The second effect is that from this point on, B sees C's clock run *slower* than A's (because the relative speed of C and B is greater than the relative speed of A and B). It turns out that this slowness wins out over the head start that C's clock had over A's. So in the end, C's clock reads less than A's. Let's be quantitative about this.

Let B's clock read t_B when C meets him. (t_B is the same as the t_A/γ in part (a), but we won't use that since we're doing things from B's point of view here.) Then when B hands off this time to C, A's clock reads only t_B/γ, because B sees A's clock run slow. We must determine how much additional time elapses on A's clock and on C's clock, by the time they meet. We'll find all times below in terms of t_B.

At time t_B (when C passes B) A is a distance vt_B from B. Let t_b be the additional time on B's clock between C passing him and C catching up with A. We can find t_b by noting that C closes the initial head start of vt_B that A had, at a relative speed of $2v/(1+v^2) - v$, as viewed by B. So

$$t_b = \frac{vt_B}{\dfrac{2v}{1+v^2} - v} \implies t_b = t_B\left(\frac{1+v^2}{1-v^2}\right). \tag{1.91}$$

During the time t_b, B sees A's and C's clocks increase by t_b divided by the relevant time-dilation factor. For A this factor is $\gamma = 1/\sqrt{1-v^2}$, and for C it is

$$\gamma_C = \frac{1}{\sqrt{1-\left(\dfrac{2v}{1+v^2}\right)^2}} = \frac{1+v^2}{1-v^2}, \tag{1.92}$$

as you can verify. Therefore, the total time shown on A's clock when A and C meet is

$$\begin{aligned} T_A &= \frac{t_B}{\gamma} + \frac{t_b}{\gamma} = t_B\sqrt{1-v^2} + t_B\left(\frac{1+v^2}{1-v^2}\right)\cdot\sqrt{1-v^2} \\ &= \frac{2t_B}{\sqrt{1-v^2}}. \end{aligned} \tag{1.93}$$

And the total time shown on C's clock when A and C meet is the handoff time of t_B plus the time elapsed on C, so

$$T_C = t_B + \frac{t_b}{\gamma_C} = t_B + t_B\left(\frac{1+v^2}{1-v^2}\right)\cdot\left(\frac{1-v^2}{1+v^2}\right) = 2t_B. \tag{1.94}$$

Therefore, $T_C = T_A\sqrt{1-v^2} = T_A/\gamma$, as desired.

(c) Let's now work in C's frame. The worldlines of A, B, and C are shown Fig. 1.74. A moves to the right at speed v, and B moves to the right at speed $2v/(1+v^2)$. As in part (b), the time-dilation factor between B and C is $\gamma_B = (1+v^2)/(1-v^2)$. Also, as in part (b), let B and C meet when B's clock reads t_B. So this is the time that B hands off to C. We'll find all times below in terms of t_B.

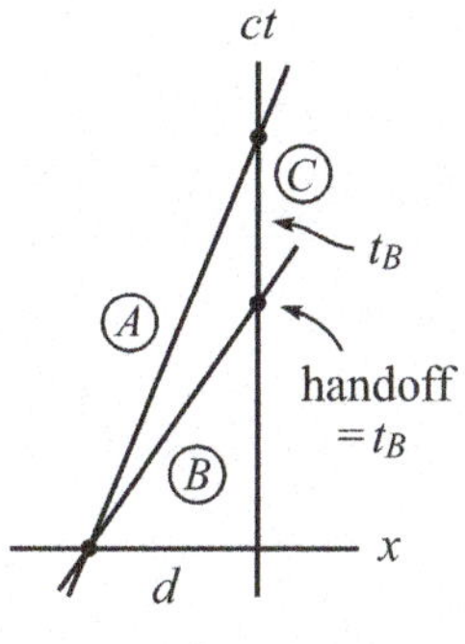

Figure 1.74

C sees B's clock running slow, so in C's frame it takes a time of $\gamma_B t_B$ for B's clock to advance by t_B, since when he met A. B therefore travels for a time of

$$t_{B\,\text{reach}\,C} = \gamma_B t_B = t_B\left(\frac{1+v^2}{1-v^2}\right) \tag{1.95}$$

between meeting A and meeting C. During this time, B covers a distance in C's frame equal to

$$d = t_B\left(\frac{1+v^2}{1-v^2}\right)\cdot\frac{2v}{1+v^2} = \frac{2vt_B}{1-v^2}. \tag{1.96}$$

A must travel this same distance (from where B passed him) to meet up with C. This allows us to find T_A. The time (as viewed by C) that it takes A to travel the distance d to reach C is

$$t_{A\,\text{reach}\,C} = \frac{d}{v} = \frac{2t_B}{1-v^2}. \tag{1.97}$$

But since C sees A's clock running slow by a factor $\sqrt{1-v^2}$, A's clock reads only

$$T_A = \frac{2t_B}{\sqrt{1-v^2}} \tag{1.98}$$

when he meets C. This agrees with Eq. (1.93), as it must, because the reading is frame-independent.

Now let's find T_C. To find T_C, we must take the handoff time of t_B and add to it the extra time it takes A to reach C, compared with the time it takes B to reach C. From Eqs. (1.95) and (1.97), this extra time is

$$t_{A\,\text{reach}\,C} - t_{B\,\text{reach}\,C} = \frac{2t_B}{1-v^2} - \frac{t_B(1+v^2)}{1-v^2} = t_B. \tag{1.99}$$

(This simple result is clear in A's frame, but not so clear in C's frame.) Therefore, C's clock reads

$$T_C = t_B + t_B = 2t_B, \tag{1.100}$$

which agrees with Eq. (1.94), as it must, because the reading is frame-independent. Hence, $T_C = T_A\sqrt{1-v^2} \equiv T_A/\gamma$, as desired.

Chapter 2

Kinematics, Part 2

In this chapter we'll continue with our treatment of kinematics. We derived all the basics in Chapter 1, but the topics in this chapter provide extensions to what we already know and offer alternative ways of looking at things. The outline is as follows.

In Section 2.1 we derive the *Lorentz transformations*. Given two events, the Lorentz transformations relate the Δx and Δt associated with the events in one frame, with the $\Delta x'$ and $\Delta t'$ in another frame. In Section 2.2 we give an alternative (and quick) derivation of the longitudinal velocity-addition formula, and then we go a step further and derive the formula for transverse velocities. Section 2.3 covers the *invariant interval*, which is an important consequence of the Lorentz transformations. Section 2.4 introduces *Minkowski diagrams*. These diagrams provide a geometrical way of thinking about the Lorentz transformations. Section 2.5 derives the relativistic *Doppler effect*, which describes how frequencies in two frames are related. Section 2.6 covers *rapidity*, which gives a useful way of expressing a relativistic velocity. In Section 2.7 we show that nearly all of special relativity (more precisely, nearly all of the information contained in the Lorentz transformations) can be derived by using only the relativity postulate (that is, ignoring the speed-of-light postulate). This is rather surprising, considering that the latter postulate is generally viewed as the one that leads to the strangeness of special relativity.

2.1 Lorentz transformations

The goal of this section is to derive the Lorentz transformations. We will provide two derivations; a third one is presented in Appendix D. As mentioned above, given two events, the Lorentz transformations relate the Δx and Δt in one frame, with the $\Delta x'$ and $\Delta t'$ in another frame. Recall that an event is something that occurs with definite space and time coordinates, for example, the clap of hands or the explosion of a paint bomb. An important point is that the Lorentz transformations deal with *two* events. They have nothing to say about a single event. A common case where the transformations are useful is when you are given a setup for which it is easy to find Δx and Δt in one frame, but hard in another. The Lorentz transformations allow you to simply plug in the easy Δ's, and the hard ones pop out.

The Lorentz transformations contain exactly the same information as the collection of fundamental effects we derived in Chapter 1: rear clock ahead, time dilation, and length contraction. So any problem that can be solved with the fundamental effects can also be solved with the Lorentz transformations, and vice versa. You therefore never *have* to use the Lorentz transformations. But sometimes they provide a quick solution to a problem.

However, note well: the Lorentz transformations are far less intuitive than the fundamental effects. Of course, "intuition" in relativity might seem like an oxymoron, but what we mean is that when using the fundamental effects, you can draw pictures to get a sense of what's going on (however strange those things may be). And you can solve a problem in various steps and check each step. You can also redo the problem in a different frame to check your answer. And so on. In contrast, when using the Lorentz transformations, all you do is plug things into some formulas. It's hard to see what's really going on. It's risky, because if you make an algebra mistake or mess up a sign, there aren't any intuitive double checks you can perform. Besides, drawing pictures and using the fundamental effects is much more fun than plugging things into formulas! I personally would be very hesitant to solve a problem using only the Lorentz transformations. But I would gladly use them as a double check of a fundamental-effect solution. That said, in more advanced physics topics that incorporate relativity, the Lorentz transformations are absolutely critical.

2.1.1 First derivation

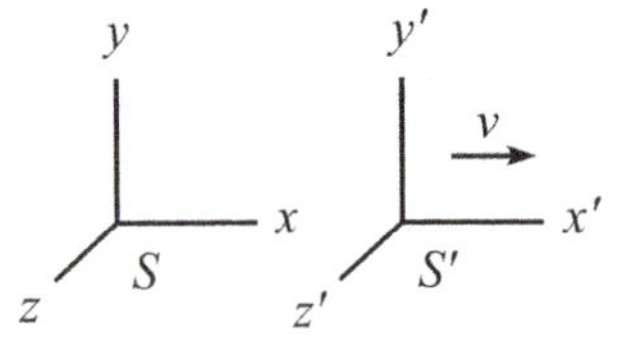

Figure 2.1

Consider a reference frame S' moving relative to another frame S, as shown in Fig. 2.1. Let the constant relative speed of the frames be v. Let the corresponding axes of S and S' point in the same direction, and let the origin of S' move along the x axis of S. Nothing exciting happens in the y and z directions (there is no length contraction or Lv/c^2 effect in the transverse direction), so we'll ignore them.

Our goal is to look at two events and relate the Δx and Δt in S to the $\Delta x'$ and $\Delta t'$ in S'. We therefore want to find the constants A, B, C, and D in the relations,

$$\begin{aligned}\Delta x &= A\,\Delta x' + B\,\Delta t',\\ \Delta t &= C\,\Delta t' + D\,\Delta x'.\end{aligned} \tag{2.1}$$

The four constants here will end up depending on v (which is constant, given the two inertial frames). But we won't explicitly write this dependence, for ease of notation.

Remarks:

1. We have assumed in Eq. (2.1) that A, B, C, and D are constants, that is, they depend at most on v, and not on x, t, x', or t'. And we have also assumed that Δx and Δt are linear functions of $\Delta x'$ and $\Delta t'$.

   The first of these assumptions (constant A, B, C, D) follows from the first of our two relativity postulates, which says that all points in (empty) space are indistinguishable. With this in mind, let's assume that we have a transformation with, say, $A = ax'$. That is, $\Delta x = (ax')\,\Delta x' + B\,\Delta t'$. The x' here implies that the absolute location in space (and not just the relative position) is important. This violates the first postulate, so such a term can't exist.

   The second assumption (linearity in the Δ's) can be justified in various ways. One is that all inertial frames should agree on what non-accelerating motion is. That is, if $\Delta x' = u'\,\Delta t'$, then we should also have $\Delta x = u\,\Delta t$, for some constant u. This is true only if $\Delta x'$ and $\Delta t'$ are raised to the first power in the transformations, as you can check. Technically the conclusion is only that the $\Delta x'$ and $\Delta t'$ terms must all be raised to the same power. But if we then invoke the fact that the transformations must reduce to $\Delta x = \Delta x'$ and $\Delta t = \Delta t'$ when $v = 0$, we see that the only power that has a chance of working is 1.

   The second assumption can also be justified by noting that any interval can be built up from a series of many infinitesimal intervals. But for infinitesimal intervals $\Delta x'$ and $\Delta t'$, any nonlinear terms such as $(\Delta t')^2$ are negligible compared with the linear terms. Therefore, if we add up all the infinitesimal intervals to obtain the given interval, we will be left with only the linear terms. Equivalently, it shouldn't matter whether we make a measurement with, say, meter sticks or half-meter sticks.

2. If the relations in Eq. (2.1) turn out to be the usual Galilean transformations (which are the ones that hold for everyday relative speeds v; see Section 1.1.1), then we will have $\Delta x = \Delta x' + v\Delta t$, and $\Delta t = \Delta t'$ (that is, $A = C = 1$, $B = v$, and $D = 0$). We will find, however, that given the postulates of special relativity, we do *not* obtain the Galilean transformations. But we will show below that the correct transformations reduce to the Galilean transformations in the limit of low speeds, as they must. ♣

The constants A, B, C, and D in Eq. (2.1) are four unknowns, and we can solve for them by using four facts that we have at our disposal, namely the three fundamental effects, along with the given fact that the relative speed of the frames is v. There are various ways to invoke the three fundamental effects. We'll find it most convenient to cast them in the forms shown in the following table:

	Effect	Condition	Result	Eq. in text
1	Time dilation	$\Delta x' = 0$	$\Delta t = \gamma\Delta t'$	(1.14)
2	Length contraction	$\Delta t' = 0$	$\Delta x' = \Delta x/\gamma$	(1.20)
3	Relative v of frames	$\Delta x = 0$	$\Delta x' = -v\Delta t'$	
4	Rear clock ahead	$\Delta t = 0$	$\Delta t' = -v\Delta x'/c^2$	(1.8)

You should pause for a moment and verify that the entries in the "result" column are in fact the proper mathematical expressions for the four effects, given the entries in the "condition" column. My advice is to keep pausing until you're comfortable with all of the entries. Note that the sign in the rear-clock-ahead effect is indeed correct, because the front clock shows less time than the rear clock. So the clock with the higher x' value is the one with the lower t' value. The above effects can be stated in other ways too, by switching the primes and unprimes. For example, time dilation can be written as, "If $\Delta x = 0$ then $\Delta t' = \gamma\Delta t$." But we've chosen the above ways of writing things because they will allow us to solve for the four constants in Eq. (2.1) in the most efficient way. We quickly find that:

Fact 1 gives $C = \gamma$.

Fact 2 gives $A = \gamma$.

Fact 3 gives $B/A = v \Longrightarrow B = \gamma v$.

Fact 4 gives $D/C = v/c^2 \Longrightarrow D = \gamma v/c^2$.

Eqs. (2.1), which are known as the *Lorentz transformations*, are therefore given by

$$\boxed{\begin{aligned}\Delta x &= \gamma(\Delta x' + v\Delta t')\\ \Delta t &= \gamma(\Delta t' + v\Delta x'/c^2)\end{aligned}} \tag{2.2}$$

where

$$\gamma \equiv \frac{1}{\sqrt{1 - v^2/c^2}}. \tag{2.3}$$

Additionally we have $\Delta y = \Delta y'$ and $\Delta z = \Delta z'$. Nothing interesting happens in the y and z directions.

We will often omit the Δ's in Eq. (2.2), but it should be understood that x really means Δx, etc. We are always concerned with the space and time *differences* between the coordinates of two events. The actual values of the coordinates are irrelevant, because there is no preferred origin.

If you want, you can invert the transformations for x and t (again, really Δx and Δt) in Eq. (2.2) and solve for x' and t' in terms of x and t. You can verify that after some algebra, the result is

$$\begin{aligned} x' &= \gamma(x - vt), \\ t' &= \gamma(t - vx/c^2). \end{aligned} \tag{2.4}$$

These are sometimes referred to as the "inverse Lorentz transformations." However, this terminology is a bit misleading, because the set of equations you label as the "inverse" transformations depends on your point of view. The S frame is no more fundamental than the S' frame, so the transformations in Eq. (2.2) are no more fundamental than the ones in Eq. (2.4).

There was actually no need to go through the algebra of inverting the equations, because the only way that the two sets of equations (writing the unprimes in terms of the primes in Eq. (2.2), or the primes in terms of the unprimes in Eq. (2.4)) can differ is in the sign of v. S' is moving forward with respect to S, whereas S is moving backward with respect to S' (given our sign conventions). So inverting the equations simply entails replacing v with $-v$.

The reason why the above derivation of Eq. (2.2) was so quick is that we already did most of the work in Section 1.3 when we derived the fundamental effects. If we wanted to derive the Lorentz transformations from scratch, that is, by starting with the two postulates in Section 1.2, then the derivation would be longer. In Appendix D we give such a derivation, where it is clear which information comes from which of the postulates. The procedure there is somewhat cumbersome, but it's worth taking a look, because we will invoke the result in a very interesting way in Section 2.7.

REMARKS:

1. In the limit $v \ll c$, we have $\gamma \approx 1$, so Eqs. (2.2) reduce to $x = x' + vt$ and $t = t'$ (dropping the Δ's), which are the familiar Galilean transformations. This makes sense, because we know from everyday experience (where $v \ll c$) that the Galilean transformations work just fine. Technically, we must also require $vx'/c^2 \ll t'$, otherwise the second equation wouldn't reduce to $t = t'$. So even if v is small, we have to be careful that x' isn't too large. In practice, however, this requirement isn't much of an issue.

2. Eq. (2.2) exhibits a nice symmetry between x and ct. With $\beta \equiv v/c$, we have

$$\begin{aligned} x &= \gamma[x' + \beta(ct')], \\ ct &= \gamma[(ct') + \beta x']. \end{aligned} \tag{2.5}$$

Equivalently, in units where $c = 1$ (for example, where one unit of distance equals $3 \cdot 10^8$ meters, or where one unit of time equals $1/(3 \cdot 10^8)$ seconds), Eq. (2.2) takes the symmetric form,

$$\begin{aligned} x &= \gamma(x' + vt'), \\ t &= \gamma(t' + vx'). \end{aligned} \tag{2.6}$$

3. In matrix form, Eq. (2.5) can be written as

$$\begin{pmatrix} x \\ ct \end{pmatrix} = \begin{pmatrix} \gamma & \gamma\beta \\ \gamma\beta & \gamma \end{pmatrix} \begin{pmatrix} x' \\ ct' \end{pmatrix}. \tag{2.7}$$

This looks similar to a rotation matrix in a 2-D plane. We'll talk more about this in Section 2.6.

4. We worked through the above derivation in terms of primed and unprimed coordinates. However, when solving problems, it's usually best to label your coordinates with subscripts such as "A" for Alice, or "t" for train. In addition to being more informative, this notation is less likely to make you think that one frame is more fundamental than the other.

5. It's easy to get confused about the sign on the righthand side of the Lorentz transformations. To figure out if it should be a plus or a minus, write down $\Delta x_A = \gamma(\Delta x_B \pm v\Delta t_B)$ with a "±" in the middle, and then imagine sitting in frame A and looking at a fixed point in frame B. This fixed point satisfies $\Delta x_B = 0$, which gives $\Delta x_A = \pm\gamma v \Delta t_B$. So if the point moves to the right (that is, if x_A increases as time increases), then pick the "+." Or if it moves to the left, then pick the "−." In other words, the sign is determined by which way A (the person associated with the coordinates on the lefthand side of the equation) sees B (the person associated with the righthand side) moving.

6. One very important thing we must check is that two successive Lorentz transformations (from S_1 to S_2 and then from S_2 to S_3) again yield a Lorentz transformation (from S_1 to S_3). This must be true because we showed that any two frames must be related by Eq. (2.2). If we composed two Lorentz transformations (L.T.'s) along the same direction and found that the transformation from S_1 to S_3 was not of the form in Eq. (2.2), for some new v, then the whole theory would be inconsistent, and we would have to drop one of our postulates.[1] You can show that the combination of an L.T. (with speed v_1) and another L.T. (with speed v_2) does indeed yield an L.T., and it has speed $(v_1 + v_2)/(1 + v_1v_2/c^2)$, which is the velocity-addition formula (you should convince yourself why this is the case). This is the task of Exercise 2.17 and also Problem 2.14, which is stated in terms of the *rapidity*, introduced in Section 2.6. ♣

2.1.2 Second derivation

We will now present a second derivation of the Lorentz transformations in Eq. (2.2). We'll consider a setup in which a train moves rightward at speed v with respect to the ground. It turns out that a simple application of length contraction will give the equation for Δx, and a simple application of time dilation will give the equation for Δt. Instead of using unprimed and primed frames, we'll work with the ground and train frames, subscripted with "g" and "t." To keep the notation uncluttered, we'll drop the Δ's. Equivalently, we'll let the first event be located at the origin in both frames.

We claim that the x equation in Eq. (2.2), namely

$$x_g = \gamma(x_t + vt_t), \tag{2.8}$$

immediately follows from Fig. 2.2, where we have shown the setup in the train frame. You should think about what the reasoning is before reading further.

The logic is as follows. In the train frame, the ground moves to the left at speed v. Let the two events be paint bombs, located at the ends of the train, that explode and leave marks on both the train and the ground. Assume that the first event is the left bomb exploding (see the top picture in the figure), and the second event is the right bomb exploding at a time t_t later (see the bottom picture). The black dots in both pictures represent the paint marks right when they are created, and the hollow dots in the bottom picture represent the marks that are already there.

As the bottom picture indicates, the distance between the marks on the ground (as measured in the *train* frame) is the separation x_t between the marks on the train, plus

[1] This statement is true only for the composition of two L.T.'s in the *same* direction. If we composed an L.T. in the x direction with an L.T. in the y direction, then the result would interestingly *not* be an L.T. along some new direction, but rather the composition of an L.T. along some direction and a rotation through some angle. This rotation results in what is known as the *Thomas precession*. See the appendix of Muller (1992) for a quick derivation of the Thomas precession. For further discussion, see Costella *et al.* (2001) and Rebilas (2002).

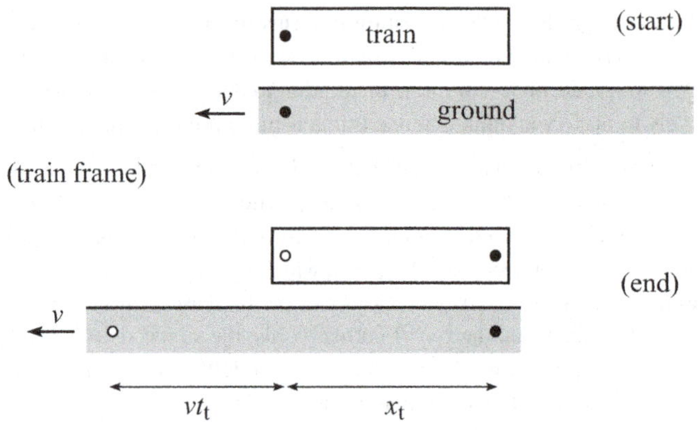

Figure 2.2

the additional distance vt_t that the ground travels while waiting for the second bomb to explode. The distance between the marks on the ground, as measured in the *ground* frame, must therefore be $x_g = \gamma(x_t + vt_t)$, because this distance is what is length contracted down to the $x_t + vt_t$ distance shown in the train frame. We have therefore proved Eq. (2.8).

Now for the t equation in Eq. (2.2),

$$t_g = \gamma(t_t + vx_t/c^2). \tag{2.9}$$

We claim that this equation immediately follows from Fig. 2.3, where we have shown the setup in the ground frame. Again, you should think about what the reasoning is before reading further.

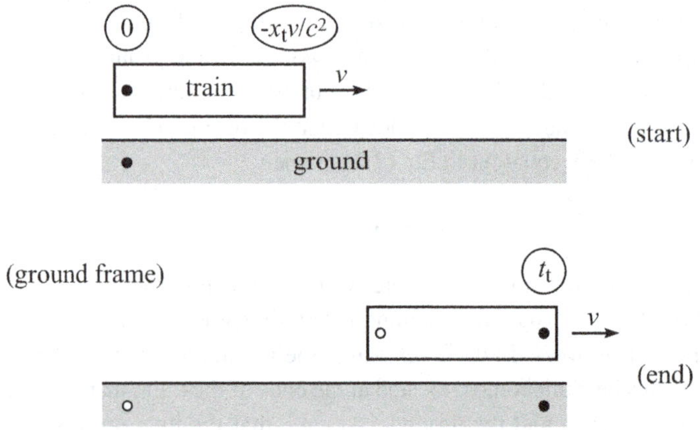

Figure 2.3

The logic is as follows. Assume that the left clock on the train reads zero when the left paint bomb explodes. Then the right clock on the train reads t_t when the right paint bomb explodes, because we are assuming that t_t is the time separation in the train frame (and the train clocks are synchronized in the train frame). In the ground frame, the front clock on the train reads $-x_t v/c^2$ at the start, due to the rear-clock-ahead effect. This is the standard $-Lv/c^2$ expression, with the proper length L of the train (which is the separation between the events in the train frame) replaced with x_t. In the ground frame, the front clock (and the back clock, too) on the train therefore advances by $t_t + x_t v/c^2$ between the explosions. Due to time dilation, a ground observer sees this clock run slow, so the time that elapses in the ground frame between the explosions is $t_g = \gamma(t_t + vx_t/c^2)$. We have therefore proved Eq. (2.9).

In the end, we used all three fundamental effects (along with the fact that the relative speed of the frames is v) in deriving the two equations, just as we did in our earlier derivation in Section 2.1.1. But this second derivation is a little clearer, because it is easier to see from the above two figures exactly how the four transformation coefficients arise.

Example: A train with proper length L moves rightward at speed $5c/13$ with respect to the ground. A ball is thrown from the back of the train to the front. The speed of the ball with respect to the train is $c/3$. As viewed by someone on the ground, how much time does the ball spend in the air, and how far does it travel?

Solution: The γ factor associated with the speed $5c/13$ is $\gamma = 13/12$. The two events we are concerned with are "ball leaving back of train" and "ball arriving at front of train." The x and t separations between these events are easy to calculate in the train frame. They are just $x_t = L$ and $t_t = L/(c/3) = 3L/c$. The Lorentz transformations therefore give the x and t separations in the ground frame as

$$\begin{aligned} x_g = \gamma(x_t + vt_t) &= \frac{13}{12}\left(L + \left(\frac{5c}{13}\right)\left(\frac{3L}{c}\right)\right) = \frac{7L}{3}, \\ t_g = \gamma(t_t + vx_t/c^2) &= \frac{13}{12}\left(\frac{3L}{c} + \frac{(5c/13)L}{c^2}\right) = \frac{11L}{3c}. \end{aligned} \tag{2.10}$$

The sign in the transformations is a "+," because the ground (associated with the lefthand side of the equation) sees the train (associated with the righthand side) moving in the positive direction.

The preceding example demonstrates the main usefulness of the Lorentz transformations: Given two events, if it is easy to calculate Δx and Δt in one frame (as it is in the train frame above), then you simply need to plug these Δ's into the Lorentz transformations to obtain the Δ's in another frame (the ground frame here), where they aren't as obvious.

In the above example, if you want to find the x and t separations in the ground frame without using the Lorentz transformations, the most reasonable strategy is the velocity-addition method in Problem 2.7(b). This method is longer, but it has the benefit of being more interesting, because you actually have to think about what's going on, instead of simply plugging things into the Lorentz transformations.

As mentioned on page 72, I personally would be wary of solving a problem using only the L.T.'s, because it's very easy to mess up a sign in the transformations or make an algebra mistake. But after solving a problem in a different way (using the fundamental effects, velocity addition, etc.), the L.T.'s provide a good opportunity to double check your answer. The other methods are generally more fun when solving a problem for the first time, while the L.T.'s are usually quick and easy to apply (perfect for a double-check).

The excitement will build in your voice,
As you rise from your seat and rejoice,
"A Lorentz transformation
Provides information,
As an alternate method of choice!"

2.1.3 The fundamental effects

Let's now see how the Lorentz transformations lead to the three fundamental effects (rear clock ahead, time dilation, and length contraction) discussed in Section 1.3. Of course, we just used these effects to *derive* the Lorentz transformations, so we know everything will work out. We'll just be going in circles. But since these fundamental effects are, well, fundamental, let's belabor the point and discuss them one more time, with the starting point being the Lorentz transformations.

Rear clock ahead

Let S' move rightward at speed v with respect to S, and let two events occur simultaneously in frame S. Then the space and time (often referred to as "spacetime") separation between them, as measured by S, is $(x, t) = (x, 0)$. (We won't bother writing the Δ's here.) Using the second of Eqs. (2.4), we see that the time between the events, as measured by S', is $t' = -\gamma v x/c^2$. This is not equal to zero (unless $x = 0$). Therefore, the events do not occur simultaneously in frame S'.

Furthermore, γx is the separation (call it L) between the events in S', due to length contraction; γx is what gets length contracted down to the x that S measures. So we see that t' takes the form of $-Lv/c^2$, which is the expected rear-clock-ahead result. The sign is negative because the event with the larger value of x' has the smaller value of t', because the front clock is behind by Lv/c^2.

Time dilation

Consider two events that occur at the same place in S', such as two ticks on a clock at rest in S'. The spacetime separation between these events takes the form $(x', t') = (0, t')$. Using the second of Eqs. (2.2), we see that the time between the events, as measured by S, is

$$t = \gamma t' \qquad (\text{if } x' = 0). \tag{2.11}$$

This time-dilation equation is simply a special case of the Lorentz transformation for t, when $x' = 0$. The γ factor is greater than or equal to 1, so $t \geq t'$. The passing of one second on S''s clock takes more than one second on S's clock.

The same strategy works if we interchange S and S'. Consider two events that occur at the same place in S. The spacetime separation between them is $(x, t) = (0, t)$. Using the second of Eqs. (2.4), we see that the time between the events, as measured by S', is

$$t' = \gamma t \qquad (\text{if } x = 0). \tag{2.12}$$

Therefore, $t' \geq t$. Another way to derive this is to set $x = 0$ in the first of Eqs. (2.2) to obtain $x' = -vt'$, and then substitute this into the second equation. However, it is much quicker to simply use the second of Eqs. (2.4), as we did. Eqs. (2.2) and (2.4) contain the same information, so either one suffices for any purpose. But depending on the task at hand, one might make things much cleaner than the other.

If we write down the above two results by themselves, $t = \gamma t'$ and $t' = \gamma t$, they appear to contradict each other. This apparent contradiction arises from the omission of the conditions that they are based on. The former equation is based on the assumption that $x' = 0$. The latter equation is based on the assumption that $x = 0$. They have nothing to do with each other. It would perhaps be better to write the equations as

$$(t = \gamma t')_{x'=0} \qquad \text{and} \qquad (t' = \gamma t)_{x=0}, \tag{2.13}$$

but this is somewhat awkward.

Length contraction

The procedure here is similar to what we did with time dilation, except that now we want to set certain time intervals equal to zero, instead of certain space intervals. We want to do this because to measure a length, we calculate the distance between two points whose positions are measured *simultaneously*. That's what a length is.

Consider a stick at rest in S', where it has length ℓ'. We want to find the length ℓ in S. Simultaneous measurements of the coordinates of the ends of the stick in S yield a spacetime separation of the form $(x, t) = (x, 0)$. Using the first of Eqs. (2.4), we find

$$x' = \gamma x \implies x = \frac{x'}{\gamma} \qquad (\text{if } t = 0). \tag{2.14}$$

But x is by definition the length ℓ in S. And x' is the length ℓ' in S', because the stick isn't moving in S'.[2] Therefore, $\ell = \ell'/\gamma$. And since $\gamma \geq 1$, we have $\ell \leq \ell'$, so S sees the stick shorter than S' sees it. Similar to the situation with time dilation, the above length-contraction equation is simply a special case of the Lorentz transformation for x', when $t = 0$.

Now interchange S and S'. Consider a stick at rest in S, where it has length ℓ. We want to find the length in S'. Measurements of the coordinates of the ends of the stick in S' yield a spacetime separation of $(x', t') = (x', 0)$. Using the first of Eqs. (2.2), we have

$$x = \gamma x' \implies x' = \frac{x}{\gamma} \qquad (\text{if } t' = 0). \tag{2.15}$$

But x' is by definition the length ℓ' in S'. And x is the length ℓ in S, because the stick isn't moving in S. Therefore, $\ell' = \ell/\gamma$, so $\ell' \leq \ell$. Another way to derive this is to set $t' = 0$ in the second of Eqs. (2.4) to obtain $t = vx/c^2$, and then substitute this into the first equation. But it is much quicker to simply use the first of Eqs. (2.2), as we did.

As with time dilation, if we write down the above two results by themselves, $\ell = \ell'/\gamma$ and $\ell' = \ell/\gamma$, they appear to contradict each other. But as before, this apparent contradiction arises from the omission of the conditions that they are based on. The former equation is based on the assumptions that $t = 0$ and the stick is at rest in S'. The latter equation is based on the assumptions that $t' = 0$ and the stick is at rest in S. They have nothing to do with each other. We should really write

$$(x = x'/\gamma)_{t=0} \qquad \text{and} \qquad (x' = x/\gamma)_{t'=0}, \tag{2.16}$$

and then identify x' in the first equation with ℓ' only after invoking the further assumption that the stick is at rest in S'. Likewise for the second equation. But this is a pain.

2.2 Velocity addition

2.2.1 Longitudinal velocity addition

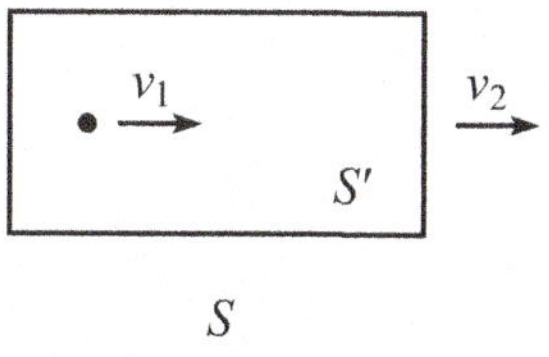

Figure 2.4

In Section 1.5 we derived the longitudinal velocity-addition formula; see Eq. (1.28). We'll now present a second (and much quicker) derivation of the formula. We'll work with frames S and S' instead of the ground and train frames, respectively. So we have the setup in Fig. 2.4. An object moves with velocity v_1 with respect to S', and S' moves

[2] The measurements of the ends made in S are *not* simultaneous in the S' frame. In the S' frame, the spacetime separation between the events is (x', t'), where both x' and t' are nonzero. This doesn't satisfy our definition of a length measurement in S', because $t' \neq 0$. However, the stick isn't moving in S', so an S' observer can measure the ends whenever he feels like it, and he will always get the same difference. So x' is indeed the length in the S' frame.

with velocity v_2 with respect to S. Our goal is to find the velocity V of the object with respect to S.

The Lorentz transformations can be used to easily determine V. The relative velocity of the frames is v_2. Consider two events along the object's path (for example, say it makes two beeps). In S', the space and time separations between these events satisfy $\Delta x'/\Delta t' = v_1$, because $\Delta x'/\Delta t'$ is by definition the velocity v_1 in S'. We want to find $\Delta x/\Delta t$, because $\Delta x/\Delta t$ is by definition the velocity V in S. The Lorentz transformations from S' to S are given in Eqs. (2.2) as

$$\Delta x = \gamma_2(\Delta x' + v_2\,\Delta t') \qquad \text{and} \qquad \Delta t = \gamma_2(\Delta t' + v_2\,\Delta x'/c^2), \tag{2.17}$$

where $\gamma_2 \equiv 1/\sqrt{1 - v_2^2/c^2}$. Note that these transformations involve only the relative velocity of the two frames, which is v_2 here; v_1 does not appear. v_1 will eventually make its appearance via $\Delta x'/\Delta t'$. The quotient of the above two equations gives the desired velocity V as (dividing both the numerator and denominator by $1/\Delta t'$ to obtain the second line here)

$$\begin{aligned} V \equiv \frac{\Delta x}{\Delta t} &= \frac{\Delta x' + v_2\,\Delta t'}{\Delta t' + v_2\,\Delta x'/c^2} \\ &= \frac{\Delta x'/\Delta t' + v_2}{1 + v_2(\Delta x'/\Delta t')/c^2} \\ &= \frac{v_1 + v_2}{1 + v_1 v_2/c^2}, \end{aligned} \tag{2.18}$$

in agreement with Eq. (1.28), with $u \to v_1$ and $v \to v_2$. As promised, this derivation was indeed quick!

2.2.2 Transverse velocity addition

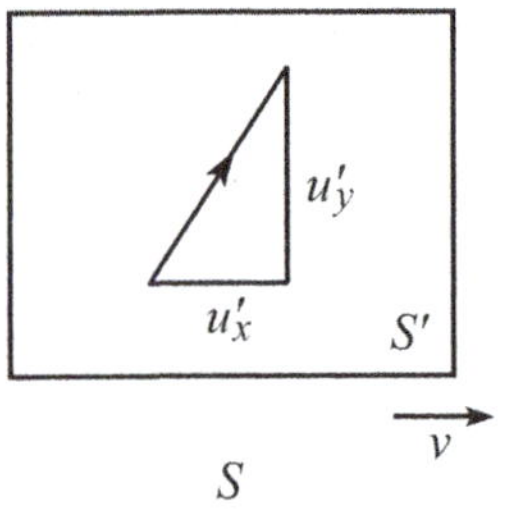

Figure 2.5

We can also use the Lorentz transformations to give a quick derivation of the *transverse* velocity-addition formula, which deals with the general two-dimensional setup shown in Fig. 2.5. An object moves with velocity (u'_x, u'_y) with respect to S', and S' moves with velocity v with respect to S, in the x direction. What is the velocity (u_x, u_y) of the object with respect to S?

The existence of motion in the y direction doesn't affect the above derivation of the velocity in the x direction (the Lorentz transformations in Eq. (2.17) are still valid), so Eq. (2.18) still holds. In the present notation, it becomes

$$\boxed{u_x = \frac{u'_x + v}{1 + u'_x v/c^2}} \tag{2.19}$$

To find u_y, we can again make easy use of the Lorentz transformations. Consider two events along the object's path. We are given that $\Delta x'/\Delta t' = u'_x$ and $\Delta y'/\Delta t' = u'_y$. Our goal is to find $u_y \equiv \Delta y/\Delta t$. The relevant Lorentz transformations from S' to S in Eq. (2.2) are

$$\Delta y = \Delta y' \qquad \text{and} \qquad \Delta t = \gamma(\Delta t' + v\,\Delta x'/c^2). \tag{2.20}$$

Therefore,

$$\begin{aligned} u_y \equiv \frac{\Delta y}{\Delta t} &= \frac{\Delta y'}{\gamma(\Delta t' + v\,\Delta x'/c^2)} \\ &= \frac{\Delta y'/\Delta t'}{\gamma(1 + v(\Delta x'/\Delta t')/c^2)}, \end{aligned} \tag{2.21}$$

which gives

$$u_y = \frac{u'_y}{\gamma_v(1 + u'_x v/c^2)} \tag{2.22}$$

This is the desired transverse velocity-addition formula. Note that the γ factor here is associated with the relative velocity v between the two frames. This γ factor has nothing to do with the velocity (u'_x, u'_y) of the object within S'. We have emphasized this by adding on the subscript v in γ_v, although we will often drop it in future discussions.

In the special case where $u'_x = 0$ (that is, the object is moving vertically in S'), Eq. (2.22) gives $u_y = u'_y/\gamma_v$. The simplicity of this result suggests that there is a quicker way of deriving it. And indeed, it can be seen as a simple consequence of time dilation, in the following way. Consider a series of equally spaced lines parallel to the x axis; see Fig. 2.6. Imagine a clock at rest in S', with the same x' value as the moving object. Let this clock tick simultaneously (as seen in S') with the object's line crossings. Then the clock also ticks simultaneously with the object's line crossings in S. (This is true because the clock and the object have the same x' value, which means that we don't have to worry about any issues with simultaneity.) But the clock ticks slower in S by the time-dilation factor $1/\gamma_v = \sqrt{1-v^2}$. Therefore the rate of the line crossings, and hence the y speed, is smaller in S by this factor, as we wanted to show. (We have used the fact that the vertical spacing between the lines is the same in the two frames, because there is no transverse length contraction.) This γ factor in the transverse speed will be very important when we deal with momentum in Chapter 3.

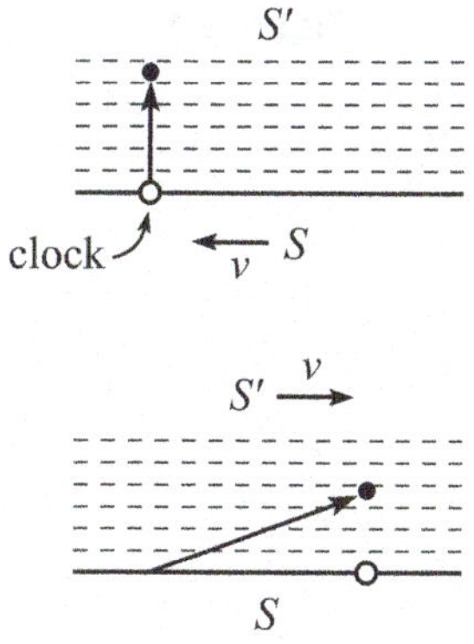

Figure 2.6

In general, if you run in the x direction relative to a frame S' in which an object has an arbitrary velocity, then its y speed in your frame may be faster or slower than in S', depending on the relative sign and size of u'_x and v, which affect the $1 + u'_x v/c^2$ factor in Eq. (2.22). Strange indeed, but no stranger than other effects we've seen.

Example (Equal transverse speeds): In the ground frame, an object moves with velocity (u_x, u_y), and you move with velocity $(v, 0)$. What should v be so that the y component of the object's velocity in your frame is also u_y? (You don't care about the x component.) One solution is of course $v = 0$. Find the other one.

Solution: From your point of view, the ground is moving with speed v in the negative x direction. And within the ground frame, the velocity of the object is (u_x, u_y). The transverse velocity-addition formula, Eq. (2.22), therefore gives the y component of the velocity in your frame as $u_y/\gamma_v(1 - u_x v)$, where we have dropped the c's to simplify the calculation. (We'll put them back in at the end, to make the units correct.) This expression is just Eq. (2.22) with $v \to -v$ and with the primes erased. Demanding that this result equals u_y gives

$$\gamma_v(1 - u_x v) = 1 \implies 1 - u_x v = \sqrt{1 - v^2}$$

$$\implies 1 - 2u_x v + u_x^2 v^2 = 1 - v^2 \implies v = \frac{2u_x}{1 + u_x^2}, \tag{2.23}$$

or $v = 0$, of course. To make the units correct with the c's, the denominator is really $1 + u_x^2/c^2$. The ground-frame view is shown in Fig. 2.7.

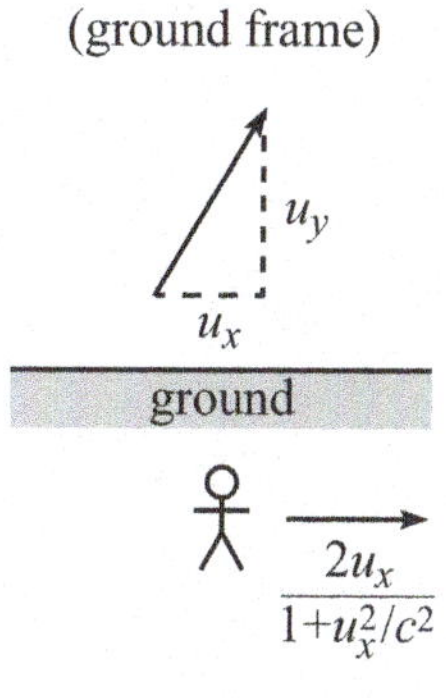

Figure 2.7

Remark: The above result for v makes sense, because this speed is the relativistic addition of u_x with itself. You are therefore moving with speed u_x relative to the speed u_x that the object has with respect to the ground. This means that both your frame and the ground frame move with speed u_x (but in opposite directions) relative to the frame moving with

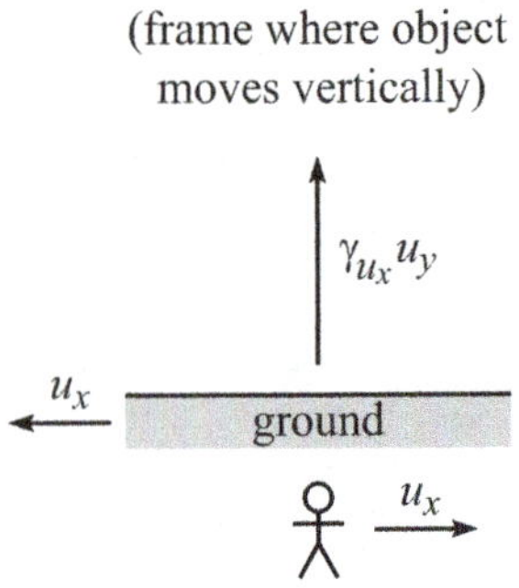

Figure 2.8

speed u_x. But that frame is the one where the object has no speed in the x direction; see Fig. 2.8. (You can quickly show that the y velocity $u_y/\gamma_v(1 - u_x v)$ equals $\gamma_{u_x} u_y$ when $v = u_x$.) By symmetry, therefore, the y velocity of the object must be the same in your frame and in the ground frame.

As an exercise, you can show that as a function of v, the object's y velocity $u_y/\gamma_v(1 - u_x v)$ is maximized when $v = u_x$ (where it takes on the value $\gamma_{u_x} u_y$). So if you imagine starting at $v = 0$ and then slowly ramping up your speed, the object's y velocity in your frame first increases and reaches a maximum of $\gamma_{u_x} u_y$ when $v = u_x$, then decreases back down to u_y when $v = 2u_x/(1 + u_x^2/c^2)$, and then continues to decrease and eventually approaches zero as v approaches c (because the γ_v in Eq. (2.22) approaches infinity). ♣

2.3 The invariant interval

Consider two events separated by Δx and Δt in one frame, and $\Delta x'$ and $\Delta t'$ in another. And consider the quantity (this is admittedly a bit out of the blue)

$$\boxed{(\Delta s)^2 \equiv c^2(\Delta t)^2 - (\Delta x)^2} \tag{2.24}$$

Technically, we should also subtract off $(\Delta y)^2$ and $(\Delta z)^2$, but nothing exciting happens in the transverse directions, so we'll ignore them. We claim that $c^2(\Delta t)^2 - (\Delta x)^2 = c^2(\Delta t')^2 - (\Delta x')^2$. That is, $(\Delta s)^2$ is *invariant*; it doesn't depend on the frame. This certainly isn't obvious, because the Δx and Δt values themselves get all messed up (via the Lorentz transformations) when going from one frame to another. But the special combination, $c^2(\Delta t)^2 - (\Delta x)^2$, of the coordinates happens to remain the same. This invariance of $(\Delta s)^2$ is a special case of more general results involving inner products and 4-vectors, which we'll discuss in Chapter 4.

We can prove that $(\Delta s)^2$ (or s^2, for short) is invariant by using the Lorentz transformations in Eq. (2.2) to write $(\Delta s)^2$ in terms of the S' coordinates. The result is (dropping the Δ's)

$$\begin{aligned}
c^2t^2 - x^2 &= c^2\gamma^2(t' + vx'/c^2)^2 - \gamma^2(x' + vt')^2 \\
&= \gamma^2\left(c^2t'^2 + \cancel{2t'vx'} + v^2x'^2/c^2\right) - \gamma^2\left(x'^2 + \cancel{2x'vt'} + v^2t'^2\right) \\
&= \gamma^2c^2t'^2\left(1 - \frac{v^2}{c^2}\right) - \gamma^2x'^2\left(1 - \frac{v^2}{c^2}\right) \\
&= c^2t'^2 - x'^2,
\end{aligned} \tag{2.25}$$

where we have used $\gamma^2 \equiv 1/(1 - v^2/c^2)$. We see that the Lorentz transformations imply that the quantity $c^2t^2 - x^2$ doesn't depend on the frame, as desired.

This result is more than we bargained for, for the following reason. Two events along the trajectory of a photon satisfy $x/t = c \implies c^2t^2 - x^2 = 0$ (again dropping the Δ's). So the speed-of-light postulate says that if $c^2t^2 - x^2 = 0$, then $c^2t'^2 - x'^2 = 0$. But Eq. (2.25) says that if $c^2t^2 - x^2 = b$, then $c^2t'^2 - x'^2 = b$, for *any* value of b, not just zero. There are enough things that change when we go from one frame to another, so it's nice to have a frame-independent quantity that we can hang on to. Given two events, all inertial observers agree on the value of $s^2 \equiv c^2t^2 - x^2$, independent of what they measure for the actual coordinates.

"Potato?! Pot*ah*to!" said she,
"And of *course* it's tom*ah*to, you see.
But the square of ct
Minus x^2 will be
Always something on which we agree."

The invariance of s^2 under the Lorentz transformations is analogous to the invariance of the square of the distance from the origin, r^2, under rotations in the x-y plane. The coordinates themselves change under the rotation, but the special combination, $r^2 = x^2 + y^2$, of the coordinates remains the same. Why is there a plus sign in $x^2 + y^2$, while there is a minus sign in $c^2t^2 - x^2$? The former is a consequence of the rotation transformation matrix, while the latter is a consequence of the Lorentz transformation matrix in Eq. (2.7). These are different matrices, so there is no reason to expect that the invariant quantities should take the same form.

A note on terminology: the separation in the coordinates, $(c\Delta t, \Delta x)$, with a c in front of the Δt, is usually referred to as the *spacetime separation*, while the quantity $(\Delta s)^2 \equiv c^2(\Delta t)^2 - (\Delta x)^2$ is referred to as the *invariant interval*, or alternatively as the *spacetime interval*. At any rate, just call it s^2, and people will know what you mean. The important part of Eq. (2.24) is the righthand side; the lefthand side is simply a convenient definition. It would be a pain to have to write out $c^2t^2 - x^2$ all the time.

Let's now look at the physical significance of $s^2 \equiv c^2t^2 - x^2$. There are three cases to consider. We'll drop the Δ's in the following discussion, but you should remember that x always means Δx, etc.

Case 1: $s^2 > 0$ (timelike separation)

If $s^2 > 0$, we say that the two events are *timelike* separated (because the ct term wins out over the x term). Since $c^2t^2 > x^2$ in this case, we have $|x/t| < c$. For concreteness, we'll assume that both x and t are positive. We can then just write $x/t < c$. Other cases can be treated similarly. Consider a frame S' moving rightward at speed v with respect to S. The Lorentz transformation for x' is

$$x' = \gamma(x - vt). \tag{2.26}$$

The minus sign here arises because S' sees S moving leftward. Since $x/t < c$, there exists a v that is less than c (namely $v = x/t$) that makes $x' = 0$. In other words, if two events are timelike separated, it is possible to find a frame S' in which the two events happen at the same place. (In short, the condition $x/t < c$ means that it is possible for an object to travel from one event to the other.) With $x' = 0$, the invariance of s^2 gives $s^2 = c^2t'^2 - x'^2 = c^2t'^2$. So we see that s/c is the time t' between the events in the frame in which the events happen at the same place. That is, s/c is the proper time.

Case 2: $s^2 < 0$ (spacelike separation)

If $s^2 < 0$, [3] we say that the two events are *spacelike* separated (because the x term wins out over the ct term). Since $c^2t^2 < x^2$ in this case, we have $|x/t| > c$. As above, we'll assume that both x and t are positive. We then have $x/t > c$. Consider again a frame S' moving rightward at speed v with respect to S. The Lorentz transformation for t' is

$$t' = \gamma(t - vx/c^2). \tag{2.27}$$

[3] It's fine if s^2 is negative, which means that s is imaginary. We can take the absolute value of s if we want to obtain a real number.

Since $x/t > c \Longrightarrow t/x < 1/c$, there exists a v that is less than c (namely $v = c^2t/x$) that makes $t' = 0$. In other words, if two events are spacelike separated, it is possible to find a frame S' in which the two events happen at the same time. (This statement isn't as easy to intuitively understand as the corresponding one in the timelike case above. But if you draw a Minkowski diagram, described in the next section, it becomes clear.) With $t' = 0$, the invariance of s^2 gives $s^2 = c^2t'^2 - x'^2 = -x'^2$. So we see that $|s|$, or equivalently $\sqrt{-s^2}$, is the distance x' between the events in the frame in which the events happen at the same time. We call this the proper distance, or proper length.

Note that if we pick v to be larger than c^2t/x, then the t' in Eq. (2.27) is negative. This means that the order of the events in S' is different from the order in S. We therefore see that if two events are spacelike separated, there exists a frame in which the temporal order of the events is reversed. This is not true for timelike separation, because $v > c^2t/x$ implies $v > c$, since $x/t < c \Longrightarrow t/x > 1/c$ for timelike events.

The analogous statement for timelike-separated events is that if we pick v to be larger than x/t, then the x' in Eq. (2.26) is negative. This means that the *spatial* order of the events in S' is different from the order in S. That is, the right event in S is the left event in S'. This is intuitively clear. If you start at event E_1 and travel rightward toward event E_2 with a large enough v so that you overshoot it, then in your frame, E_2 will happen to the left of where E_1 happened (which was at your location). This is not possible for spacelike separation, because $v > x/t$ implies $v > c$, since $x/t > c$ for spacelike events. In short, if $x/t > c$ then you can't overshoot the second event.

Case 3: $s^2 = 0$ (lightlike separation)

If $s^2 = 0$, we say that the two events are *lightlike* separated. Since $c^2t^2 = x^2$ in this case, we have $|x/t| = c$. This relation is frame independent due to the invariance of s^2. So in every frame, a photon emitted at one of the events will arrive at the other (if aimed in the right direction, of course). This is simply the statement that the speed of light is c in every inertial frame. Note that it is not possible to find a frame S' in which the two events happen at the same place or the same time, because such a frame would need to travel at the speed of light. This is true because the necessary speeds are x/t and c^2t/x, respectively (from above), and these are both equal to c when $x/t = c$.

Example 1 (Time dilation): An illustration of the usefulness of the invariance of s^2 is a derivation of time dilation. Let frame S' move at speed v with respect to frame S. Consider two events at a fixed location in S', separated by time t'. This fixed location moves a distance vt in S during a time t, so the separations between the events are

$$\begin{aligned} \text{in } S': &\quad (x', t') = (0, t'), \\ \text{in } S: &\quad (x, t) = (vt, t). \end{aligned} \tag{2.28}$$

The invariance of s^2 implies $c^2t'^2 - 0 = c^2t^2 - v^2t^2$, which gives

$$t = \frac{t'}{\sqrt{1 - v^2/c^2}} \equiv \gamma t', \tag{2.29}$$

as desired. This method makes it clear that the $t = \gamma t'$ time-dilation result rests on the assumption that $x' = 0$.

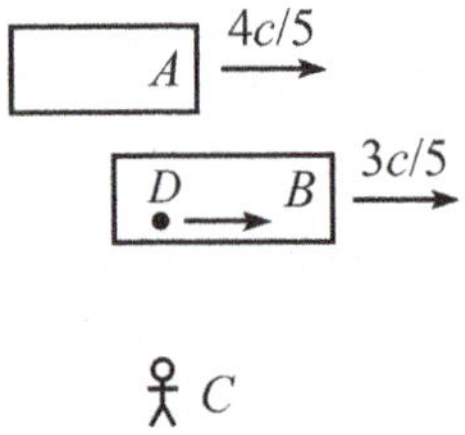

Figure 2.9

Example 2 (Passing trains, yet again): Consider again the scenario in the "Passing trains" examples in Sections 1.3.3 and 1.5; see Fig. 2.9. Verify that $(\Delta s)^2$ between the events E_1 and E_2 is the same in all of the frames, A, B, C, and D.

Solution: The only quantity that we'll need that we haven't already found in the two earlier examples is the distance between E_1 and E_2 in C's frame (the ground frame). In this frame, train A travels at speed $4c/5$ for a time $t_C = 7L/c$, covering a distance of $28L/5$. But event E_2 occurs at the back of train A, which is a distance $3L/5$ behind the front end (this is the contracted length in the ground frame). Therefore, the distance between events E_1 and E_2 in the ground frame is $28L/5 - 3L/5 = 5L$. You can also apply the same line of reasoning using train B, in which the $5L$ result takes the form of $(3c/5)(7L/c) + 4L/5$. B's length gets added instead of subtracted here.

Putting all of the earlier results together, we have the following separations between the events in the various frames:

	A	B	C	D
Δt	$5L/c$	$5L/c$	$7L/c$	$2\sqrt{6}L/c$
Δx	$-L$	L	$5L$	0

From the table, we see that $(\Delta s)^2 \equiv c^2(\Delta t)^2 - (\Delta x)^2 = 24L^2$ in all four frames. So $(\Delta s)^2$ is indeed invariant, as desired. We could have worked backwards, of course, and used the $(\Delta s)^2 = 24L^2$ result from frames A, B, or D, to deduce that $\Delta x = 5L$ in frame C. Verifying that $(\Delta s)^2$ is in fact invariant, as we have just done, is a very quick and easy double check you can often perform after solving a problem. If you have generated Δt and Δx values in two different frames, you should always take a few seconds and check that the $(\Delta s)^2$ values agree.

In Problem 2.1 you are asked to perform the tedious task of checking that the values in the above table satisfy the Lorentz transformations between the six different pairs of frames.

Causality

We have mentioned on a few occasions that it is impossible for a (massive) object to travel at the speed of light. This can be justified in a number of ways (using energy, momentum, force, or velocity addition); see the discussion at the end of Section 3.5.1. A related fact is that it is impossible for a *signal* (that is, any kind of information) to travel *faster* than the speed of light. If a signal *could* travel faster than c, we would be able to generate setups that violate causality. *Causality* is the principle that in any given frame, causes must precede effects. If a ball is thrown and ends up crashing through a window, then in all reference frames it must be the case that the ball is thrown before the window breaks. Let's see how faster-than-light signals lead to a violation of causality.

Assume that a given signal travels faster than c from one event to another. For example, let's say A flips a coin and then immediately sends a signal to B with the result. If the signal travels faster than c, then the "A flips coin" and "B receives signal" events are spacelike separated, because $\Delta x > c\Delta t$. Therefore, as we noted above in the discussion of spacelike separation, there exists a frame in which the order of events is reversed. That is, there exists a frame in which B knows the result of the coin flip *before* the coin has actually been flipped. In this new frame, since the order of events is reversed, the signal actually goes *from B* back *to A*. This is a violation of causality. If Event 1 causes Event 2, then it is nonsensical for Event 2 to occur before Event 1 in a given frame. We therefore conclude that signals cannot travel faster than light.

However, you might be thinking that since so many other obvious "truths" from everyday experience get thrown out the window in relativity, what's the harm in having a little causality violation? Can we actually produce a genuine contradiction if we assume that signals are able to travel faster than light? Indeed we can, in the following way. The upshot of the scenario below is that faster-than-light signals imply that you can be 100% certain what a coin flip will be, *before* you flip the coin. Note that this statement involves only *one* person instead of the two people in the preceding paragraph. This makes the violation of causality more transparent. In the following discussion, we'll work with a lottery machine instead of a coin flip, since the lower odds make the effect more striking.

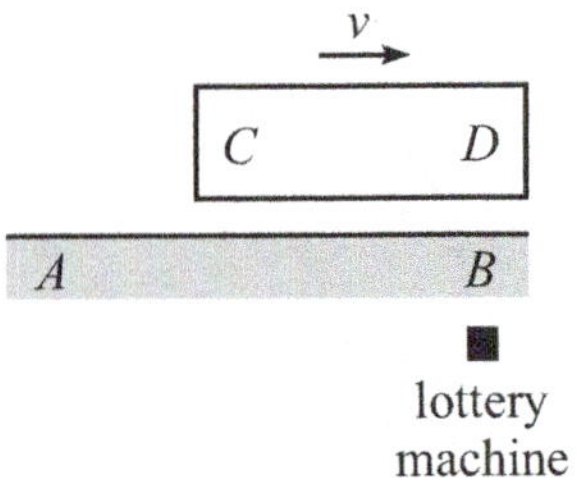

Figure 2.10

Imagine that A and B stand some distance apart along an east-west set of train tracks, with B to the east. A train carrying C and D moves eastward with speed v. C is at the back of the train, and D is at the front; see Fig. 2.10. B is standing next to a lottery machine that is about to produce the winning numbers. The numbers are produced right when D passes by, and B immediately gives the information to D. In the frame of the train, D then sends a signal instantaneously to C, telling her what the numbers are.[4] C then immediately gives the information to A, who happens to be right next to her at this instant (although this isn't the case in Fig. 2.10; we'll explain why below). A then immediately sends a signal instantaneously (as measured in the ground frame) to B, letting him know what the winning numbers are.

We claim that in this scenario, B receives the signal containing the winning numbers *before* the lottery machine picks them. This can be seen as follows. Assume that D's clock reads zero when B hands off the numbers to him. Then C's clock reads zero when she receives the signal, because we are assuming that the relay is instantaneous in the train frame. But here is the critical point: in the ground frame, when D's clock reads zero, the rear-clock-ahead effect tells us that C's clock reads Lv/c^2, where L is the proper length of the train. See the bottom (later) picture in Fig. 2.11. Now, C's clock reads zero *before* it reads Lv/c^2 (of course), so C must have received the signal *before* (as measured in the ground frame) D sent it (which was when B handed it off, and D's clock read zero). Therefore, when A gets the signal from C and instantaneously sends it to B, B will receive it (see the top, earlier, picture in Fig. 2.11) *before* his original interaction with D (which occurs right when the numbers are picked). In the ground frame, the time between the two pictures is $\gamma Lv/c^2$, because each train clock advances by Lv/c^2, and these clocks run slow. But this exact time isn't important here. All that matters is that it is nonzero.

The above conclusion (that B receives the signal containing the winning numbers before the lottery machine picks them) is clearly absurd. But if absurdity isn't enough and you want to generate a concrete contradiction, you can make the rule that B gives the information to D if and only if he (B) doesn't know what the numbers are before they are drawn. Then if we assume that B *doesn't* know the numbers before they are drawn, then he *does* tell D what they are right when they are drawn, in which case the above reasoning tells us that the signal gets back to him (B) before the numbers are drawn, which means that he *does* know the numbers before they are drawn. This contradicts our initial assumption. Conversely, if we assume that B *does* know the numbers before they are drawn, then he *doesn't* tell D what they are right when they are drawn, in which case there is no signal in the first place, which means that he *doesn't* know the numbers before they are drawn. This again contradicts our initial assumption. We end up with a contradiction no matter what route we take. We therefore conclude that it is impossible for signals to travel instantaneously. This conclusion arises for any (nonzero) value of the train's speed v. Even a slow speed like 10 mph is theoretically enough.

If we consider a *superluminal* (faster than c) signal instead of an *instantaneous* (infinitely fast) signal as we have been doing, then the contradiction requires that v take

[4]By instantaneous, we mean that the signal takes zero time (as measured in the train frame) to get to C. That is, the signal travels infinitely fast in the train frame. There is actually no need to have the signal be instantaneous, in order to generate the contradiction we're aiming for. But we'll assume it is, because this simplifies the discussion.

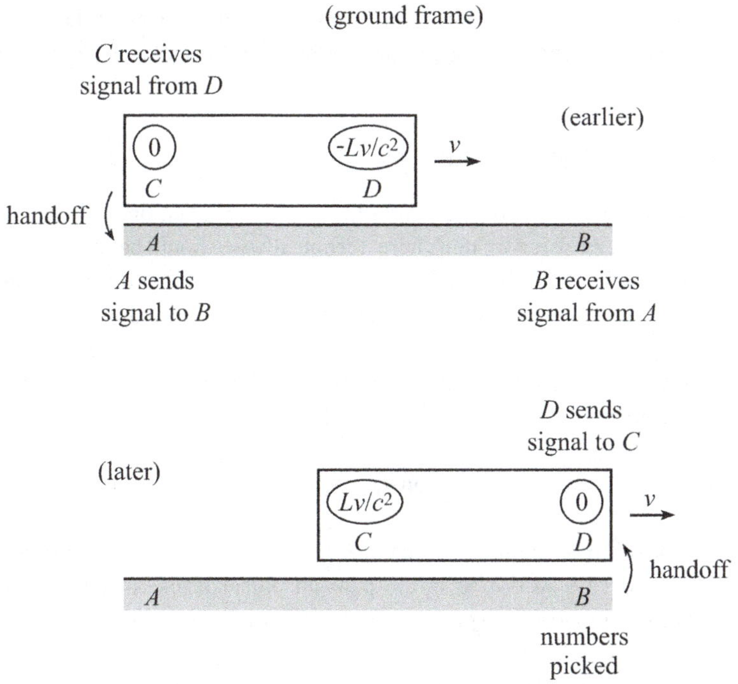

Figure 2.11

on a minimum value, dependent on the given superluminal speed. (The easiest way to see this is via a Minkowski diagram, where a simple picture explains concisely what's going on. We'll discuss this in the next section.) Since we are free to pick v as large as we want (but less than c, of course, because we're talking about the speed of a train), the contradiction is still realized. The existence of one contradiction is all we need, so we conclude that superluminal signal speeds are not possible.

2.4 Minkowski diagrams

Minkowski diagrams (also called *spacetime diagrams*) are extremely useful in seeing how coordinates (or technically coordinate differences) transform between different reference frames. If you want to produce exact numbers in a problem, you'll probably need to use one of the strategies we've encountered so far (fundamental effects, Lorentz transformations, etc.). But when it comes to getting an overall intuitive picture of a setup, there is no better tool than a Minkowski diagram. Here's how you make one.

Let frame S' move at speed v with respect to frame S (along the x axis, as usual, and ignore the y and z components). Draw the x and t axes of frame S. We choose to plot x on the horizontal axis and t (or actually ct) on the vertical axis, which is the opposite of what is done in a standard kinematics plot. We choose to plot ct instead of t, so that the trajectory of a light beam lies at a nice 45° angle. Alternatively, we could work with units where $c = 1$.

What do the x' and ct' axes of S' look like, superimposed on the axes of S? That is, at what angles are the axes inclined, and what is the size of one unit on these axes? (There is no reason why one unit on the x' and ct' axes should have the same length on the paper as one unit on the x and ct axes.) We can answer these questions by using the Lorentz transformations, Eqs. (2.2). We'll look first at the ct' axis, and then at the x' axis. We'll assume that the origins of the two frames coincide. Equivalently, we'll let a given event define the origin of both frames, so that the differences of the coordinates (Δx, etc.) to a second event are simply equal to the coordinates (x, etc.) of the second event.

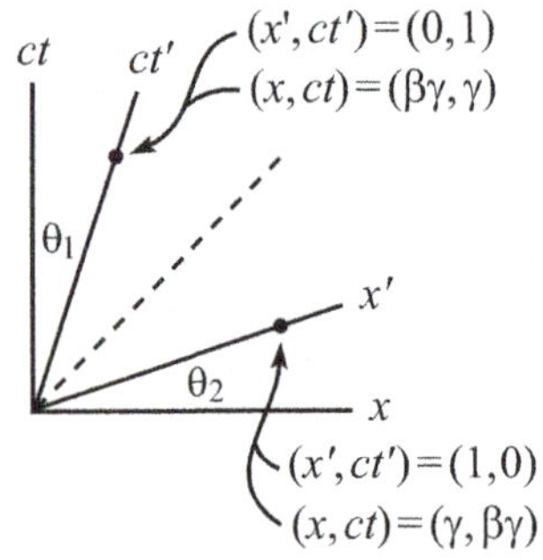

Figure 2.12

ct'-axis angle and unit size

Consider the point $(x', ct') = (0, 1)$, which lies on the ct' axis, one ct' unit from the origin; see Fig. 2.12. We'll ignore units here; technically we should be writing $ct' = 1$ m, instead of just $ct' = 1$. We've drawn both the ct' and x' axes titled inward because this will turn out to be how they behave. But at the moment this certainly isn't obvious.

With $x' = 0$ and $t' = 1/c$, Eq. (2.2) tells us that the point $(x', ct') = (0, 1)$ is the same as the point $(x, ct) = (\gamma v/c, \gamma)$. From Fig. 2.12 the angle between the ct' and ct axes is therefore given by $\tan\theta_1 = x/ct = v/c$. With $\beta \equiv v/c$, we have

$$\boxed{\tan\theta_1 = \beta} \tag{2.30}$$

Alternatively, the ct' axis is simply the *worldline* of the origin of S', because the origin always has $x' = 0$. (A worldline is the path an object takes as it travels through spacetime.) The origin moves at speed v with respect to S. Therefore, points on the ct' axis satisfy $x/t = v$, or $x/ct = v/c$.

On the paper, the point $(x', ct') = (0, 1)$, which we just found to be the same as the point $(x, ct) = (\gamma v/c, \gamma)$, is a distance $\gamma\sqrt{1 + v^2/c^2}$ from the origin, by the Pythagorean theorem. (We're assuming that the x and ct unit size on the paper is 1.) Therefore, using the definitions of β and γ, we see that

$$\boxed{\frac{\text{one } ct' \text{ unit}}{\text{one } ct \text{ unit}} = \sqrt{\frac{1+\beta^2}{1-\beta^2}}} \tag{2.31}$$

as measured on a grid where the x and ct axes are orthogonal. This ratio is greater than or equal to 1, so the ct' unit size is always at least as large as the ct unit size. The ratio approaches infinity as $\beta \to 1$. And it correctly equals 1 if $\beta = 0$, because S' is the same as S in that case.

x'-axis angle and unit size

The same basic reasoning holds for the x' axis. Consider the point $(x', ct') = (1, 0)$, which lies on the x' axis, one x' unit from the origin; see Fig. 2.12. With $x' = 1$ and $t' = 0$, Eq. (2.2) tells us that the point $(x', ct') = (1, 0)$ is the same as the point $(x, ct) = (\gamma, \gamma v/c)$. From Fig. 2.12 the angle between the x' and x axes is therefore given by $\tan\theta_2 = ct/x = v/c$. So, as in the ct'-axis case,

$$\boxed{\tan\theta_2 = \beta} \tag{2.32}$$

On the paper, the point $(x', ct') = (1, 0)$, which we just found to be the same as the point $(x, ct) = (\gamma, \gamma v/c)$, is a distance $\gamma\sqrt{1 + v^2/c^2}$ from the origin. So again, as in the ct'-axis case,

$$\boxed{\frac{\text{one } x' \text{ unit}}{\text{one } x \text{ unit}} = \sqrt{\frac{1+\beta^2}{1-\beta^2}}} \tag{2.33}$$

as measured on a grid where the x and ct axes are orthogonal. The x' and ct' axes are therefore stretched by the same factor, and tilted inward by the same angle, relative to the x and ct axes. In retrospect, this symmetry is clear from the symmetry in x and ct exhibited by the Lorentz transformation in Eq. (2.5). Consistent with this, the 45° worldline of a photon bisects the angle between the x' and ct' axis. This means that the photon's worldline is described by the relation $x' = ct'$. In other words, the photon has speed c in S', as we know it must.

Note that the "squeezing in" of the S' axes in a Minkowski diagram (associated with a given Lorentz transformation) is different from what happens in a rotation, where both axes rotate in the same direction. Also note that if the velocity of S' with respect to S is negative, then the S' axes actually fan out instead of squeeze in, as you can quickly check. The ct' axis lies to the left of the ct axis, and the x' axis lies below the x axis.

A few limits of the above results: If $\beta \equiv v/c = 0$, then $\theta_1 = \theta_2 = 0$. So the ct' and x' axes coincide with the ct and x axes, as they should. And the unit-size ratio in Eqs. (2.31) and (2.33) equals 1, as it should. If β is very close to 1, then $\theta_1 = \theta_2 \to 45°$, which means that the ct' and x' axes are both very close to the 45° light-ray line. The unit-size ratio in Eqs. (2.31) and (2.33) is very large when $\beta \to 1$, which means that the S' axes are stretched by a large factor.

In the above discussion, we chose one of our events to be the origin, for simplicity. But now that we know what the ct' and x' axes look like, we can let our two events be any two points in the diagram. Given these two points, we can just read off the Δx, $c\,\Delta t$, $\Delta x'$, and $c\,\Delta t'$ values that our two observers measure, assuming that our graph is accurate enough. Although these quantities are of course related by the Lorentz transformations, the advantage of a Minkowski diagram is that you can actually see geometrically what's going on.

There are very useful physical interpretations of the ct' and x' axes. If you stand at the origin of S', then the ct' axis is the "here" axis, and the x' axis is the "now" axis (the line of simultaneity). That is, all events on the ct' axis take place at your position (the ct' axis is your $x' = 0$ worldline, after all), and all events on the x' axis take place simultaneously (they all have $t' = 0$).

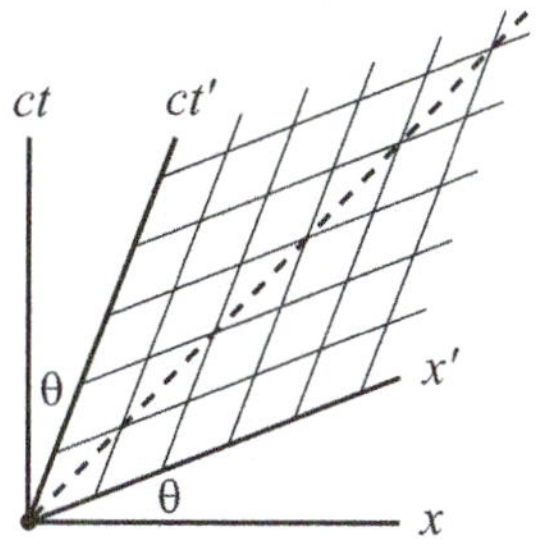

Figure 2.13

Likewise, lines of constant x' are parallel to the ct' axis, and lines of constant t' are parallel to the x' axis, as shown in Fig. 2.13. (We've drawn only the lines in the first "quadrant," but they fill up the entire plane, of course.) These statements are true because we could simply repeat the above derivations with a new origin, having the value of x' or t' that we're concerned with. All the angles would come out the same. Note that lines of constant x' (or t') are *not perpendicular* to the x' (or ct') axis. The lines *are* perpendicular to the axes in the original S frame, but only because the ct and x axes are perpendicular. The *parallelness* (if that's a word) described in the first sentence of this paragraph is what characterizes lines of constant x' or t'.

Example (Length contraction): For both parts of this problem, use a Minkowski diagram where the axes in frame S are orthogonal.

(a) S' moves at speed v with respect to S, in the positive x direction. A meter stick lies along the x axis and is at rest in S. If S' measures its length, what is the result?

(b) Now let the meter stick lie along the x' axis and be at rest in S'. If S measures its length, what is the result?

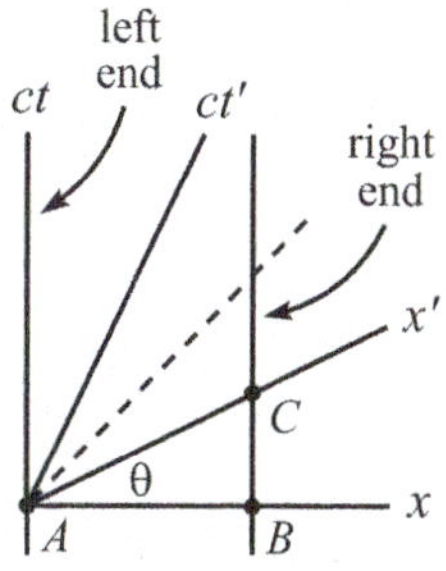

Figure 2.14

Solution:

(a) Pick the left end of the stick to be at the origin in S. Since the stick (in particular, the ends) is at rest in S, the worldlines of the two ends are shown in Fig. 2.14. The distance AB is one meter in the S frame, because A and B are the endpoints of the meter stick at simultaneous times (namely $t = 0$) in S; this is how a length is measured. So the length of AB on the paper is 1.

How does S' measure the length of the stick? He writes down the x' coordinates of the ends at simultaneous times (as measured by him, of course) and takes the difference. Let the time he makes the measurements be $t' = 0$. Then he measures the ends to be at the points A and C.

We now need to do a little geometry. The length of AC equals $(AB)/\cos\theta = 1/\cos\theta = \sqrt{1+\beta^2}$, where we have used the fact that the $\tan\theta = \beta$ relation in Eq. (2.32) implies $\cos\theta = 1/\sqrt{1+\beta^2}$. To determine how many x' units this length corresponds to, recall that Eq. (2.33) tells us that one unit on the x' axis has length $\sqrt{1+\beta^2}/\sqrt{1-\beta^2}$ on the paper. The $\sqrt{1+\beta^2}$ length of AC is $\sqrt{1-\beta^2}$ times this S' unit length. So S' measures the meter stick to have length $\sqrt{1-\beta^2} \equiv 1/\gamma$, which is the standard length-contraction result.

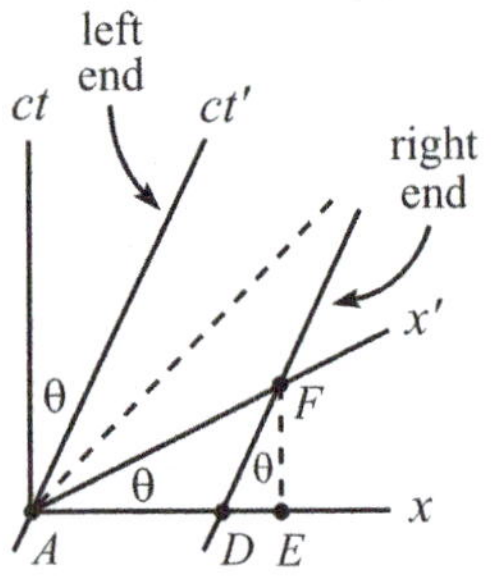

Figure 2.15

(b) The stick is now at rest in S', and we want to find the length that S measures. Pick the left end of the stick to be at the origin in S'. Then the worldlines of the two ends are shown in Fig. 2.15. The distance AF is one meter in the S' frame, because A and F are the endpoints of the meter stick at simultaneous times (namely $t' = 0$) in S'. And since Eq. (2.33) tells us that one unit on the x' axis has length $\sqrt{1+\beta^2}/\sqrt{1-\beta^2}$, this is the length on the paper of the segment AF.

In measuring the length of the stick, S writes down the x coordinates of the ends at simultaneous times (as measured by him) and takes the difference. Let the time he makes the measurements be $t = 0$. Then he measures the ends to be at the points A and D.

We now need to do the geometry, which is a little more involved in this case. Our goal is to find the length of segment AD, given that segment AF has length $\sqrt{1+\beta^2}/\sqrt{1-\beta^2}$. We know that the primed axes are tilted at an angle θ, where $\tan\theta = \beta$. Hence, $EF = (AF)\sin\theta$. And since $\angle DFE = \theta$ (it's the same as the angle the ct' axis makes with the vertical), we have $DE = (EF)\tan\theta = (AF)\sin\theta\cdot\tan\theta$. Therefore,

$$\begin{aligned} AD = AE - DE &= (AF)\cos\theta - (AF)\sin\theta\tan\theta \\ &= (AF)\cos\theta\,(1-\tan^2\theta) \\ &= \sqrt{\frac{1+\beta^2}{1-\beta^2}}\,\frac{1}{\sqrt{1+\beta^2}}(1-\beta^2) \\ &= \sqrt{1-\beta^2}\,. \end{aligned} \tag{2.34}$$

So S measures the meter stick to have length $\sqrt{1-\beta^2} \equiv 1/\gamma$, which again is the standard length-contraction result.

The analysis used in the above example also works for time intervals. The derivation of time dilation, using a Minkowski diagram, is the task of Exercise 2.26. Additionally, the derivation of the Lv/c^2 rear-clock-ahead result is the task of Exercise 2.27.

Causality revisited

In the last subsection of Section 2.3, we talked about causality. In particular, we discussed a setup involving lottery numbers and a train. We found that if instantaneous signals were possible, then person B would be able to know what the winning numbers were, *before* they were drawn. The setup involved B handing off the numbers to D, who sends an instantaneous signal (in the train frame) to C, who hands off the numbers to A, who sends an instantaneous signal (in the ground frame) to B. This signal arrives at B *before* the numbers are drawn.

What does all this look like in terms of a Minkowski diagram? Let the ground frame be S and the train frame be S'. Then the setup is summarized in Fig. 2.16, where the worldlines of the four people are indicated by the circled letters. At the upper right dot, three things happen: the lottery numbers are drawn, B immediately hands them off to D, and D sends an instantaneous signal (in the train frame) to C. At the left dot, three other things happen: C receives the signal, C immediately hands off the information to A, and A sends an instantaneous signal (in the ground frame) to B. At the lower right dot, B receives the signal.

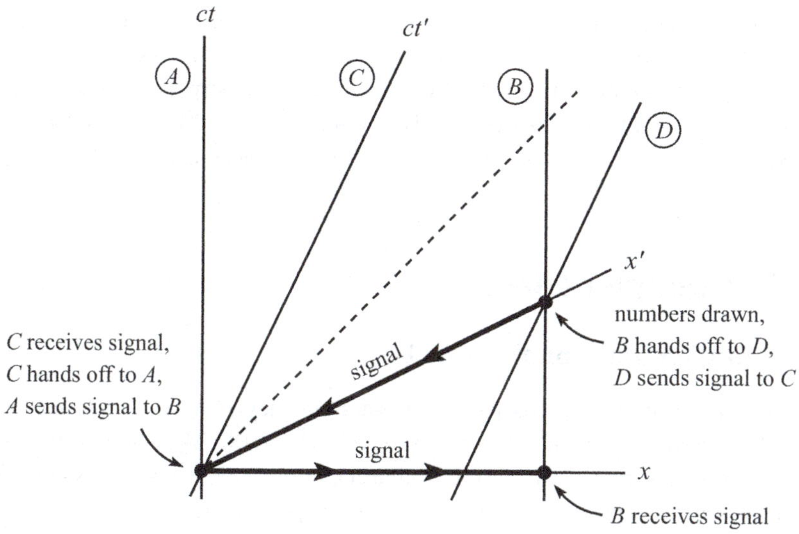

Figure 2.16

The worldline of the instantaneous signal from D to C lies along the x' axis, because this axis is a line of simultaneity in the train frame. (All points on the x' axis have the same value of t', namely $t' = 0$.) Likewise, the worldline of the instantaneous signal from A to B lies along the x axis, because this axis is a line of simultaneity in the ground frame (with $t = 0$). The Minkowski diagram makes it clear that B receives the signal *before* the numbers are drawn, as we found in Section 2.3 by using the rear-clock-ahead effect. We therefore conclude that instantaneous signals are not possible. The Minkowski diagram also makes it clear that in the ground frame, the signal from D to C actually goes "backward" *from C to D*, because at the given events, D has a larger t value than C.

As we mentioned in the last paragraph of Section 2.3, it isn't actually necessary to have the signals be instantaneous in order to generate the contradiction where B knows the winning numbers before they are drawn. Even if the signal from D to C isn't sloped downward quite as much as the x' axis (meaning that the speed of the signal in the train frame is fast, but not infinitely fast), and even if the signal from A to B is sloped slightly upward (likewise meaning that the speed of the signal in the ground frame is fast, but not infinitely fast), it is still possible for the "B receives signal" dot in Fig. 2.16 to be lower (that is, earlier) than the "numbers drawn" dot. This contradiction shows that the particular superluminal speeds that we have just chosen are not possible.

Can we show that *any* superluminal speed is impossible, even one that is only slightly larger than c? Yes indeed. If we pick the speed v of S' (the train) with respect to S (the ground) to be very close to c, then the S' axes in the Minkowski diagram are squeezed in very close to the 45° dotted line. As an exercise, you can convince yourself that superluminal signals with speeds only slightly faster than c will (assuming the axes are squeezed in enough) yield the contradiction where B knows the winning numbers before they are drawn. Therefore, by choosing v to be sufficiently close to c, we can generate a contradiction for any proposed speed of a superluminal signal.

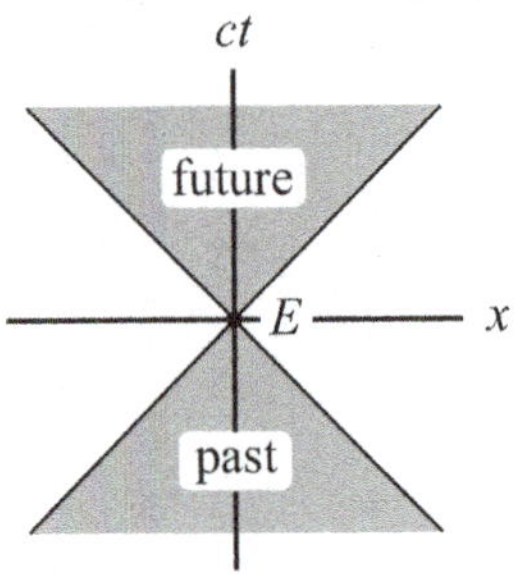

Figure 2.17

Given that signals can't travel faster than c, consider an event E located at the origin in Fig. 2.17. What other events can E possibly influence? Due to the speed limit of c, the worldline of a signal sent from E must lie above (or on) the $\pm 45°$ worldlines of photons sent from E. So the possible events that the signal can reach are ones in the upper shaded region shown. Similarly, the events that E can possibly be influenced *by* are ones in the lower shaded region. Events outside the shaded regions are ones that E can neither influence nor be influenced by; they are causally disconnected from E. As far as E is concerned, the future is limited to the upper shaded region, and the past is limited to the lower shaded region. If we add on a y axis perpendicular to the page, then the shaded triangles become cones. The boundary of the cone (the worldlines of photons) is called the *light cone*. For 3-D space, we would need to draw a 4-D diagram.

2.5 The Doppler effect

2.5.1 Longitudinal Doppler effect

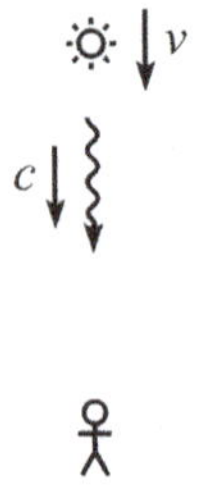

Figure 2.18

Consider a light source that emits short flashes at frequency f' (in its own frame) while moving directly toward you at speed v, as shown in Fig. 2.18. Alternatively, you could simply have someone running toward you while clapping at frequency f'. By definition, the frequency is the number of flashes (or whatever) per second; the units are s^{-1}. So a frequency of, say, $5\,s^{-1}$ means that five flashes happen each second. In our setup, we're not concerned with the actual frequency of the light (which determines its color), although we could just as well have the setup consist of a continuous beam of light with a given frequency. We've chosen instead to work with short flashes, because those are easy to visualize.

The question we will answer here is: What is the frequency at which the flashes hit your eye? In these Doppler-effect problems, we must be careful to distinguish between the time at which an event *occurs* in your frame, and the time at which you *see* the event occur. This is one of the few situations where we are concerned with the latter.

There are two effects contributing to the longitudinal Doppler effect. The first is relativistic time dilation. Since the source's clock runs slow in your frame, there is more time between the flashes. This effect decreases the frequency at which the flashes hit your eye. The second is the everyday Doppler effect (as with sound), arising from the motion of the source. Successive flashes have a smaller (or larger, if v is negative) distance to travel to reach your eye. This effect increases (or decreases, if v is negative) the frequency at which the flashes hit your eye.

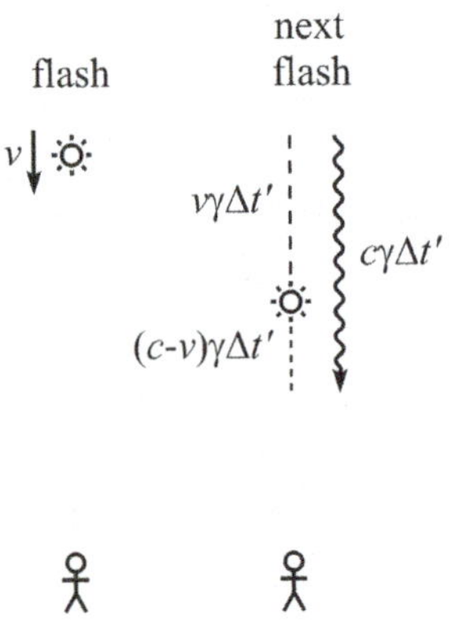

Figure 2.19

Let's now be quantitative and find the observed frequency. The time between flashes in the source's frame is $\Delta t' = 1/f'$. (For example, a frequency of $5\,s^{-1}$ implies a time of 1/5 of a second between flashes.) The time between flashes in your frame is then $\Delta t = \gamma\Delta t'$, due to time dilation. The photons of one flash have therefore traveled a distance (in your frame) of $c\,\Delta t = c\gamma\,\Delta t'$ by the time the next flash occurs. During this time between flashes, the source has traveled a distance $v\,\Delta t = v\gamma\,\Delta t'$ toward you in your frame; see Fig. 2.19. So at the instant the next flash occurs, the photons of this next flash are a distance (in your frame) of $c\,\Delta t - v\,\Delta t = (c - v)\gamma\,\Delta t'$ behind the photons of the previous flash. This result holds for all adjacent flashes. Therefore, after a given flash hits you, the next flash has to travel a remaining distance of $(c - v)\gamma \Delta t'$. It does this at speed c, so the time ΔT between the arrivals of the flashes at your eye is

$$\begin{aligned}\Delta T &= \frac{1}{c} \cdot (c - v)\gamma \Delta t' = \frac{1-\beta}{\sqrt{1-\beta^2}}\Delta t' \\ &= \sqrt{\frac{1-\beta}{1+\beta}}\Delta t' = \sqrt{\frac{1-\beta}{1+\beta}} \cdot \frac{1}{f'}. \end{aligned} \tag{2.35}$$

The frequency you see is therefore

$$f = \frac{1}{\Delta T} \quad \Longrightarrow \quad \boxed{f = \sqrt{\frac{1+\beta}{1-\beta}}\, f'} \tag{2.36}$$

If $\beta > 0$ (that is, the source is moving toward you), then $f > f'$. The everyday Doppler effect (which in this case serves to increase the frequency) wins out over the time-dilation effect (which always serves to decrease the frequency, when doing the calculation in your frame). In this case we say that the light is "blueshifted," because blue light is at the high-frequency end of the visible spectrum. The light need not have anything to do with the color blue, of course; by "blue" we just mean that the frequency is increased. If $\beta < 0$ (that is, the source is moving away from you), then $f < f'$. Both the everyday Doppler effect and the time-dilation effect serve to decrease the frequency. In this case we say that the light is "redshifted," because red light is at the low-frequency end of the visible spectrum.

We can also derive Eq. (2.36) by working in the frame of the source. (Never pass up a chance to redo a problem in different frame!) In this frame the source is at rest, so the distance between the photons in successive flashes is $c\Delta t'$; see Fig. 2.20. Since you are moving toward the source at speed v, the relative speed of you and the photons (as measured in the frame of the source) is $c + v$. This is the rate at which the $c\Delta t'$ gap between successive flashes is closed. So the time between your running into successive flashes is $c\Delta t'/(c + v) = \Delta t'/(1 + \beta)$, as measured in the frame of the source. But your clock runs slow in this frame, so a time of only $\Delta T = (1/\gamma)\Delta t'/(1 + \beta)$ elapses on your watch. Since $\gamma \equiv 1/\sqrt{1-\beta^2}$, we find that $\Delta T = \Delta t'\sqrt{(1-\beta)/(1+\beta)}$, in agreement with the ΔT in Eq. (2.35).

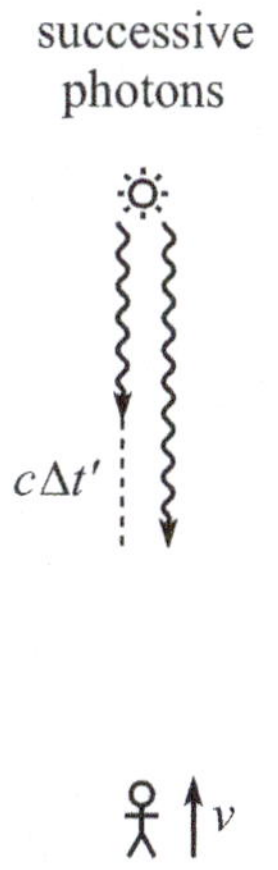

Figure 2.20

We certainly needed to obtain the same result (for the frequency f at which the flashes hit your eye, as measured by you) by working in the frame of the source, because the principle of relativity states that the result can't depend on which object we consider to be the one at rest; there is no preferred reference frame. This is different from the standard nonrelativistic Doppler effect (relevant to a siren moving toward you), because the frequency there *does* depend on whether you or the source is the one that is moving. The reason for this is that when we say "moving" here, we mean with respect to the rest frame of the air, which is the medium that sound travels in. We therefore do in fact have a preferred reference frame, unlike in relativity where there is no "ether." Using the arguments given above for the two different frames, but without the γ factors, it follows that the two nonrelativistic Doppler results are $f = f'/(1 - \beta)$ if a source is moving toward a stationary you, and $f = (1+\beta)f'$ if you are moving toward a stationary source. ("Stationary" means with respect to the air.) Here β is the ratio of the speed of the moving object to the speed of sound. In view of the fact that we have two different frequencies in the nonrelativistic case, the relativistic Doppler effect can be considered to be a simpler effect, in the sense that there is only one frequency to remember.

Example: Person A is at rest in the ground frame and emits flashes at frequency f_A. Person B runs with speed v toward A. When B passes a tree that is a distance L from A (as measured in the ground frame), B starts counting the number of flashes he receives. What is the total number of flashes B receives, by the time he arrives at A? Answer this by:

(a) using the Doppler effect.

(b) working in the ground frame and counting the number of flashes that B runs into.

Solution:

(a) As viewed by B, the time of the process is $L/\gamma v$. This follows from using time dilation in A's frame, which gives B's time as $(L/v)/\gamma$. Alternatively, it follows from using length contraction in B's frame, which gives B's time as $(L/\gamma)/v$. During this time, B receives the flashes at frequency $f_B = f_A\sqrt{(1+\beta)/(1-\beta)}$, from Eq. (2.36). The total number of flashes that B receives is therefore

$$N = f_B t_B = \sqrt{\frac{1+\beta}{1-\beta}}\, f_A \cdot \frac{L}{\gamma v} = \sqrt{\frac{1+\beta}{1-\beta}}\, f_A \cdot \frac{L\sqrt{1-\beta^2}}{v} = \frac{L(1+\beta)}{v} f_A. \quad (2.37)$$

(b) The first flash that B runs into is the one he receives when he passes the tree. Let this occur at time $t = 0$ in the ground frame. Since the tree is a distance L from A, this flash must have been emitted by A at time $t = -L/c$. The last flash that B runs into is the one he receives right when he arrives at A. Since B travels at speed v, it takes him a time of L/v to reach A. So the last flash that B runs into is emitted by A at time $t = L/v$. B therefore runs into the flashes that are emitted from $t = -L/c$ to $t = L/v$. The number of flashes that A emits in this time of $L/c + L/v$ is

$$N = \left(\frac{L}{v} + \frac{L}{c}\right) f_A = \frac{L}{v}\left(1 + \frac{v}{c}\right) f_A = \frac{L(1+\beta)}{v} f_A, \quad (2.38)$$

in agreement with the result in part (a). Note that if you forgot to use relativity in part (a), then t_B would be larger by a factor γ, while f_B would be smaller by a factor γ. (Recall how time dilation entered into the discussion of Fig. 2.20.) The relativistic effects would therefore cancel in Eq. (2.37), and the result for N would be the same. This means that N would be the same for, say, a (nonrelativistic) sound wave, given the same f_A. This makes sense, because the method in part (b) didn't involve relativity. (Using the speed c doesn't count as using relativity, because in a different setup, c is simply replaced by the speed of sound, etc.)

2.5.2 Transverse Doppler effect

Let's now consider a two-dimensional setup. Consider a source that emits short flashes at frequency f' (in its own frame), while moving across your field of vision at speed v. There are two reasonable questions we can ask about the frequency you observe:

- CASE 1: At the instant the source is at its closest approach to you, with what frequency do the flashes hit your eye?
- CASE 2: When you *see* the source at its closest approach to you, with what frequency do the flashes hit your eye?

The difference between these two scenarios is shown in Fig. 2.21 and Fig. 2.22, where the source's motion is taken to be parallel to the x axis. In the first case, the photons you see must have been emitted at an earlier time, because the source moves during the nonzero time it takes the light to reach you. In this scenario, we are dealing with photons that hit your eye *when* the source crosses the y axis. You therefore see the photons come in at an angle with respect to the y axis. It *looks* like the source is off to the side, at the given moment when it crosses the y axis and is at its closest approach to you.

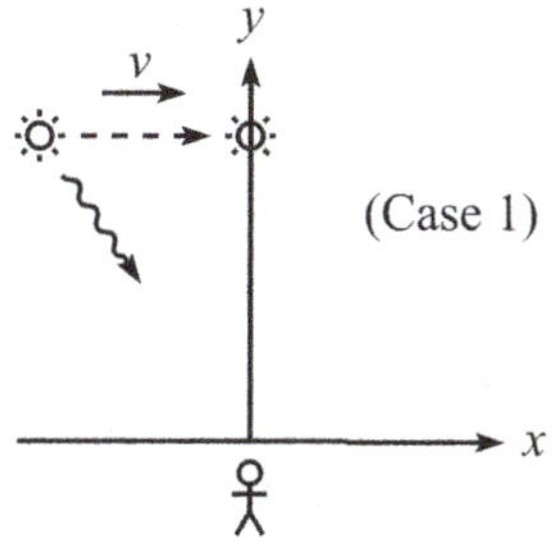

Figure 2.21

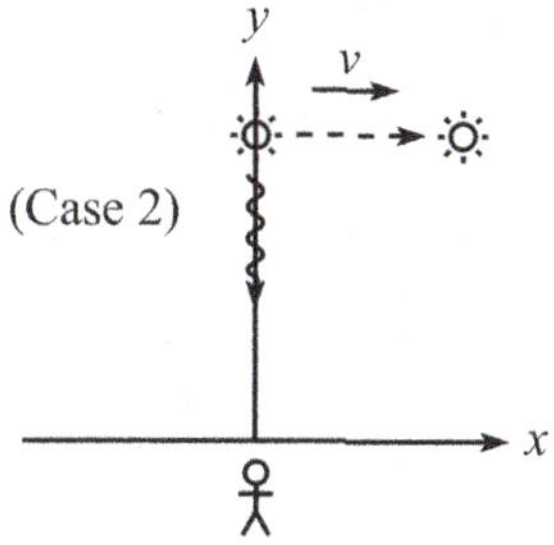

Figure 2.22

In the second case, you see the photons come in along the y axis (by the definition of this scenario). At the instant you observe one of these photons, the source is at a position past the y axis (which you don't care about). Let's find the observed frequencies in these two cases.

Case 1: Let your frame be S, and let the source's frame be S'. Consider the situation from S''s point of view. An S' person sees you moving across his field of vision at speed v. The relevant photons hit your eye when you cross the y' axis of the S' frame, defined to be the vertical axis that passes through the source. This is true because you can imagine the source carrying a long vertical stick. You are concerned with the photons that hit your eye when the stick brushes by you. This is a frame-independent statement.

We claim that S' sees you get hit by a flash every $\Delta t' = 1/f'$ seconds in its frame. This follows from the fact that when you are very close to the y' axis, the relevant photons are received at points that are essentially equidistant from the source. So they all travel the same distance, and we don't have to worry about any longitudinal effects, in the S' frame. Therefore, as observed in the S' frame, the flashes hit you at the same rate they are emitted (which is one flash every $\Delta t' = 1/f'$ seconds). But S' sees your clock running slow, so you get hit by a flash every $\Delta T = \Delta t'/\gamma = 1/(f'\gamma)$ seconds on your clock. The frequency in your frame is therefore

$$f = \frac{1}{\Delta T} = \gamma f' \quad \Longrightarrow \quad \boxed{f = \frac{f'}{\sqrt{1-\beta^2}}} \qquad (2.39)$$

Hence, f is greater than f'. You see the flashes at a higher frequency than the source emits them.

Case 2: Again, let your frame be S, and let the source's frame be S'. Consider the situation from your point of view. Because of time dilation, a clock on the source runs slow (in your frame) by a factor of γ. So you get hit by a flash every $\Delta T = \gamma \Delta t' = \gamma/f'$ seconds in your frame. This follows from the fact that when the source is very close to the y axis, the relevant photons are emitted from points that are essentially equidistant from you. So they all travel the same distance, and we don't have to worry about any longitudinal effects, in your frame. Therefore, as observed in your frame, the flashes hit you at the same rate they are emitted (which is one flash every $\gamma \Delta t' = \gamma/f'$ seconds). Since we are working in your frame here, this time γ/f' is the time ΔT between flashes as measured on your clock. When you see the source cross the y axis, you therefore observe a frequency of

$$f = \frac{1}{\Delta T} = \frac{f'}{\gamma} \quad \Longrightarrow \quad \boxed{f = f'\sqrt{1-\beta^2}} \qquad (2.40)$$

Hence, f is smaller than f'. You see the flashes at a lower frequency than the source emits them.

When people talk about the "transverse Doppler effect," they sometimes mean Case 1, and they sometimes mean Case 2. The name "transverse Doppler" is therefore ambiguous, so you should remember to state exactly which scenario you are talking about. Other cases that are "in between" these two can also be considered. But they get a bit messy.

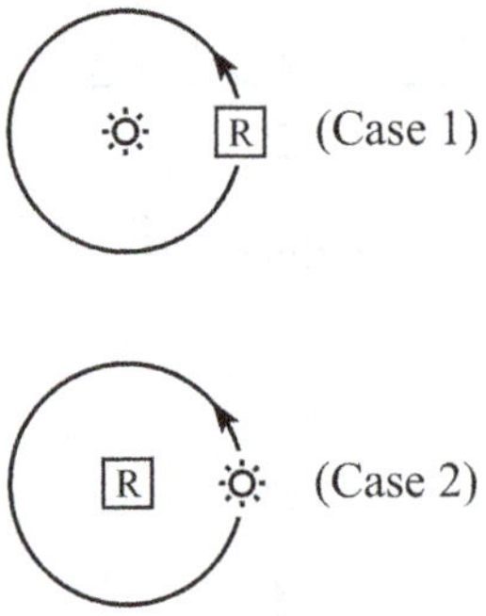

Figure 2.23

REMARKS:

1. The above two cases may alternatively be described, respectively (as you can convince yourself), in the following ways; see Fig. 2.23.

   - CASE 1: A receiver moves with speed v in a circle around a source. What frequency does the receiver register?
   - CASE 2: A source moves with speed v in a circle around a receiver. What frequency does the receiver register?

   These two setups make it clear that the results in Eqs. (2.39) and (2.40) arise from a simple time-dilation argument used by the inertial object at the center of each circle. In Case 1, the receiver's clock runs slow, so it gets hit by a larger number of flashes during one tick of its clock. In Case 2, the source's clock runs slow, so the flashes are emitted (and hence received by the receiver) at a lower rate.

   These circular setups involve accelerating objects. We must therefore invoke the fact (which has plenty of experimental verification) that if an inertial observer looks at a moving clock, only the instantaneous speed of the clock is important in computing the time-dilation factor. The acceleration is irrelevant. (See the third remark in the solution to Problem 2.12.) The acceleration is, however, very important if things are considered from an accelerating object's point of view. We'll discuss this in our treatment of general relativity in Chapter 5.

2. Beware of the following incorrect reasoning for Case 1 (the original Case 1 in Fig. 2.21, not the circular setup in Fig. 2.23), leading to an incorrect version of Eq. (2.39). "You (in frame S) see things in S' slowed down by a factor γ (that is, $\Delta t = \gamma \Delta t'$), by the usual time-dilation effect. Hence, you see the light flashing at a slower rate. Therefore, $f = f'/\gamma$." This reasoning puts the γ in the wrong place. Where is the error? The error lies in confusing the time at which an event *occurs* in your frame, with the time at which you *see* (with your eyes) the event occur. The flashes certainly *occur* at a lower rate in S, due to time dilation. But there is an additional effect: due to the motion of the source relative to you, it turns out that the flashes meet your eye at a faster rate, because the source is moving slightly toward you while it is emitting the relevant photons. So there is a partial longitudinal effect. You can work out the details from your (S's) point of view in Exercise 2.31. Although this is a fun exercise, it should convince you that it is much easier to look at things in the frame in which there are no longitudinal effects, as we did in the original solutions above. ♣

2.6 Rapidity

2.6.1 Definition

The *rapidity*, ϕ, is defined by

$$\boxed{\tanh\phi \equiv \beta \equiv \frac{v}{c}}. \tag{2.41}$$

where tanh is the "hyperbolic tangent function," given by $\tanh\phi \equiv (e^{\phi}-e^{-\phi})/(e^{\phi}+e^{-\phi})$. A few properties of the hyperbolic trig functions are listed in Appendix H. These functions are similar in many ways to the normal trig functions.

The rapidity defined in Eq. (2.41) is very useful in relativity because many expressions take a particularly nice form when written in terms of it. Consider, for example, the velocity-addition formula. Let $\beta_1 = \tanh\phi_1$ and $\beta_2 = \tanh\phi_2$. Then if we add β_1

and β_2 using the velocity-addition formula, Eq. (1.29), we obtain a new β value given by

$$\beta = \frac{\beta_1 + \beta_2}{1 + \beta_1\beta_2} = \frac{\tanh\phi_1 + \tanh\phi_2}{1 + \tanh\phi_1 \tanh\phi_2} = \tanh(\phi_1 + \phi_2), \tag{2.42}$$

where we have used the addition formula for $\tanh\phi$, which you can prove by writing things in terms of the $e^{\pm\phi}$ exponentials. Therefore, while the velocities add in the strange manner of Eq. (1.29), the rapidities add by standard addition. That is, the relativistic addition of velocities described by the rapidities ϕ_1 and ϕ_2 is the velocity described by the rapidity $\phi_1 + \phi_2$.

The Lorentz transformations also take a nice form when written in terms of the rapidity. The γ factor can be written as

$$\gamma \equiv \frac{1}{\sqrt{1-\beta^2}} = \frac{1}{\sqrt{1-\tanh^2\phi}} = \cosh\phi, \tag{2.43}$$

which you can again prove by writing things in terms of the $e^{\pm\phi}$ exponentials. Also,

$$\gamma\beta \equiv \frac{\beta}{\sqrt{1-\beta^2}} = \frac{\tanh\phi}{\sqrt{1-\tanh^2\phi}} = \sinh\phi. \tag{2.44}$$

Therefore, the Lorentz transformations in matrix form, Eqs. (2.7), become

$$\begin{pmatrix} x \\ ct \end{pmatrix} = \begin{pmatrix} \cosh\phi & \sinh\phi \\ \sinh\phi & \cosh\phi \end{pmatrix} \begin{pmatrix} x' \\ ct' \end{pmatrix}. \tag{2.45}$$

This transformation looks similar to a rotation in a plane, which is given by

$$\begin{pmatrix} x \\ y \end{pmatrix} = \begin{pmatrix} \cos\theta & -\sin\theta \\ \sin\theta & \cos\theta \end{pmatrix} \begin{pmatrix} x' \\ y' \end{pmatrix}. \tag{2.46}$$

(Depending on the direction of rotation, the minus sign may instead appear in the lower-left entry.) To switch from rotations to Lorentz transformations, we just need to replace the standard trig functions with hyperbolic trig functions, and get rid of the minus sign. The fact that $s^2 \equiv c^2t^2 - x^2$ doesn't depend on the reference frame follows quickly from Eq. (2.45). The cross terms in the squares cancel, and $\cosh^2\phi - \sinh^2\phi = 1$ (provable with the $e^{\pm\phi}$ exponentials), so we have

$$\begin{aligned} c^2t^2 - x^2 &= \big((\sinh\phi)x' + (\cosh\phi)ct'\big)^2 - \big((\cosh\phi)x' + (\sinh\phi)ct'\big)^2 \\ &= (\cosh^2\phi - \sinh^2\phi)c^2t'^2 - (\cosh^2\phi - \sinh^2\phi)x'^2 \\ &= c^2t'^2 - x'^2, \end{aligned} \tag{2.47}$$

as desired. This should be compared with the invariance of $r^2 \equiv x^2 + y^2$ for rotations in a plane, where the cross terms arising from Eq. (2.46) likewise cancel, and $\cos^2\theta + \sin^2\theta = 1$.

The quantities associated with a Minkowski diagram also take a nice form when written in terms of the rapidity. The angle between the S and S' axes satisfies

$$\tan\theta = \beta = \tanh\phi. \tag{2.48}$$

And the size of one unit on the ct' and x' axes is, from Eqs. (2.31) and (2.33),

$$\sqrt{\frac{1+\beta^2}{1-\beta^2}} = \sqrt{\frac{1+\tanh^2\phi}{1-\tanh^2\phi}} = \sqrt{\frac{\cosh^2\phi + \sinh^2\phi}{\cosh^2\phi - \sinh^2\phi}} = \sqrt{\cosh 2\phi}, \tag{2.49}$$

where we have used $\cosh^2\phi - \sinh^2\phi = 1$ and also the double-angle formula for cosh (provable with the $e^{\pm\phi}$ exponentials). For large ϕ, $\sqrt{\cosh 2\phi}$ is approximately equal to $e^\phi/\sqrt{2}$.

2.6.2 Physical meaning

The fact that the rapidity makes many of our formulas look nice is reason enough to consider it. But in addition, it turns out to have a very meaningful physical interpretation. Consider the following setup. A spaceship is initially at rest in the lab frame. At a given instant, it starts to accelerate. Let a be the *proper acceleration*, which is defined as follows. Let t be the time in the spaceship's frame. (This frame is of course changing as time goes by, because the spaceship is accelerating. The time t is simply the spaceship's proper time.) If the proper acceleration is a, then at time $t + dt$ (where dt is small), the spaceship is moving at speed $a\,dt$ relative to the frame it was just in at time t. An equivalent definition is that the astronaut feels a force of ma applied to his body by the spaceship. If he is standing on a scale, the scale shows a reading of $F = ma$.

What is the speed $v(t)$ of the spaceship with respect to the lab frame at (the spaceship's) time t? We can answer this question by considering two nearby times and using the velocity-addition formula. From the above definition of a, at time $t + dt$ we have the same setup as in Fig. 2.4. The spaceship is moving at speed $v_1 \equiv a\,dt$ with respect to the earlier frame of the spaceship, which was moving at speed $v_2 \equiv v(t)$ with respect to the lab frame. So Eq. (2.18) gives

$$v(t + dt) = \frac{a\,dt + v(t)}{1 + a\,dt\,v(t)/c^2}. \tag{2.50}$$

The lefthand side can be written as $v(t) + dv$, by the definition of dv. If we multiply the righthand side by 1 in the form of $(1 - a\,dt\,v(t)/c^2)/(1 - a\,dt\,v(t)/c^2)$, and if we ignore the (very small) terms of order dt^2 in both the numerator and denominator, then Eq. (2.50) becomes (with $v(t) \to v$ for simplicity)

$$v + dv = a\,dt + v - a\,dt\,v^2/c^2. \tag{2.51}$$

The v's cancel, so we end up with

$$dv = a\left(1 - \frac{v^2}{c^2}\right) dt \quad \Longrightarrow \quad \int_0^v \frac{dv}{1 - v^2/c^2} = \int_0^t a\,dt, \tag{2.52}$$

where we have separated variables and integrated. Using $\int dz/(1 - z^2) = \tanh^{-1} z$ and assuming that a is constant, we obtain the speed of the spaceship as a function of the spaceship's time t,

$$v(t) = c\tanh(at/c), \tag{2.53}$$

as you can verify. Alternatively, you can use $1/(1 - z^2) = (1/(1 - z) + 1/(1 + z))/2$, and then integrate to obtain some logs, which in turn yield the tanh.

For small a or small t (more precisely, for $at/c \ll 1$), we can apply the $\tanh z \approx z$ approximation to Eq. (2.53). (You can derive this by applying the Taylor series $e^{\pm\phi} \approx 1 \pm \phi$ to the definition of $\tanh\phi$ in terms of exponentials.) Eq. (2.53) then correctly reduces to the nonrelativistic result, $v(t) \approx at$. And for $at/c \gg 1$, we can use $\tanh z \approx 1$, for large z. This gives $v(t) \approx c$, as it should, because if the object accelerates for a long enough time, eventually its speed will approach c.

If a is a function of time, $a(t)$, then we can't take the a outside the integral in Eq. (2.52). You can quickly show that in this case we end up with the general formula,

$$\boxed{v(t) = c\tanh\left(\frac{1}{c}\int_0^t a(t)\,dt\right)} \tag{2.54}$$

which correctly yields Eq. (2.53) if a is constant. Since the rapidity ϕ was defined in Eq. (2.41) by $v = c \tanh \phi$, Eq. (2.54) therefore tells us that the rapidity is given by

$$\boxed{\phi(t) = \frac{1}{c}\int_0^t a(t)\,dt} \tag{2.55}$$

Remember that t is the spaceship's time. Note that whereas v has c as a limiting value, ϕ can become arbitrarily large. Looking at Eq. (2.55), and recalling that $F = ma \Longrightarrow a(t) = F(t)/m$, we see that $\phi(t) = (1/mc)\int_0^t F(t)\,dt$. In words, this says that the ϕ associated with a given v is $1/mc$ times the time integral of the force (felt by the astronaut) that is needed to bring the astronaut up to speed v. By applying a force for an arbitrarily long time, we can make ϕ be arbitrarily large.

The integral $\int a(t)\,dt$ may be described as the naive, incorrect speed. That is, it is the speed that the astronaut might think he has, if he has his eyes closed and knows nothing about the theory of relativity. And in fact his thinking would be essentially correct for small speeds, where Newtonian physics applies. For relativistic speeds, the quantity $\int a(t)\,dt = \int F(t)\,dt/m$ looks like a reasonably physical thing, so it seems like it should have *some* meaning. And indeed, although it doesn't equal v, all you have to do to get v is take a tanh and throw in some factors of c.

Eq. (2.54) is the reason why rapidities add via simple addition when using the velocity-addition formula, as we saw in Eq. (2.42). There is really nothing more going on here than the fact that

$$\int_{t_0}^{t_2} a(t)\,dt = \int_{t_0}^{t_1} a(t)\,dt + \int_{t_1}^{t_2} a(t)\,dt. \tag{2.56}$$

To be explicit, let a force (not necessarily constant) be applied from t_0 to t_1 that brings a mass up to a speed of $\beta_1 = \tanh\phi_1 = \tanh\big(\int_{t_0}^{t_1} a(t)\,dt\big)$. And then let an additional force be applied from t_1 to t_2 that adds on an additional speed of $\beta_2 = \tanh\phi_2 = \tanh\big(\int_{t_1}^{t_2} a(t)\,dt\big)$, relative to the mass's frame at t_1. Then the resulting speed may be looked at in two ways: (1) it is the result of relativistically adding the speeds $\beta_1 = \tanh\phi_1$ and $\beta_2 = \tanh\phi_2$, and (2) it is the result of applying the force from t_0 to t_2. You get the same final speed, of course, whether or not you bother to record the speed along the way at t_1. This latter result is $\beta = \tanh\big(\int_{t_0}^{t_2} a(t)\,dt\big) = \tanh(\phi_1 + \phi_2)$, where the second equality here comes from the statement, Eq. (2.56), that integrals simply add. Therefore, the relativistic addition of $\tanh\phi_1$ and $\tanh\phi_2$ gives $\tanh(\phi_1 + \phi_2)$, as we wanted to show.

2.7 Relativity without c

In Section 1.2 we introduced the two postulates of special relativity, namely the relativity postulate and the speed-of-light postulate. In subsequent sections, we then derived many counterintuitive and strange consequences of these postulates. But you might wonder – does the strangeness come more from one of them than the other? Is one the real "culprit" that makes Einstein's world so much different from Newton's?

Appendix D gives a derivation of the Lorentz transformations, working directly from the two postulates and not using the three fundamental effects, which were the basis of the derivations in Section 2.1. Let's see what happens if we omit the speed-of-light postulate, which is the one that most people would say produces the strangeness of relativity. It's hard to imagine a reasonable (empty) universe where the relativity postulate doesn't hold, but it's easy to imagine a universe where the speed of light depends on the reference frame. Light could behave like a baseball, for example. So we'll drop the speed-of-light postulate and see what we can say about the coordinate transformations between frames, using only the relativity postulate. For further discussion of this topic, see Lee and Kalotas (1975) and references therein.

In Appendix D the form of the transformations, just prior to invoking the speed-of-light postulate, is given in Eq. (6.11) as (assuming S' is moving rightward at speed v with respect to S)

$$x = A_v(x' + vt'),$$
$$t = A_v\left(t' + \frac{1}{v}\left(1 - \frac{1}{A_v^2}\right)x'\right). \tag{2.57}$$

We have put a subscript on A here, to remind us of the v dependence. Can we say anything about A_v without invoking the speed-of-light postulate? Indeed we can. Define V_v by

$$\frac{1}{V_v^2} \equiv \frac{1}{v^2}\left(1 - \frac{1}{A_v^2}\right) \implies A_v = \frac{1}{\sqrt{1 - v^2/V_v^2}}. \tag{2.58}$$

Eq. (2.57) then becomes

$$x = \frac{1}{\sqrt{1 - v^2/V_v^2}}(x' + vt'),$$
$$t = \frac{1}{\sqrt{1 - v^2/V_v^2}}\left(\frac{v}{V_v^2}x' + t'\right). \tag{2.59}$$

We picked the positive square root in Eq. (2.58) because when $v = 0$ we should have $x = x'$ and $t = t'$. All we've done so far is make a change of variables from A_v to V_v, but we now make the following claim.

Claim 2.1 *V_v^2 is independent of v.*

Proof: As stated in the last remark in Section 2.1.1, we know that two successive applications of the transformations in Eq. (2.59) must again yield a transformation of the same form, otherwise the theory wouldn't be consistent. Consider a transformation characterized by velocity v_1, and another one characterized by velocity v_2. For simplicity, define

$$V_1 \equiv V_{v_1}, \qquad V_2 \equiv V_{v_2},$$
$$\gamma_1 \equiv \frac{1}{\sqrt{1 - v_1^2/V_1^2}}, \qquad \gamma_2 \equiv \frac{1}{\sqrt{1 - v_2^2/V_2^2}}. \tag{2.60}$$

To calculate the composite transformation, it is easiest to use matrix notation. Looking at Eq. (2.59), we see that the composite transformation is given by the matrix

$$\begin{pmatrix} \gamma_2 & \gamma_2 v_2 \\ \gamma_2\dfrac{v_2}{V_2^2} & \gamma_2 \end{pmatrix}\begin{pmatrix} \gamma_1 & \gamma_1 v_1 \\ \gamma_1\dfrac{v_1}{V_1^2} & \gamma_1 \end{pmatrix} = \gamma_1\gamma_2\begin{pmatrix} 1 + \dfrac{v_1 v_2}{V_1^2} & v_2 + v_2 \\ \dfrac{v_1}{V_1^2} + \dfrac{v_2}{V_2^2} & 1 + \dfrac{v_1 v_2}{V_2^2} \end{pmatrix}. \tag{2.61}$$

The composite transformation must still be of the form in Eq. (2.59). But this implies that the upper-left and lower-right entries of the composite matrix must be equal. Therefore, $V_1^2 = V_2^2$. Since this holds for arbitrary v_1 and v_2, we see that V_v^2 must be a constant, independent of v. ■

Denote the constant value of V_v^2 by V^2. Then the coordinate transformations in Eq. (2.59) become

$$x = \frac{1}{\sqrt{1 - v^2/V^2}}(x' + vt'),$$
$$t = \frac{1}{\sqrt{1 - v^2/V^2}}\left(t' + \frac{v}{V^2}x'\right). \tag{2.62}$$

We have obtained this result using only the relativity postulate. These transformations have the same form as the Lorentz transformations in Eq. (2.2). The only extra information in Eq. (2.2) is that V equals the speed of light, c. It is remarkable that we were able to prove so much by using only the relativity postulate. In this sense, we see that most of the strangeness of relativity can be surprisingly thought of as arising from the relativity postulate, as opposed to the speed-of-light postulate.

We can say a few more things. There are four basic possibilities for the value of V^2. However, two of these are unphysical.

- $V^2 = \infty$: This gives the Galilean transformations, $x = x' + vt'$ and $t = t'$.
- $0 < V^2 < \infty$: This gives transformations of the Lorentz type. V is the limiting speed of an object. Experiments show that this case is the one that corresponds to the universe we live in.
- $V^2 = 0$: This case isn't physical, because any nonzero value of v makes the $\gamma \equiv 1/\sqrt{1 - v^2/V^2}$ factor be zero (technically the imaginary number $0i$). So in Eq. (2.62), x is always zero. And t is always infinite if $x' \neq 0$ (because x' is divided by V^2). Since any nonzero value of v yields these nonsensical results, nothing can ever move.
- $V^2 < 0$: It turns out that this case is also unphysical. You might be concerned that the square of V is less than zero, but this is fine because V appears in the transformations (2.62) only through its square. There is no need for V to actually be the speed of anything. The trouble is that if $V^2 < 0$, the nature of Eq. (2.62) implies the possibility of time reversal. This opens the door for causality violation and all the other problems associated with time reversal. We therefore reject this case.

  To be more explicit, define $b^2 \equiv -V^2$, where b is a positive quantity. If we then additionally define θ by $\tan\theta \equiv v/b$, you can verify that Eq. (2.62) can be rewritten in the form,

$$x = x'\cos\theta + (bt')\sin\theta,$$
$$bt = -x'\sin\theta + (bt')\cos\theta. \tag{2.63}$$

  θ must lie in the range $-\pi/2 \leq \theta \leq \pi/2$, because the coefficients of the x' and t' terms in Eq. (2.62) that aren't multiplied by v are necessarily positive (or zero, if $v \to \infty$), which implies that the $\cos\theta$ coefficients in Eq. (2.63) must satisfy $\cos\theta \geq 0$.

  Since we have normal trig functions in Eq. (2.63) instead of the hyperbolic trig functions that appeared in the Lorentz transformation in Eq. (2.45), the transformation in Eq. (2.63) is simply a rotation of the axes through an angle θ in the plane. The S axes are rotated counterclockwise by an angle θ relative to the S' axes. (The direction is indeed counterclockwise, because the point $(x', bt') = (0, 1)$ yields the point $(x, bt) = (\sin\theta, \cos\theta)$, which lies in the first

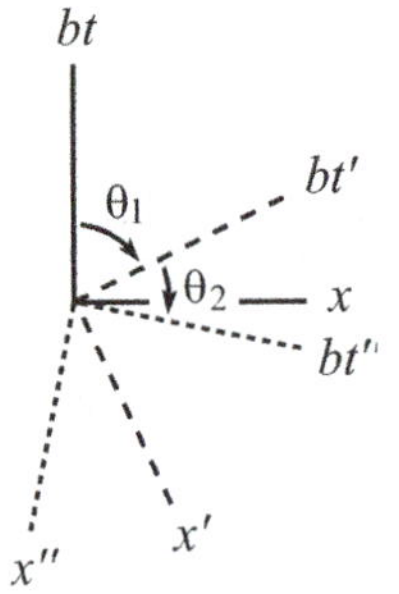

Figure 2.24

quadrant of S.) Equivalently, the S' axes are rotated clockwise by an angle θ relative to the S axes.

Eq. (2.63) satisfies the requirement that the composition of two transformations is again a transformation of the same form. Rotation by θ_1, and then by θ_2, yields a rotation by $\theta_1 + \theta_2$. However, if θ_1 and θ_2 are positive, and if the resulting rotation angle of $\theta \equiv \theta_1 + \theta_2$ is greater than 90°, then we have a problem. The tangent of such an angle is negative. Therefore, $\tan\theta = v/b$ implies that v is negative.

This situation is shown in Fig. 2.24. Frame S'' moves at velocity $v_2 > 0$ with respect to frame S', which moves at velocity $v_1 > 0$ with respect to frame S. From the figure, we see that someone standing at rest in frame S'' (that is, someone whose worldline is the bt'' axis) is going to have some serious issues in frame S. For one, the bt'' axis has a negative slope in frame S, which means that if we imagine moving upward in the figure, then as t increases for points on this axis, x decreases. The person is therefore moving with a *negative* velocity with respect to S. Adding two positive velocities and obtaining a negative one is absurd. But even worse, someone standing at rest in S'' is moving in the positive direction along the bt'' axis, which means that he is traveling *backward* in time in S. That is, he will die before he is born. This is not good.

An equivalent method of dismissing this case, given in Lee and Kalotas (1977), but one that doesn't specifically refer to causality violation, is to note that the transformations in Eq. (2.63) don't form a closed group. In other words, successive applications of the transformations can eventually yield a transformation that isn't of the form in Eq. (2.63), due to the $-\pi/2 \le \theta \le \pi/2$ restriction. (In contrast, the rotations in a plane form a closed group, because there are no restrictions on what θ can be.) This argument is equivalent to the time-reversal argument above, because $-\pi/2 \le \theta \le \pi/2$ is equivalent to $\cos\theta \ge 0$, which is equivalent to the statement that the coefficients of t and t' in Eq. (2.63) have the same sign.

Note that all of the finite $0 < V^2 < \infty$ possibilities (the ones that yield transformations of the Lorentz type) are essentially the same. Any difference in the numerical value of V can be absorbed into the definitions of the unit sizes of x and t. Given that V is finite, it has to be *something*, so it doesn't make sense to put much importance on its particular numerical value. There is therefore only one decision to be made when constructing the spacetime structure of an (empty) universe. You just have to say whether V is finite or infinite, that is, whether the universe is Lorentzian or Galilean. Equivalently, all you have to say is whether or not there is an upper limit on the speed of any object. If there is, then you can simply postulate the existence of something that moves with this limiting speed. In other words, to create your universe, you simply have to say, "Let there be light."

2.8 Summary

In this chapter we studied additional kinematics topics by building on the material we learned in Chapter 1. In particular, we learned:

- The Lorentz transformations,

$$\begin{aligned} \Delta x &= \gamma(\Delta x' + v\,\Delta t'), \\ \Delta t &= \gamma(\Delta t' + v\,\Delta x'/c^2), \end{aligned} \tag{2.64}$$

tell us how the x and t separations between two events transform between frames. These transformations contain the same information as the collection of the three fundamental effects in Section 1.3. The sign on the righthand side is determined by which way S (associated with the coordinates on the lefthand side) sees S' (associated with the righthand side) moving.

- The Lorentz transformations provide quick derivations of both the longitudinal and transverse velocity-additional formulas. These formulas are, respectively,

$$u_x = \frac{u'_x + v}{1 + u'_x v/c^2} \quad \text{and} \quad u_y = \frac{u'_y}{\gamma(1 + u'_x v/c^2)}, \tag{2.65}$$

where the object moves with velocity (u'_x, u'_y) with respect to S', and S' moves with velocity v with respect to S, in the x direction.

- The Lorentz transformations imply that the quantity

$$(\Delta s)^2 \equiv c^2(\Delta t)^2 - (\Delta x)^2 \tag{2.66}$$

is invariant. That is, it has the same value in all frames. If the events are timelike separated ($(\Delta s)^2 > 0$), then $(\Delta s)/c$ is the proper time between the events. If the events are spacelike separated ($(\Delta s)^2 < 0$), then $|\Delta s|$ is the proper distance between the events.

- Minkowski diagrams (or spacetime diagrams) contain the same information as the Lorentz transformations, but in geometric form. The x' and ct' axes are swung inward by the same angle (given by $\tan\theta = \beta$), and the unit size on these axes is longer than on the x and ct axes by the factor $\sqrt{(1+\beta^2)/(1-\beta^2)}$.

- Longitudinal Doppler effect: If a source and an observer are moving in 1-D with relative velocity β (with positive β defined as toward each other), then the observed frequency is $\sqrt{(1+\beta)/(1-\beta)}$ times the emitted frequency.

  Transverse Doppler effect: If a source is moving across your vision, then there are two fundamental cases to consider, described in Figs. 2.21/2.22 and in Fig. 2.23. The observed frequencies in these two cases are, respectively, γ and $1/\gamma$ times the emitted frequency.

- The rapidity ϕ, defined by $\tanh\phi = \beta$, has many nice properties. It is given by $\phi(t) = (1/c)\int_0^t a(t)\,dt$, assuming that an object starts at rest and accelerates with proper acceleration $a(t)$, where t is the object's time.

- In Section 2.7 we showed that the Lorentz transformations can be almost completely obtained by ignoring the speed-of-light postulate and using only the relativity postulate. The speed-of-light postulate serves only to specify one remaining parameter in the transformations.

2.9 Problems

Section 2.1: The Lorentz transformations

2.1. **A bunch of L.T.'s** **

Verify that the values of Δx and Δt in the table in the "Passing trains" example in Section 2.3 satisfy the Lorentz transformations between the six pairs of frames, namely AB, AC, AD, BC, BD, and CD. See Fig. 2.25.

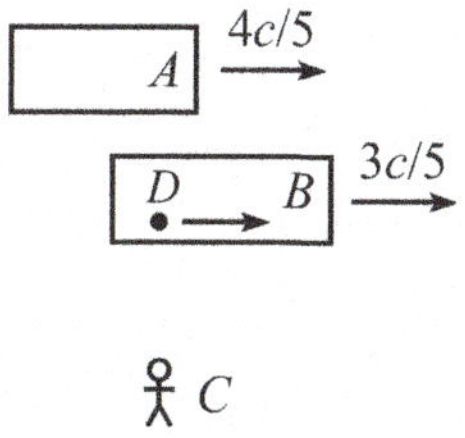

Figure 2.25

Section 2.2: Velocity addition

2.2. **Vertical to diagonal** $*$

A photon moves vertically in the ground frame. With what speed v should you travel to the left, so that the photon moves rightward and upward at a 45° angle in your frame?

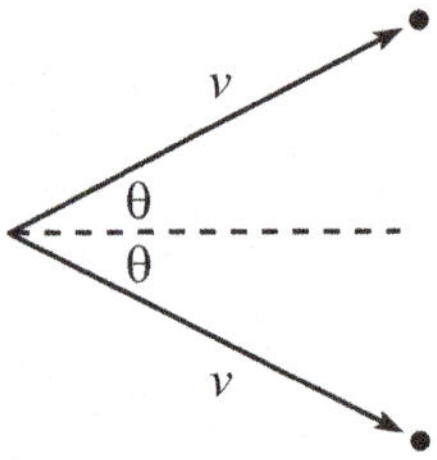

Figure 2.26

2.3. **Relative speed** $**$

In the lab frame, two particles move with speed v along the paths shown in Fig. 2.26. The angle between the trajectories is 2θ. What is the speed of one particle, as viewed by the other? (Note: This problem is posed again in Chapter 4, where it can be solved in a much simpler way, using 4-vectors.)

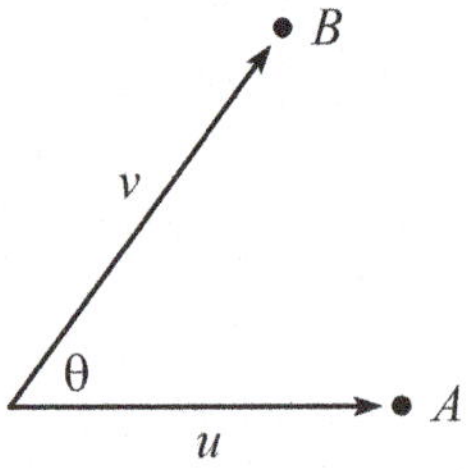

Figure 2.27

2.4. **Another relative speed** $**$

In the lab frame, particles A and B move with speeds u and v along the paths shown in Fig. 2.27. The angle between the trajectories is θ. What is the speed of one particle, as viewed by the other? (Note: This problem is posed again in Chapter 4, where it can be solved in a much simpler way, using 4-vectors.)

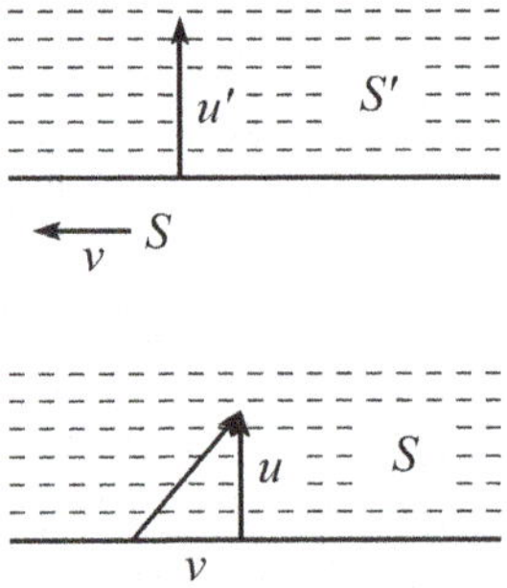

Figure 2.28

2.5. **Transverse velocity addition** $**$

In the special case where $u'_x = 0$, the transverse velocity-addition formula, Eq. (2.22), yields $u_y = u'_y/\gamma_v$. Derive this in the following way: In frame S', an object moves with velocity $(0, u')$, as shown in the first picture in Fig. 2.28. Frame S moves to the left with speed v, so the situation in S is shown in the second picture in Fig. 2.28, with the y speed now u. Consider a series of equally spaced horizontal dotted lines, as shown. The ratio of the times between successive passes of the dotted lines in frames S and S' is $T_S/T_{S'} = u'/u$, because there is no transverse length contraction. Produce another expression for this ratio by using time-dilation arguments applied to a clock on the object, and then equate the two expressions to solve for u in terms of u' and v. (This argument is more involved than the simple one we gave in Section 2.2.2, where we applied time dilation to a clock at rest in S'; see the discussion of Fig. 2.6.)

Section 2.3: The invariant interval

2.6. **Passing trains** $**$

Train A with proper length L moves eastward at speed v, while train B with proper length $2L$ moves westward also at speed v. How much time does it take for the trains to pass each other (defined as the time between the fronts coinciding and the backs coinciding):

(a) In A's frame?

(b) In B's frame?

(c) In the ground frame?

(d) Verify that the invariant interval is the same in all three frames.

2.7. **Throwing on a train** $**$

A train with proper length L moves at speed $c/2$ with respect to the ground. A ball is thrown from the back to the front, at speed $c/3$ with respect to the train. How much time does this take, and what distance does the ball cover, in:

(a) The train frame?

(b) The ground frame? Solve this by:

i. Using a velocity-addition argument.

ii. Using the Lorentz transformations to go from the train frame to the ground frame.

(c) The ball frame?

(d) Verify that the invariant interval is the same in all three frames.

(e) Show that the times in the ground frame and ball frame are related by the relevant γ factor.

(f) Likewise for the train frame and ball frame.

(g) Show that the times in the ground frame and train frame are *not* related by the relevant γ factor. Why not?

(h) Show that the time in the ground frame *is* related by the relevant γ factor to the time that elapses on *a given clock* on the train, as viewed from the ground.

Section 2.4: Minkowski diagrams

2.8. **A new frame** $*$

In a given reference frame, Event 1 happens at $x = 0$, $ct = 0$, and Event 2 happens at $x = 2$, $ct = 1$ (in units of some specified length). Find a frame in which the two events are simultaneous.

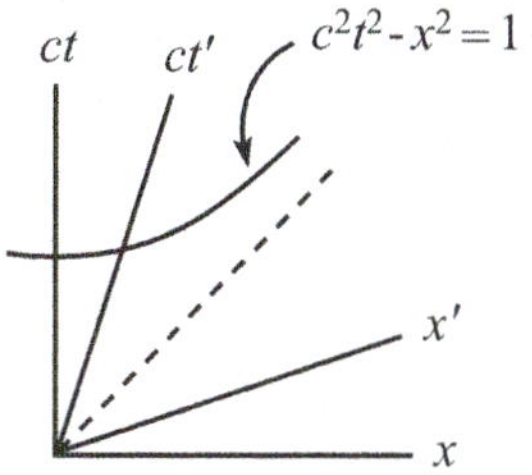

Figure 2.29

2.9. **Minkowski diagram units** $*$

Consider the Minkowski diagram in Fig. 2.29. In frame S, the hyperbola $c^2t^2 - x^2 = 1$ is drawn. Also drawn are the axes of frame S', which moves at speed v with respect to S. Use the invariance of the interval $s^2 = c^2t^2 - x^2$, along with the fact that the angle θ_1 in Fig. 2.12 satisfies $\tan\theta_1 = \beta$, to derive the ratio of the unit sizes on the ct' and ct axes, and check your result with Eq. (2.31).

2.10. **Velocity addition via Minkowski** $**$

An object moves at speed v_1 with respect to frame S'. Frame S' moves at speed v_2 with respect to frame S (in the same direction as the motion of the object). What is the speed V of the object with respect to frame S? Solve this problem (that is, derive the velocity-addition formula) by drawing a Minkowski diagram with frames S and S', drawing the worldline of the object, and doing some geometry.

2.11. **Clapping both ways** $**$

Twin A stays on the earth, and twin B flies with speed v to a distant star and back. For both of the following setups, draw a Minkowski diagram that explains what is happening.

(a) Throughout the trip, B claps in such a way that his claps occur at equal time intervals Δt in A's frame. At what time intervals do the claps occur in B's frame?

(b) Now let A clap in such a way that his claps occur at equal time intervals $\Delta t'$ in B's frame. At what time intervals do the claps occur in A's frame? (Be careful on this one. The sum of all the time intervals must equal the increase in A's age, which is greater than the increase in B's age, as we know from the usual twin paradox.)

2.12. Acceleration and redshift ***

Use a Minkowski diagram to solve the following problem: Two people stand a distance d apart. They simultaneously start accelerating in the same direction (along the line between them) with the same proper acceleration a. At the instant they start to move, how fast does each person's clock tick in the (changing) frame of the other person?

2.13. Break or not break? ***

Figure 2.30

Two spaceships float in space and are at rest relative to each other. They are connected by a string; see Fig. 2.30. The string is strong, but it cannot withstand an arbitrary amount of stretching. At a given instant, the spaceships simultaneously (with respect to their initial inertial frame) start accelerating in the same direction (along the line between them) with the same constant proper acceleration. In other words, assume they bought identical engines from the same store, and they put them on the same setting. Will the string eventually break? (This is a classic, so don't look at the solution until you've settled on a final answer!)

Section 2.6: Rapidity

2.14. Successive Lorentz transformations *

Eq. (2.45) gives the Lorentz transformation in matrix form as

$$\begin{pmatrix} x \\ ct \end{pmatrix} = \begin{pmatrix} \cosh\phi & \sinh\phi \\ \sinh\phi & \cosh\phi \end{pmatrix} \begin{pmatrix} x' \\ ct' \end{pmatrix}. \tag{2.67}$$

Show that if you apply an L.T. with $v_1 = \tanh\phi_1$, and then another one with $v_2 = \tanh\phi_2$, the result is an L.T. with $v = \tanh(\phi_1 + \phi_2)$.

2.15. Accelerator's time **

A spaceship is initially at rest in the lab frame. At a given instant, it starts to accelerate. Let this happen when the lab clock reads $t = 0$ and the spaceship clock reads $t' = 0$. The proper acceleration has the constant value a. (That is, at time $t' + dt'$, the spaceship is moving at speed $a\,dt'$ relative to the frame it was just in at time t'.) Later on, a person in the lab frame records the reading t' that he observes at a given time t. What is the relation between t and t'?

2.10 Exercises

Section 2.1: The Lorentz transformations

2.16. Still speed c *

In frame S' a photon moves at speed c (of course) in the x' direction. S' moves at speed v with respect to frame S, in the x direction. Use the Lorentz transformations to show that the speed of the photon in frame S is also c, as it must be. *Hint*: Pick two events located along the path of the photon. You know how $\Delta x'$ is related to $\Delta t'$, and you are trying to show how Δx is related to Δt.

2.17. Successive L.T.'s **

Show that the combination of an L.T. (with speed v_1) and another L.T. (with speed v_2) yields an L.T. with speed $u = (v_1 + v_2)/(1 + v_1 v_2/c^2)$.

2.18. **Loss of simultaneity** $**$

A train moves eastward at speed v with respect to the ground. Two events occur simultaneously, a distance L apart, in the train frame. What are the time and space separations in the ground frame? Solve this by:

(a) Using the Lorentz transformations.

(b) Using only the results in Section 1.3. Do this by working in the ground frame, and then again in the train frame.

Section 2.2: Velocity addition

2.19. **Slanted time dilation** $*$

A clock moves vertically with speed u in a given frame, and you run horizontally with speed v with respect to this frame. Show that you see the clock run slow by the nice simple factor $\gamma_u \gamma_v$.

2.20. **45-degree photon** $*$

With respect to a train, a photon moves leftward and upward at a 45° angle. The train moves rightward at speed $c/\sqrt{2}$ with respect to the ground. By using the velocity-addition formulas, find the x and y components of the photon's velocity in the ground frame. Then explain why your results make sense.

2.21. **Angled photon** $**$

In frame S' a photon moves at an angle θ with respect to the x' axis. S' moves at speed v with respect to frame S, along the x axis. Calculate the components of the photon's velocity in S, and verify that the speed is c.

Section 2.3: The invariant interval

2.22. **Passing a train** $**$

Person A stands on the ground, train B with proper length L moves to the right at speed $3c/5$, and person C runs to the right at speed $4c/5$. C starts behind the train and eventually passes it. Let event E_1 be "C coincides with the back of the train," and let event E_2 be "C coincides with the front of the train." Find Δt and Δx between the events E_1 and E_2 in the frames of A, B, and C, and verify that $c^2(\Delta t)^2 - (\Delta x)^2$ is the same in all three frames.

2.23. **Jousting stick and trees** $**$

A person runs to the right at speed $4c/5$ while carrying a jousting stick with proper length L. There are two trees a distance $2L$ apart on the ground. Let Event 1 be the back of the stick coinciding with the left tree, and let Event 2 be the front of the stick coinciding with the right tree.

(a) Using length contraction, find the time separation between the two events in the ground frame.

(b) Using the Lorentz transformations, find the time and space separations in the person's frame (the stick's frame).

(c) Use a length-contraction argument in the person's frame to check your answers to part (b).

(d) Check your answers again by verifying that the invariant interval between the two events is the same in both frames.

2.24. **Throwing on a train** $**$

A train with proper length L moves at speed $3c/5$ with respect to the ground. A ball is thrown from the back to the front, at speed $c/2$ with respect to the train. How much time does this take, and what distance does the ball cover, in:

(a) The train frame?

(b) The ground frame? Solve this by:

i. Using a velocity-addition argument.

ii. Using the Lorentz transformations to go from the train frame to the ground frame.

(c) The ball frame?

(d) Verify that the invariant interval is the same in all three frames.

(e) Show that the times in the ground frame and ball frame are related by the relevant γ factor.

(f) Likewise for the train frame and ball frame.

(g) Show that the times in the ground frame and train frame are *not* related by the relevant γ factor. Why not?

(h) Show that the time in the ground frame *is* related by the relevant γ factor to the time that elapses on *a given clock* on the train, as viewed from the ground.

Section 2.4: Minkowski diagrams

2.25. **Pole in barn** $*$

(a) A pole with proper length L moves rightward at speed $3c/5$ through a barn with proper length L. In the barn frame, what are the Δx and Δt separations between the "left end of pole passing left end of barn" and "right end of pole passing right end of barn" events?

(b) If you analyze the setup in the pole frame, you will quickly see that the order of the above events is reversed from the order in the barn frame. By continuity, there should therefore be a frame in which the two events happen at the same time. Use a Minkowski diagram to find this frame.

2.26. **Time dilation via Minkowski** $**$

In the spirit of the example in Section 2.4, use a Minkowski diagram to derive the time-dilation result between frames S and S' (in both directions, as in the example).

2.27. Lv/c^2 **via Minkowski** $**$

In the spirit of the example in Section 2.4, use a Minkowski diagram to derive the Lv/c^2 rear-clock-ahead result for frames S and S' (in both directions, as in the example).

2.28. **Simultaneous waves again** $**$

Solve Exercise 1.24 by using a Minkowski diagram from the point of view of an observer who sees Alice and Bob moving with equal and opposite speeds.

2.29. Short train in a tunnel again ***

Solve Exercise 1.39 by using a Minkowski diagram from the point of view of the train, and also of the tunnel.

Section 2.5: The Doppler effect

2.30. Running toward each other **

A and B move toward each other. They each have speed v with respect to the ground. A sends out flashes at frequency f (as measured by her). With what frequency does B receive the signals (as measured by him)? Solve this in two ways:

(a) Imagine a stationary intermediate person C who receives the flashes and then immediately emits flashes. So the flashes effectively pass right through her. See Fig. 2.31(a).

(b) Work in B's frame, where A approaches at a certain speed. See Fig. 2.31(b).

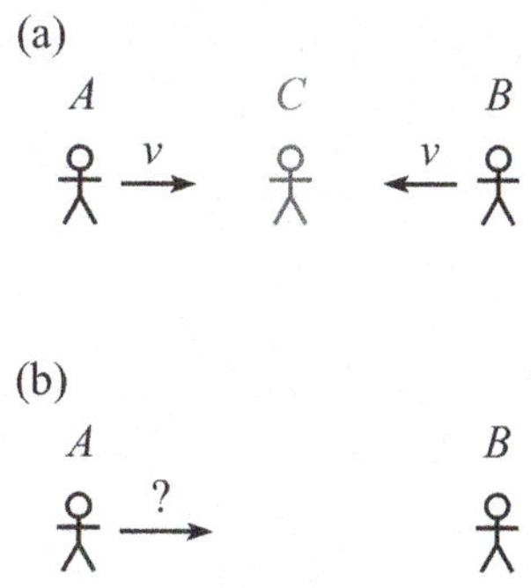

Figure 2.31

2.31. Transverse Doppler **

As mentioned in the second remark in Section 2.5.2, it is possible to derive the transverse Doppler effect for Case 1 by working in the frame of the observer, provided that you account for the longitudinal component of the source's motion. Analyze the setup in this way and reproduce Eq. (2.39).

2.32. Twin paradox via Doppler ***

Twin A stays on the earth, and twin B flies at speed v to a distant star and back. The star is a distance L from the earth in the earth-star frame. Use the longitudinal Doppler effect to show that B is younger by a factor γ when she returns. (Don't use any time-dilation or length-contraction results.) Do this in the following two ways; both are doable by working in either A's frame or B's frame(s), so take your pick.

(a) A sends out flashes at intervals of t seconds (as measured in his frame). By considering the numbers of redshifted and blueshifted flashes that B receives, show that $T_B = T_A/\gamma$.

(b) B sends out flashes at intervals of t seconds (as measured in her frame). By considering the numbers of redshifted and blueshifted flashes that A receives, show again that $T_B = T_A/\gamma$.

Section 2.6: Rapidity

2.33. Adding rapidities *

Rapidities $\phi_1 = 1/2$ and $\phi_2 = 2/3$ are added to obtain $\phi = 7/6$. Show (with a calculator) that the associated speeds are correctly related via the velocity-addition formula.

2.34. Time of travel ***

Consider the setup in Problem 2.15 (and feel free to use the results from that problem in this exercise). Let the spaceship travel to a planet a distance L from the earth.

(a) By working in the frame of the earth, find the time of the journey, as measured by the earth. Check the large and small L (compared with c^2/a) limits.

(b) By working in the (changing) frame of the spaceship, find the time of the journey, as measured by the spaceship (an implicit equation is fine). Check the small L limit. How does the time behave for large L?

2.11 Solutions

2.1. **A bunch of L.T.'s**

Using the results from the "Passing trains" example in Section 1.5, the relative speeds and the associated γ factors for the six pairs of frames are

	AB	AC	AD	BC	BD	CD
v	$5c/13$	$4c/5$	$c/5$	$3c/5$	$c/5$	$5c/7$
γ	$13/12$	$5/3$	$5/2\sqrt{6}$	$5/4$	$5/2\sqrt{6}$	$7/2\sqrt{6}$

From the example in Section 2.3, the separations between the two events in the four frames are

	A	B	C	D
Δx	$-L$	L	$5L$	0
Δt	$5L/c$	$5L/c$	$7L/c$	$2\sqrt{6}L/c$

The Lorentz transformations are

$$\begin{aligned}\Delta x &= \gamma(\Delta x' + v\Delta t'),\\ \Delta t &= \gamma(\Delta t' + v\Delta x'/c^2).\end{aligned} \tag{2.68}$$

For each of the six pairs, we'll transform from the faster frame to the slower frame. That is, when using Eq. (2.68) the faster frame will be S' and the slower frame will be S. Since S sees S' moving rightward, the sign on the righthand side of the L.T.'s will therefore always be a "+." In the AB case, for example, we'll write "Frames B and A," in that order, to signify that the B coordinates are on the lefthand side and the A coordinates are on the righthand side. We'll simply list the L.T.'s for the six cases, and you can check that they all do indeed work out.

Frames B and A:

$$\begin{aligned}L &= \frac{13}{12}\left(-L + \left(\frac{5c}{13}\right)\left(\frac{5L}{c}\right)\right)\\ \frac{5L}{c} &= \frac{13}{12}\left(\frac{5L}{c} + \frac{\frac{5c}{13}(-L)}{c^2}\right)\end{aligned}$$

Frames C and A:

$$\begin{aligned}5L &= \frac{5}{3}\left(-L + \left(\frac{4c}{5}\right)\left(\frac{5L}{c}\right)\right)\\ \frac{7L}{c} &= \frac{5}{3}\left(\frac{5L}{c} + \frac{\frac{4c}{5}(-L)}{c^2}\right)\end{aligned}$$

Frames D and A:

$$\begin{aligned}0 &= \frac{5}{2\sqrt{6}}\left(-L + \left(\frac{c}{5}\right)\left(\frac{5L}{c}\right)\right)\\ \frac{2\sqrt{6}L}{c} &= \frac{5}{2\sqrt{6}}\left(\frac{5L}{c} + \frac{\frac{c}{5}(-L)}{c^2}\right)\end{aligned}$$

Frames C and B :

$$5L = \frac{5}{4}\left(L + \left(\frac{3c}{5}\right)\left(\frac{5L}{c}\right)\right)$$
$$\frac{7L}{c} = \frac{5}{4}\left(\frac{5L}{c} + \frac{\frac{3c}{5}L}{c^2}\right)$$

Frames B and D :

$$L = \frac{5}{2\sqrt{6}}\left(0 + \left(\frac{c}{5}\right)\left(\frac{2\sqrt{6}L}{c}\right)\right)$$
$$\frac{5L}{c} = \frac{5}{2\sqrt{6}}\left(\frac{2\sqrt{6}L}{c} + \frac{\frac{c}{5}(0)}{c^2}\right)$$

Frames C and D :

$$5L = \frac{7}{2\sqrt{6}}\left(0 + \left(\frac{5c}{7}\right)\left(\frac{2\sqrt{6}L}{c}\right)\right)$$
$$\frac{7L}{c} = \frac{7}{2\sqrt{6}}\left(\frac{2\sqrt{6}L}{c} + \frac{\frac{5c}{7}(0)}{c^2}\right) \qquad (2.69)$$

2.2. **Vertical to diagonal**

The 45° angle implies that our goal is to have the u_x and u_y velocity components in your frame be equal. Since you see the ground moving to the right, the velocity-addition formulas will contain plus signs. With the ground frame as S', and with $u'_x = 0$ since the photon is moving vertically in the ground frame, Eq. (2.19) gives the horizontal velocity in your frame S as

$$u_x = \frac{u'_x + v}{1 + u'_x v/c^2} = \frac{0 + v}{1 + 0} = v, \qquad (2.70)$$

which is no surprise. With $u'_y = c$, Eq. (2.22) gives the vertical velocity in your frame as

$$u_y = \frac{u'_y}{\gamma(1 + u'_x v/c^2)} = \frac{c}{\gamma(1 + 0)} = \frac{c}{\gamma}. \qquad (2.71)$$

Setting $u_x = u_y$ yields

$$v = \frac{c}{\gamma} \implies \frac{v}{c} = \frac{1}{\gamma} \implies \frac{v^2}{c^2} = 1 - \frac{v^2}{c^2} \implies v = \frac{c}{\sqrt{2}}. \qquad (2.72)$$

Remarks:

1. In retrospect, this answer makes sense because we know that the photon must still move with speed c in your frame. So if it moves at a 45° angle, its u_x component in your frame must be $c/\sqrt{2}$. You must therefore move with speed $c/\sqrt{2}$ leftward with respect to the ground, to give the photon a speed $c/\sqrt{2}$ rightward in your frame (since u'_x is zero in the ground frame).

2. Let's check that the speed of the photon in your frame is c (as it must be), for any arbitrary value of v. Applying the Pythagorean theorem to the velocity components in Eqs. (2.70) and (2.71), we find that the square of the speed in your frame is

$$v^2 + \frac{c^2}{\gamma^2} = v^2 + c^2\left(1 - \frac{v^2}{c^2}\right) = c^2, \qquad (2.73)$$

as desired. This calculation is a special case of the more involved calculation in Exercise 2.21.

3. If this problem were instead posed as a nonrelativistic problem with a baseball moving vertically with speed V in the ground frame, then the answer would simply be that you should run leftward with speed $v = V$. This is true because the vertical component of the baseball's velocity doesn't change (nonrelativistically) when going

from one frame to another, in contrast with what happens with the transverse velocity-addition formula. Note that the baseball's resulting overall speed of $\sqrt{2}V$ in your frame is larger than the speed V in the ground frame, in contrast with the invariance of the speed of light in the original problem. ♣

2.3. **Relative speed**

Let S be the lab frame, and consider the frame S' that travels along with the point P midway between the particles. S' moves at speed $v\cos\theta$, so the γ factor relating it to the lab frame is (dropping the c's)

$$\gamma = \frac{1}{\sqrt{1-v^2\cos^2\theta}}. \qquad (2.74)$$

In S' the particles move vertically. Let's find these vertical speeds. Since the particles have $u'_x = 0$, the transverse velocity-addition formula, Eq. (2.22), gives the vertical speed in S (which we know is $v\sin\theta$) as $v\sin\theta = u'_y/\gamma$. Therefore, in S' each particle moves along the vertical axis away from the point P with speed

$$u'_y = \gamma v\sin\theta = \frac{v\sin\theta}{\sqrt{1-v^2\cos^2\theta}}. \qquad (2.75)$$

The speed of one particle as viewed by the other can now be found via the longitudinal (along the y' direction) velocity-addition formula,

$$V = \frac{2u'_y}{1+u'^2_y} = \frac{\dfrac{2v\sin\theta}{\sqrt{1-v^2\cos^2\theta}}}{1+\dfrac{v^2\sin^2\theta}{1-v^2\cos^2\theta}} = \frac{2v\sin\theta\sqrt{1-v^2\cos^2\theta}}{1-v^2\cos 2\theta}, \qquad (2.76)$$

where we have used the double-angle formula $\cos^2\theta - \sin^2\theta = \cos 2\theta$. You can verify that V can be written as (for future reference in Chapter 4)

$$V = \sqrt{1-\frac{(1-v^2)^2}{(1-v^2\cos 2\theta)^2}}. \qquad (2.77)$$

Remark: If $2\theta = 180°$ (which means that the particles are moving in opposite directions in 1-D), then Eq. (2.76) gives $V = 2v/(1+v^2)$; this is correctly the velocity addition of v with itself. If $\theta = 0$, then $V = 0$, which is correct. If θ is very small but nonzero, then Eq. (2.76) reduces to $V \approx 2v\sin\theta/\sqrt{1-v^2}$, because the cosine terms can be replaced with 1. This result is simply the nonrelativistic addition of the speed in Eq. (2.75) with itself (with $\cos\theta \approx 1$), as it should be. ♣

2.4. **Another relative speed**

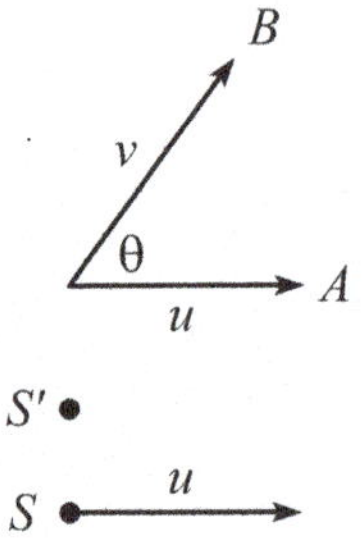

Figure 2.32

In Fig. 2.32 the velocity of A points in the x direction. Let S' be the lab frame, and let S be A's frame. Then S' moves at velocity $-u$ with respect to S. The x and y speeds of B in S' are $v\cos\theta$ and $v\sin\theta$. Therefore, the longitudinal and transverse velocity-addition formulas, Eqs. (2.19) and (2.22), give the components of B's velocity in S as

$$V_x = \frac{v\cos\theta - u}{1-uv\cos\theta},$$
$$V_y = \frac{v\sin\theta}{\gamma_u(1-uv\cos\theta)} = \frac{\sqrt{1-u^2}\,v\sin\theta}{1-uv\cos\theta}. \qquad (2.78)$$

The total speed of B in frame S (that is, from A's point of view) is therefore

$$V = \sqrt{V_x^2+V_y^2} = \sqrt{\left(\frac{v\cos\theta-u}{1-uv\cos\theta}\right)^2 + \left(\frac{\sqrt{1-u^2}\,v\sin\theta}{1-uv\cos\theta}\right)^2}$$
$$= \frac{\sqrt{u^2+v^2-2uv\cos\theta-u^2v^2\sin^2\theta}}{1-uv\cos\theta}, \qquad (2.79)$$

where we have used $\cos^2\theta + \sin^2\theta = 1$. You can verify that V can be written as

$$V = \sqrt{1 - \frac{(1-u^2)(1-v^2)}{(1-uv\cos\theta)^2}}. \qquad (2.80)$$

The reason why this takes such an organized form will become clear in Chapter 4.

REMARK: If $u = v$, then Eq. (2.80) reduces to Eq. (2.77) in the preceding problem (if we replace θ by 2θ). If $\theta = 180°$, then you can show that $V = (u+v)/(1+uv)$, as it should; we have 1-D motion with the particles moving in opposite directions, so the longitudinal velocity-addition formula applies. And if $\theta = 0$, then $V = |v-u|/(1-uv)$, as it should. we have 1-D motion with the particles moving in the same direction. ♣

2.5. **Transverse velocity addition**

Assume that a clock on the object ticks off a time T between successive passes of the dotted lines. In frame S' the speed of the object is u', so the time-dilation factor is $\gamma' = 1/\sqrt{1-u'^2}$. The time in S' between successive passes of the dotted lines is therefore $T_{S'} = \gamma' T$.

In frame S, the speed of the object is $\sqrt{v^2+u^2}$. (Yes, the Pythagorean theorem holds for these velocity components, because both components are measured with respect to the same frame.) Hence, the time-dilation factor is $\gamma = 1/\sqrt{1-v^2-u^2}$. The time in S between successive passes of the dotted lines is therefore $T_S = \gamma T$.

The ratio $T_S/T_{S'}$ therefore equals $\gamma T/\gamma' T = \gamma/\gamma'$. Equating this with the $T_S/T_{S'} = u'/u$ result mentioned in the statement of the problem yields

$$\frac{u'}{u} = \frac{\gamma}{\gamma'} \implies \frac{u'}{u} = \frac{\sqrt{1-u'^2}}{\sqrt{1-v^2-u^2}}. \qquad (2.81)$$

Squaring and cross multiplying gives

$$u'^2(1-v^2-u^2) = u^2(1-u'^2) \implies u = u'\sqrt{1-v^2} \equiv \frac{u'}{\gamma_v}, \qquad (2.82)$$

as desired.

2.6. **Passing trains**

(a) We'll drop the c's in this solution, for ease of computation. From the velocity-addition formula, A sees B moving at speed $2v/(1+v^2)$, which has an associated γ factor of $(1+v^2)/(1-v^2)$, as you can verify. In A's frame, B must travel the sum of the lengths of the two trains at speed $2v/(1+v^2)$. Since B is length contracted, the time is therefore

$$\Delta t_A = \frac{L + \frac{1-v^2}{1+v^2}\cdot 2L}{\frac{2v}{1+v^2}} = \frac{L(3-v^2)}{2v}. \qquad (2.83)$$

(b) Similar reasoning holds in B's frame, so the time is

$$\Delta t_B = \frac{\frac{1-v^2}{1+v^2}\cdot L + 2L}{\frac{2v}{1+v^2}} = \frac{L(3+v^2)}{2v}. \qquad (2.84)$$

(c) In the ground frame, the back ends are initially $3L\sqrt{1-v^2}$ apart due to length contraction. They move toward each other at a relative speed of $2v$ (yes, simple addition works fine here), so the time is

$$\Delta t_g = \frac{3L\sqrt{1-v^2}}{2v}. \qquad (2.85)$$

(d) For A, we have $\Delta x_A = L$, because the two events are located at the ends of the train. (Remember, it is the distance between the events that we are concerned with, not the distance that the other train travels.) Therefore (ignoring the c's),

$$\Delta t_A^2 - \Delta x_A^2 = \left(\frac{L(3-v^2)}{2v}\right)^2 - L^2 = \frac{L^2}{4v^2}\left(9 - 10v^2 + v^4\right). \tag{2.86}$$

For B, we have $\Delta x_B = 2L$, so

$$\Delta t_B^2 - \Delta x_B^2 = \left(\frac{L(3+v^2)}{2v}\right)^2 - (2L)^2 = \frac{L^2}{4v^2}\left(9 - 10v^2 + v^4\right). \tag{2.87}$$

For the ground, we have $\Delta x_g = (L/2)\sqrt{1-v^2}$, because the meeting of the back ends is midway between the initial positions of the backs (because the trains have the same speed), which means that it is $(3L/2)\sqrt{1-v^2}$ from each, or $(L/2)\sqrt{1-v^2}$ from the initial position of the fronts; see Fig. 2.33. Therefore,

$$\Delta t_g^2 - \Delta x_g^2 = \left(\frac{3L\sqrt{1-v^2}}{2v}\right)^2 - \left(\frac{L\sqrt{1-v^2}}{2}\right)^2 = \frac{L^2}{4v^2}\left(9 - 10v^2 + v^4\right). \tag{2.88}$$

The above three results are the same, as desired.

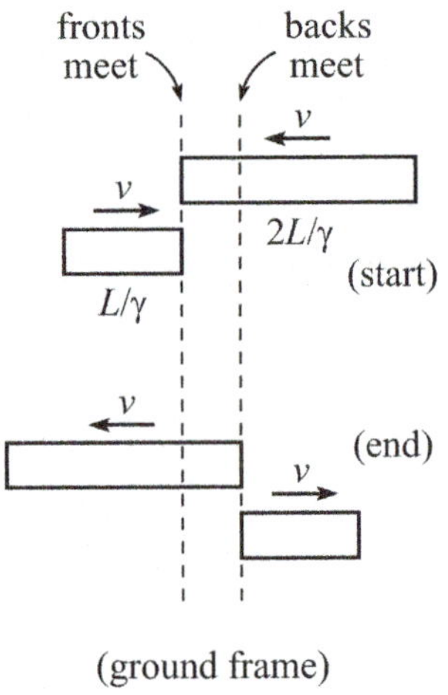

Figure 2.33

2.7. **Throwing on a train**

(a) In the train frame, the distance is simply $\Delta x_t = L$. And the time is $\Delta t_t = L/(c/3) = 3L/c$.

(b) i. The velocity-addition formula gives the velocity of the ball with respect to the ground as (with $u = c/3$ and $v = c/2$)

$$V_g = \frac{u+v}{1+\frac{uv}{c^2}} = \frac{\frac{c}{3}+\frac{c}{2}}{1+\frac{1}{3}\cdot\frac{1}{2}} = \frac{5c}{7}. \tag{2.89}$$

The length of the train in the ground frame is $L/\gamma_{1/2} = \sqrt{3}L/2$. At time t in the ground frame, the position of the ball is $V_g t$ (relative to the initial location of the back of the train). And the position of the front of the train is $\sqrt{3}L/2 + vt$. These two positions are equal when

$$V_g t = \frac{\sqrt{3}L}{2} + vt \implies \Delta t_g = \frac{\frac{\sqrt{3}L}{2}}{V_g - v} = \frac{\frac{\sqrt{3}L}{2}}{\frac{5c}{7}-\frac{c}{2}} = \frac{7L}{\sqrt{3}c}. \tag{2.90}$$

Equivalently, this time is obtained by noting that the ball closes the initial head start of $\sqrt{3}L/2$ that the front of the train had, at a relative speed of $V_g - v$, as viewed from the ground.

The distance the ball travels is $\Delta x_g = V_g t = (5c/7)(7L/\sqrt{3}c) = 5L/\sqrt{3}$.

ii. From part (a), the space and time intervals in the train frame are $\Delta x_t = L$ and $\Delta t_t = 3L/c$. The γ factor between the frames is $\gamma_{1/2} = 2/\sqrt{3}$, so the Lorentz transformations give the coordinates in the ground frame as (the sign here is a "+" because the ground sees the train moving to the right)

$$\begin{aligned}\Delta x_g &= \gamma(\Delta x_t + v\Delta t_t) = \frac{2}{\sqrt{3}}\left(L + \frac{c}{2}\left(\frac{3L}{c}\right)\right) = \frac{5L}{\sqrt{3}},\\ \Delta t_g &= \gamma(\Delta t_t + v\Delta x_t/c^2) = \frac{2}{\sqrt{3}}\left(\frac{3L}{c} + \frac{(c/2)L}{c^2}\right) = \frac{7L}{\sqrt{3}c},\end{aligned} \tag{2.91}$$

in agreement with the above results.

(c) In the ball frame, the train has length $L/\gamma_{1/3} = 2\sqrt{2}L/3$. So the time it takes the train to fly past the ball at speed $c/3$ is $\Delta t_b = (2\sqrt{2}L/3)/(c/3) = 2\sqrt{2}L/c$. And the distance is $\Delta x_b = 0$, of course, because the ball doesn't move in the ball frame.

(d) The values of $c^2(\Delta t)^2 - (\Delta x)^2$ in the three frames are:

Train frame: $c^2(\Delta t_t)^2 - (\Delta x_t)^2 = c^2(3L/c)^2 - L^2 = 8L^2.$

Ground frame: $c^2(\Delta t_g)^2 - (\Delta x_g)^2 = c^2(7L/\sqrt{3}c)^2 - (5L/\sqrt{3})^2 = 8L^2.$

Ball frame: $c^2(\Delta t_b)^2 - (\Delta x_b)^2 = c^2(2\sqrt{2}L/c)^2 - (0)^2 = 8L^2.$

These are all equal, as they should be. Note that because of the "0" value for Δx_b in the ball frame, the invariance of $c^2(\Delta t)^2 - (\Delta x)^2$ implies that Δt takes on the smallest possible value in the ball frame; the Δt is larger in any other frame. This is consistent with time dilation, because (as we'll see below) the time in any other frame is obtained by multiplying the time in the ball frame by a γ factor, which is larger than 1.

(e) The relative speed of the ball frame and the ground frame is $5c/7$. Therefore, since $\gamma_{5/7} = 7/2\sqrt{6}$, the times are indeed related by

$$\Delta t_g = \gamma \Delta t_b \quad \Longleftrightarrow \quad \frac{7L}{\sqrt{3}c} = \frac{7}{2\sqrt{6}}\left(\frac{2\sqrt{2}L}{c}\right) \quad \text{(true)} \tag{2.92}$$

(f) The relative speed of the ball frame and the train frame is $c/3$. Therefore, since $\gamma_{1/3} = 3/2\sqrt{2}$, the times are indeed related by

$$\Delta t_t = \gamma \Delta t_b \quad \Longleftrightarrow \quad \frac{3L}{c} = \frac{3}{2\sqrt{2}}\left(\frac{2\sqrt{2}L}{c}\right) \quad \text{(true)} \tag{2.93}$$

(g) The relative speed of the train frame and the ground frame is $c/2$. Therefore, since $\gamma_{1/2} = 2/\sqrt{3}$, we see that the times are *not* related by a simple time-dilation factor, because

$$\Delta t_g \neq \gamma \Delta t_t \quad \Longleftrightarrow \quad \frac{7L}{\sqrt{3}c} \neq \frac{2}{\sqrt{3}}\left(\frac{3L}{c}\right). \tag{2.94}$$

We don't obtain an equality here, because time dilation is legal to use only if the two events happen at the *same place* in one of the frames. Mathematically, the Lorentz transformation $\Delta t = \gamma(\Delta t' + (v/c^2)\Delta x')$ leads to $\Delta t = \gamma \Delta t'$ only if $\Delta x' = 0$. In this problem, the "ball leaving back" and "ball hitting front" events happen at the same place in the ball frame, but not at the same place in the train frame or the ground frame. Equivalently, neither the train frame nor the ground frame is any more special than the other, as far as these two events are concerned. So if someone insisted on trying to use time dilation, he would have a hard time deciding which side of the equation the γ should go on. When used properly, the γ goes on the side of the equation associated with the frame in which the two events happen at the same place. This frame is "special" (at least with regard to the two events).

(h) Assume that the ball leaves the back of the train when a clock there says zero. Then we know from part (a) that the ball hits the front when a clock there says $3L/c$. These are frame-independent statements; everyone agrees on these readings when the ball leaves the back and when it hits the front. When viewed from the ground, we must incorporate the rear-clock-ahead effect, which tells us that the front clock starts the process not at zero, but at $-Lv/c^2 = -L/2c$. See Fig. 2.34.

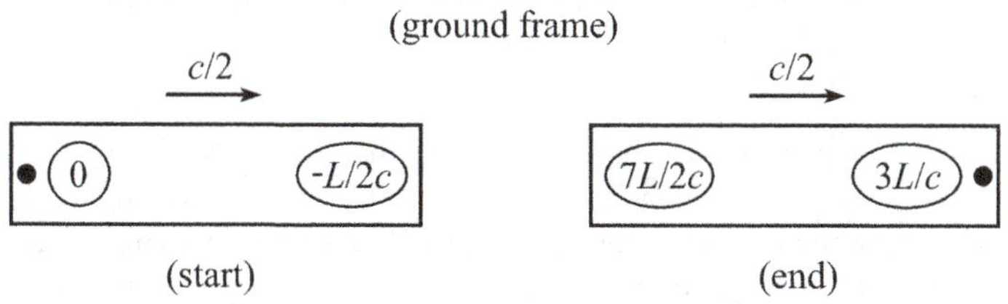

Figure 2.34

If we look at this *one clock*, we see that it advances by $3L/c - (-L/2c) = 7L/2c$ during the process. (We could just as well look at the rear clock; it reads zero at the start and $3L/c + Lv/c^2 = 7L/2c$ at the end.) Applying time dilation to this *one clock* yields the proper correction to Eq. (2.94):

$$\Delta t_g = \gamma \Delta t_{\text{one clock on train}} \quad \Longleftrightarrow \quad \frac{7L}{\sqrt{3}c} = \frac{2}{\sqrt{3}}\left(\frac{7L}{2c}\right) \qquad \text{(true)} \qquad (2.95)$$

The error in the application of time dilation in part (g) is that we tried to apply it by taking the difference in readings of *two different clocks* at the front and back of the train. But all that time dilation says is, "If you look at a single clock flying by, you see it run slow."

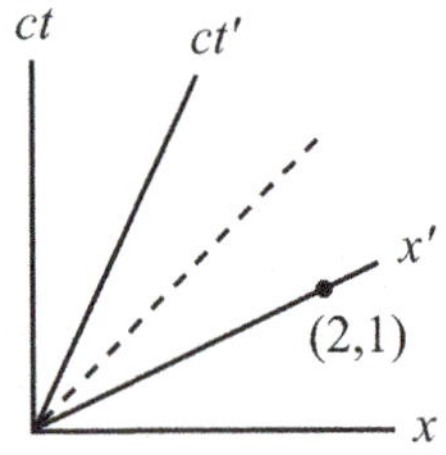

Figure 2.35

2.8. **A new frame**

First solution: Consider the Minkowski diagram shown in Fig. 2.35. In frame S, Event 1 is at the origin, and Event 2 is at the point $(2, 1)$. Consider the frame S' whose x' axis passes through the point $(2, 1)$. Since all points on the x' axis are simultaneous in the frame S' (they all have $t' = 0$), we see that S' is the desired frame. From Eq. (2.32), the slope of the x' axis equals $\beta \equiv v/c$. Since the slope is $1/2$, the desired value of v is $v = c/2$. Note that by looking at the Minkowski diagram, it is clear that if the relative speed of S and S' is less than $c/2$, then Event 2 occurs after Event 1 in S', because the x' axis lies below the point $(2, 1)$. Conversely, if the relative speed is greater than $c/2$, then Event 2 occurs before Event 1 in S'.

Second solution: Let the original frame be S, and let the desired frame be S'. Let S' move at speed v with respect to S, in the positive x direction. Our goal is to find the v that makes $\Delta t' = 0$. The $\Delta t'$ Lorentz transformation from S to S' is

$$\Delta t' = \gamma(\Delta t - v\Delta x/c^2). \qquad (2.96)$$

The sign here is "–" because S' sees S moving to the left. With $\Delta t'$ equal to zero, we obtain $\Delta t - v\Delta x/c^2 = 0 \implies v = c^2\Delta t/\Delta x$. We are given $\Delta x = 2$ and $\Delta t = 1/c$, so the desired value of v is $c/2$.

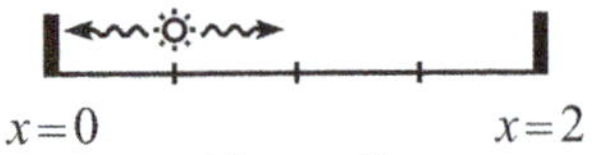

(frame S)

Figure 2.36

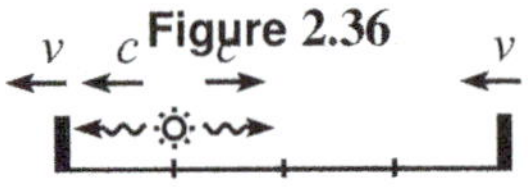

(frame S')

Figure 2.37

Third solution: Consider the setup in Fig. 2.36, which explicitly constructs the two given events, as observed in frame S. Receivers are located at $x = 0$ and $x = 2$, and a light source is located at $x = 1/2$. The source emits a flash, and when the light hits a receiver, we will say that an event has occurred. So the left event happens at $x = 0$, $ct = 1/2$. And the right event happens at $x = 2$, $ct = 3/2$. If we want, we can shift our clocks by $-1/(2c)$ in order to make the events happen at $ct = 0$ and $ct = 1$, but this shift is irrelevant because all we are concerned with are differences in time.

Now consider an observer (frame S') moving to the right at speed v. S' sees the apparatus moving to the left at speed v; see Fig. 2.37. Our goal is to find the v for which the photons hit the receivers at the same time in S'. Consider the photon moving to the left. S' sees it moving at speed c (as always), but the left receiver is receding at speed v. So the relative speed (as measured in S') of the photon and the left receiver is $c - v$. By similar reasoning, the relative speed of the photon and the right receiver is $c + v$. The light source is three times as far from the right receiver as it is from the left receiver. (This is the same ratio as in S, because the length-contraction effect is uniform in space.) If the light is to reach the two receivers at the same time in S', the relative speeds must be in the same ratio as the distances. That is, $(c + v)/(c - v) = 3/1 \implies v = c/2$.

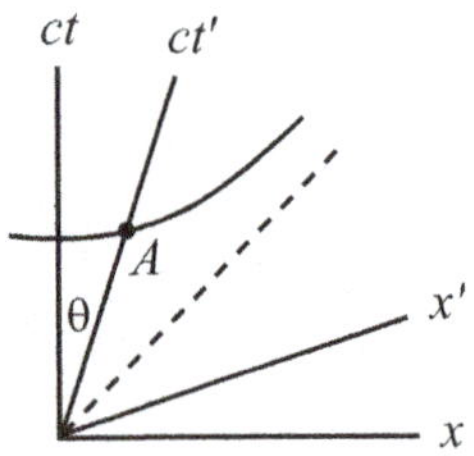

Figure 2.38

2.9. **Minkowski diagram units**

All points on the ct' axis have the property that $x' = 0$. All points on the given hyperbola have the property that $c^2t'^2 - x'^2 = 1$, due to the invariance of s^2. So the ct' value at the intersection point A in Fig. 2.38 equals 1. Therefore, we simply have to determine the distance on the paper from A to the origin. We'll do this by finding the (x, ct) coordinates of A. We know that $\tan\theta = \beta$. But $\tan\theta = x/ct$. Therefore, $x = \beta(ct)$. (This is correctly just the statement that $x = vt$. Point A lies on the worldline of the $x' = 0$

point in S', after all.) Plugging $x = \beta(ct)$ into the given equation for the hyperbola, $c^2t^2 - x^2 = 1$, we find $ct = 1/\sqrt{1-\beta^2}$. The distance from A to the origin, which equals $\sqrt{c^2t^2 + x^2} = \sqrt{c^2t^2 + (\beta ct)^2}$, is then

$$ct\sqrt{1+\beta^2} = \sqrt{\frac{1+\beta^2}{1-\beta^2}}. \tag{2.97}$$

This quantity is the desired ratio of the unit sizes on the ct' and ct axes, because we are assuming that the unit size on the ct axis has length 1 on the paper. This result agrees with Eq. (2.31). The same analysis holds for the x-axis unit size ratio, using the hyperbola $x^2 - c^2t^2 = 1$.

2.10. **Velocity Addition via Minkowski**

Pick a point P on the object's worldline, as shown in Fig. 2.39. (Assume that the object and the origins of S and S' all coincide at a given earlier moment.) Let the coordinates of P in frame S be (x, ct). Our goal is to find the speed $V = x/t$. Throughout this problem, it will be easier to work with the quantities $\beta \equiv v/c$. So our goal is to find $\beta_V \equiv x/(ct)$.

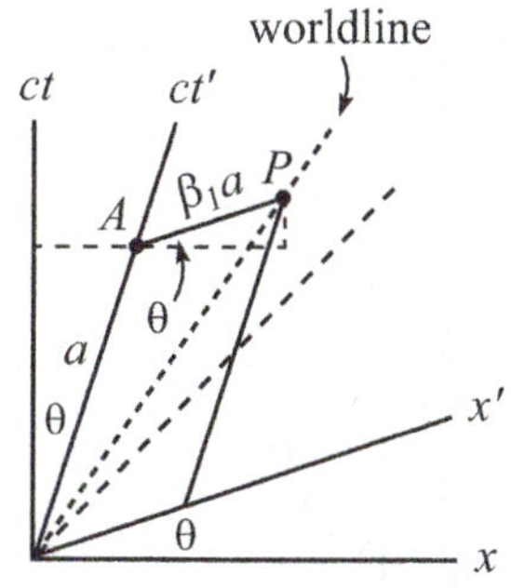

Figure 2.39

The coordinates of P in S', namely (x', ct'), are indicated by the parallelogram in the figure. For convenience, let ct' have length a on the paper. Then from the given information, we have $x' = v_1t' \equiv \beta_1(ct')$. Therefore, since x' corresponds to the segment AP, this segment has length $\beta_1 a$ on the paper. In terms of a, we can now determine the coordinates (x, ct) of P. Assuming that the unit sizes of x and ct have length 1 on the paper, the coordinates of point A are

$$(x, ct)_A = (a\sin\theta, a\cos\theta). \tag{2.98}$$

The coordinates of P, relative to A, are

$$(x, ct)_{P-A} = (\beta_1 a\cos\theta, \beta_1 a\sin\theta). \tag{2.99}$$

Adding these two sets of coordinates gives the complete coordinates of point P as

$$(x, ct)_P = (a\sin\theta + \beta_1 a\cos\theta, a\cos\theta + \beta_1 a\sin\theta). \tag{2.100}$$

The ratio of x to ct at point P is therefore

$$\beta_V \equiv \frac{x}{ct} = \frac{\sin\theta + \beta_1\cos\theta}{\cos\theta + \beta_1\sin\theta} = \frac{\tan\theta + \beta_1}{1 + \beta_1\tan\theta} = \frac{\beta_2 + \beta_1}{1 + \beta_1\beta_2}, \tag{2.101}$$

where we have used $\tan\theta = v_2/c \equiv \beta_2$, because S' moves at speed v_2 with respect to S. If we change from the β's back to the v's, the result is $V = (v_2 + v_1)/(1 + v_1v_2/c^2)$, as desired.

2.11. **Clapping both ways**

(a) From the usual time-dilation result, A sees B's clock run slow, so B must clap his hands at time intervals of $\Delta t/\gamma$ (according to him) in order for the intervals to be Δt in A's frame. Since B's $\Delta t/\gamma$ intervals are shorter than A's Δt intervals, and since A and B must agree on the total number of claps, B therefore ages less than A (as we well know). That's the quick explanation. Let's now rederive this by using a Minkowski diagram.

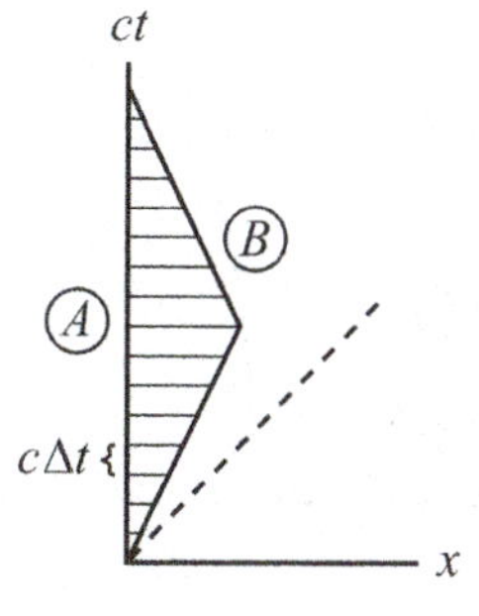

Figure 2.40

The relevant Minkowski diagram is shown in Fig. 2.40. Let B clap at the spacetime locations where the horizontal lines (lines of simultaneity in A's frame) intersect B's worldline. From the given information, the vertical spacing between the lines is $c\Delta t$. Since the slope of B's worldline is $1/\beta$, the horizontal spacing d between adjacent claps is given by $c\Delta t/d = 1/\beta \Longrightarrow \beta(c\Delta t)$, which is just $v\Delta t$, of course. The Pythagorean theorem then gives the spacing between adjacent claps (distance on the paper) along B's tilted worldline as $\sqrt{(c\Delta t)^2 + (\beta c\Delta t)^2} = \sqrt{1+\beta^2}\,c\Delta t$. But we know from Eq. (2.31) that the unit size of B's ct' axis on the paper is

$\sqrt{(1+\beta^2)/(1-\beta^2)}$ times the unit size of A's ct axis. Therefore, the time interval between claps in B's frame is

$$\frac{\sqrt{1+\beta^2}\,\Delta t}{\sqrt{(1+\beta^2)/(1-\beta^2)}} = \sqrt{1-\beta^2}\,\Delta t \equiv \frac{\Delta t}{\gamma}, \tag{2.102}$$

as we found above. We've basically just rederived time dilation.

(b) From the usual time-dilation result, B sees A's clock run slow, so A must clap his hands at time intervals of $\Delta t'/\gamma$ (according to him) in order for the intervals to be $\Delta t'$ in B's frame. Since A's time intervals are shorter than B's, we seem to be heading toward the (incorrect) conclusion that A ages less than B. But as we have seen on a few occasions, the resolution to the paradox is that B can invoke the above time-dilation reasoning only during the parts of the trip where he is in an inertial frame. That is, he cannot invoke it during the turnaround period. So we've given only a partial solution here. Let's use a Minkowski diagram to give a full solution.

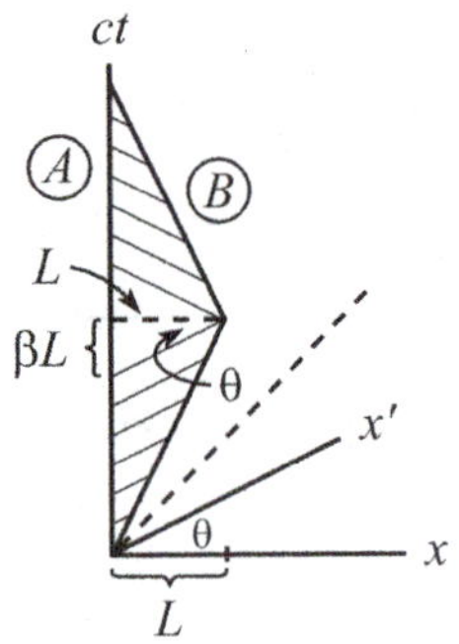

Figure 2.41

The relevant Minkowski diagram is shown in Fig. 2.41. Let A clap at the spacetime locations where the tilted lines (lines of simultaneity in B's frame) intersect A's worldline. From the given information, the tilted spacing between the lines along B's worldline is $c\,\Delta t'$ time units, which has a length $\sqrt{(1+\beta^2)/(1-\beta^2)}\,c\,\Delta t'$ on the paper, from Eq. (2.31). But then by exactly the same type of geometry reasoning that led to Eq. (2.34), you can verify that the vertical spacing between the lines along A's worldline is $\sqrt{1-\beta^2}\,c\,\Delta t'$. This means that the time interval in A's frame is $\Delta t'/\gamma$, as we found above.

But the critical point here is that the slope of the tilted lines changes abruptly when B turns around, that is, when B changes inertial frames. There is therefore an interval of time in the middle of A's worldline where the tilted lines don't hit it. The result is that A claps frequently for a while, then doesn't clap at all for a while, then claps frequently again. The overall result, as we will now show, is that more time elapses in A's frame (it will turn out to be $2L/v$, of course, where L is the distance to the star in A's frame) than in B's frame ($2L/\gamma v$, from the usual time-dilation result).

Since the time between claps is shorter in A's frame than in B's frame (except in the middle region), the time elapsed on A's clock while he is clapping is $1/\gamma$ times the total time elapsed on B's clock, which gives $(2L/\gamma v)/\gamma = 2L/\gamma^2 v$. But as shown in Fig. 2.41, the length of the region on A's ct axis where there is no clapping is $2\beta L$ (because the slope of the x' axis is β, from Eq. (2.32)). This corresponds to a time of $ct = 2\beta L \Longrightarrow t = 2vL/c^2$. The total time elapsed on A's clock during the entire process is therefore

$$\frac{2L}{\gamma^2 v} + \frac{2Lv}{c^2} = \frac{2L}{v}\left(\frac{1}{\gamma^2} + \frac{v^2}{c^2}\right) = \frac{2L}{v}\left(\left(1 - \frac{v^2}{c^2}\right) + \frac{v^2}{c^2}\right) = \frac{2L}{v}, \tag{2.103}$$

as expected. The $2Lv/c^2$ no-clapping interval in A's frame here is equivalent to the $2Lv/c^2$ whipping-ahead effect (which is a consequence of the rear-clock-ahead effect) in Exercise 1.30.

2.12. **Acceleration and redshift**

There are various ways to solve this problem, for example, by sending photons between the people, or by invoking the Equivalence Principle in general relativity. As suggested, we'll solve it here by using a Minkowski diagram. This strategy will demonstrate that the "redshift" result can be derived perfectly well by using only basic special relativity.

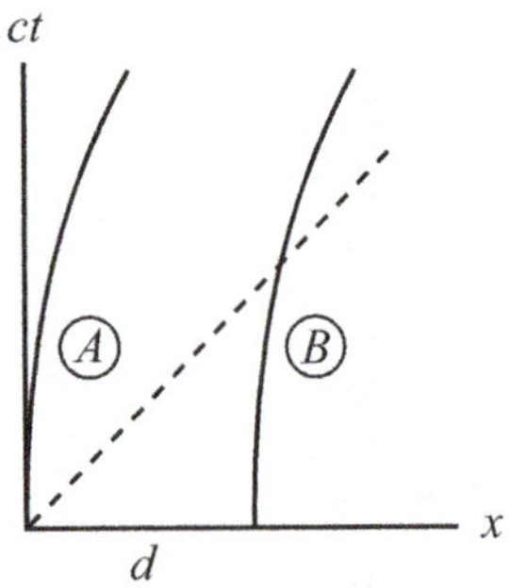

Figure 2.42

Draw the worldlines of the two people, A and B, as seen by an observer, C, in the frame where they are initially at rest. The diagram is shown in Fig. 2.42. Consider an infinitesimal time Δt, as measured by C. At this time (in C's frame), A and B are both moving at (essentially) speed $a\Delta t$. The axes of A's frame are shown in Fig. 2.43. Both A and B have moved a distance $a(\Delta t)^2/2$, which can be neglected because Δt is small. (It will turn out that the leading-order terms in the result are of order Δt, so any $(\Delta t)^2$ terms can be ignored.) Also, the special-relativistic time-dilation factor between any of the A, B, and C frames can be neglected, because the relative speeds are at most $v = a\Delta t$, which means that the time-dilation factors differ from 1 by order $(\Delta t)^2$, since γ involves the square of v. If we let A make a little explosion (call this event E_1) at time Δt in C's frame, then Δt is also the time of the explosion as measured by A, up to an error of order $(\Delta t)^2$.

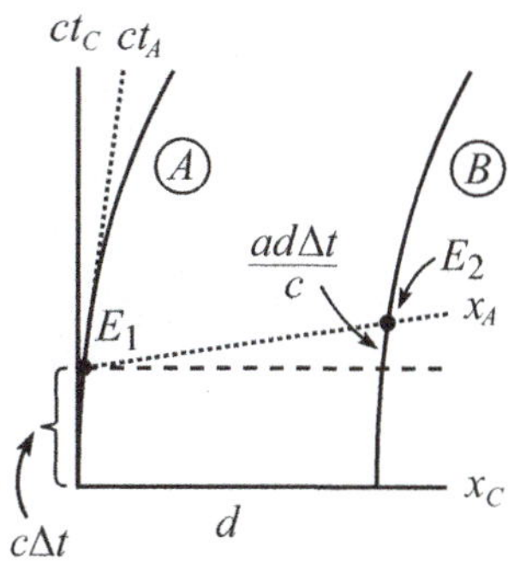

Figure 2.43

Let's figure out where A's x axis (that is, the "now" axis in A's frame) meets B's worldline. From Eq. (2.32) the slope of A's x axis in Fig. 2.43 is $\beta \equiv v/c = a\Delta t/c$. So the axis starts at a height $c\Delta t$ and then climbs up by the amount $\beta d = ad\Delta t/c$, over the distance d. (The distance is indeed d, up to corrections of order $(\Delta t)^2$.) Therefore, the axis meets B's worldline at a height $c\Delta t + ad\Delta t/c$ as viewed by C, that is, at a time $\Delta t + ad\Delta t/c^2$ as viewed by C. But C's time is the same as B's time (up to order $(\Delta t)^2$), so B's clock reads $(1 + ad/c^2)\Delta t$. Let's say that B makes a little explosion (event E_2) at this time.

Events E_1 and E_2 both occur at the same time in A's frame, because they both lie along a line of constant time in A's frame (the x axis). This means that in A's frame, B's clock reads $(1 + ad/c^2)\Delta t$ when A's clock reads Δt. Since this holds for all small values of Δt, we see that in A's (changing) frame, B's clock is sped up by the factor,

$$\frac{\Delta t_B}{\Delta t_A} = 1 + \frac{ad}{c^2}. \tag{2.104}$$

We can perform the same procedure to see how A's clock behaves in B's frame. If we draw B's x axis at time Δt, the picture is basically the same. But the relevant fact now is that B's x axis *decreases* in height when heading *leftward* toward A's worldline. So the only change is a minus sign in front of the ad/c^2 term. Therefore, in B's (changing) frame, A's clock is slowed down by the factor,

$$\frac{\Delta t_A}{\Delta t_B} = 1 - \frac{ad}{c^2}. \tag{2.105}$$

We'll discuss this type of setup in more detail in Chapter 5, including the case where there is a nonzero relative speed between A and B; see Exercise 5.7.

REMARKS:

1. In the usual special-relativity situation where two observers fly past each other with relative speed v, they *both* see the other person's clock slowed down by the same factor. This had better be the case, because the situation is symmetric between the observers. But in the present problem, A sees B's clock sped up, whereas B sees A's clock slowed down. This difference is possible because the situation is *not* symmetric between A and B. The acceleration vector determines a direction in space, and one person (namely B) is farther along this direction than the other person.

2. Another derivation of the above ad/c^2 result is the following. Consider the setup a short time after the start. An outside observer C sees A's and B's clocks showing the same time (as measured by C), because both A and B have exactly the same motion in C's frame. Therefore, by the usual vd/c^2 rear-clock-ahead result in special relativity, B's clock (which is the front clock) must read vd/c^2 more than A's in the A/B frame moving with speed v (because the front-clock-behind effect then lowers it so that C sees equal clock readings). The increase in B's clock reading per unit time, as viewed by A, is therefore $(vd/c^2)/t = (v/t)d/c^2 = ad/c^2$, in agreement with Eq. (2.104). Note that any special-relativistic time-dilation or length-contraction effects are of second order in v/c, and hence negligible because v is small here. At any later time, we can repeat (roughly) this derivation in the instantaneous rest frame of A. There are, however, a few subtleties to consider; we'll discuss these in Chapter 5, in particular in Section 5.3.

3. We showed in this problem that if an accelerating observer looks at a clock, then the acceleration a matters; it appears in Eq. (2.104). However, as mentioned in the fifth remark on page 21, if an inertial observer looks at an accelerating clock, then the acceleration *doesn't* matter. The standard special-relativity time-dilation result is the only thing we need. The clock simply runs slow by a factor of γ; all that matters is the instantaneous value of v.

   To see why this is true, imagine an inertial clock C whose velocity (and position) coincides with that of an accelerating clock B at a given instant. After a very short time interval t as measured by C, the relative speed of B and C is at, which is essentially zero. So C sees B run at essentially the same rate. An inertial observer A, who sees B and C move with speed v (or slightly more, for B), therefore sees the clocks run at essentially the same rate, which means that they both run slow by a factor γ, because C's does so. We have used the fact that the distance $at^2/2$ between B and C is negligible. This implies that we don't have to worry about any loss of simultaneity from A's point of view.

   To be rigorous, we must justify why C sees B run at the same rate if B is accelerating. Consider an (inherently nonrelativistic) setup where C sits at rest, and where B starts at rest next to C and accelerates with a large acceleration (say, $100g$). For all we know, right at the start when B's speed is still small, C might see B's clock run at a different rate, due to some new acceleration effect outside the scope of special relativity. But it is an experimental fact that this is not the case. So we were correct in saying above that C sees B run at essentially the same rate. In any case, whatever worries you might have in this nonrelativistic setup are independent of relativity. What we showed in the preceding paragraph is that the relativistic and nonrelativistic scenarios behave the same way. If we lived in a world where a relativistic accelerating clock didn't run slow by a simple factor of γ, then we would necessarily also live in a world where a nonrelativistic clock accelerating from rest didn't run at the same rate as a clock at rest next to it. (But we know it does.)

   Although the above reasoning made no use of a Minkowski diagram, a diagram is very helpful in illustrating why there is an asymmetry with regard to who is doing the accelerating (clock or observer). For the setup we have been discussing in this remark, the Minkowski diagram from the inertial observer A's point of view is shown in Fig. 2.44. The worldlines of B (the accelerating clock) and C (moving with constant v) are essentially the same during a small time interval. The curvature in B's worldline is inconsequential (for a short time). This should be contrasted with the highly consequential curvature in A's worldline in Fig. 2.43 above, in the case where the observer A is the one who is accelerating. This causes the x_A axis to tilt upward in Fig. 2.43, leading to the ad/c^2 term in Eq. (2.104). Returning to the case in Fig. 2.44 where B is the one who is accelerating, although it is certainly true that B's x axis becomes tilted upward, this isn't relevant, because B isn't the one doing the observing. ♣

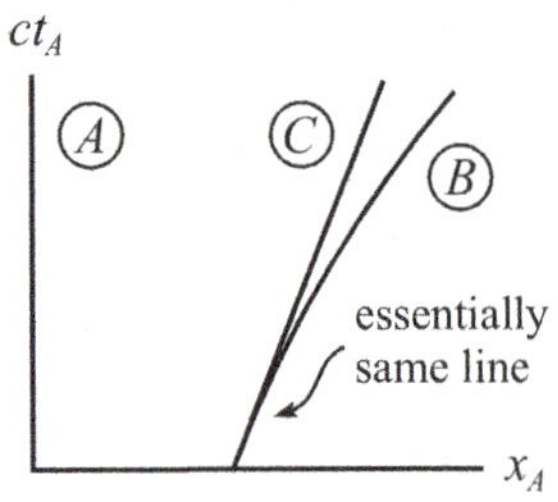

Figure 2.44

2.13. **Break or not break?**

There are two possible reasonings, so we seem to have a paradox:

- To an observer in the original rest frame, the spaceships stay the same distance d apart. Therefore, in the frame of the spaceships, the distance d' between them must equal γd. This is true because d' is the distance that gets length contracted down to d. After a long enough time, γ will differ appreciably from 1, so the string will be stretched by a large factor. Therefore, it will break.

- Let A be the rear spaceship, and let B be the front spaceship. From A's point of view, it looks like B is doing exactly what A is doing (and vice versa). A says that B has the same acceleration that he has. So B should stay the same distance ahead of him. Therefore, the string shouldn't break.

The first reasoning is correct (or mostly correct; see the first remark below). The string *will* break. So that's the answer to the problem. But as with any good relativity paradox, we shouldn't feel at ease until we've explained what is wrong with the incorrect reasoning.

The problem with the second reasoning is that A does *not* see B doing exactly what he is doing. Rather, we know from Problem 2.12 that A sees B's clock running fast (and B sees A's clock running slow). A therefore sees B's engine running faster, so B pulls away from A. Therefore, the string eventually breaks.[5]

Things become more clear if we draw a Minkowski diagram. Fig. 2.45 shows the x' and ct' axes of A's frame. The x' axis is tilted up, so it meets B's worldline farther to the right than you might think. The distance PQ along the x' axis is the distance that A measures the string to be. Although it isn't obvious that this distance in A's frame is larger than d (because the unit size on the x' axis is larger than the unit size in the original frame), we can demonstrate this as follows. In A's frame, the distance PQ is greater than the distance PQ'. But PQ' is the length of something in A's frame that is length contracted down to d in the original frame (the dashed length d in Fig. 2.45; see the discussion of Fig. 2.15). So PQ' is γd in A's frame. And since $PQ > PQ' = \gamma d > d$ in A's frame, the string breaks.

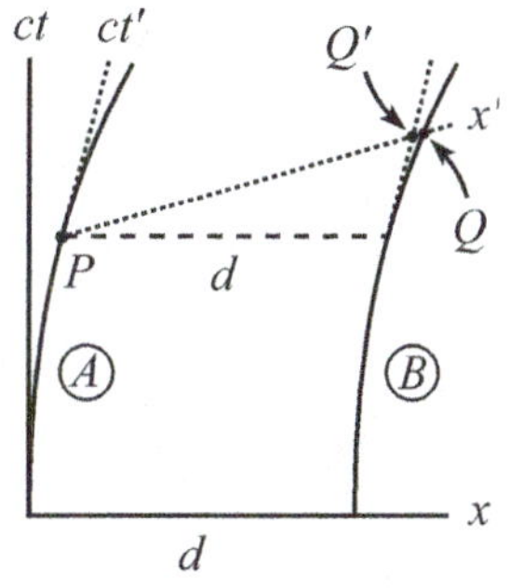

Figure 2.45

REMARKS:

1. There is one slight (inconsequential) flaw in the first reasoning above. There isn't one "frame of the spaceships." Their frames differ, because they measure a relative speed between them (the string gets longer). Therefore, it isn't clear exactly what is meant by the "length" of the string, because it isn't clear what frame the measurement should take place in. This ambiguity, however, does not change the fact that A and B observe their separation to be (roughly) γd.

   If we want there to eventually be a well-defined "frame of the spaceships," we can modify the problem by stating that after a while, the spaceships stop accelerating simultaneously, as measured in the original inertial frame. Equivalently, A and B turn off their engines after equal proper times. What A sees is the following. B moves faster than A, so B pulls away from A. B then turns off his engine. The gap continues to widen. But A continues to fire his engine until be reaches B's speed. They then sail onward, in a common frame, keeping a constant separation, which is greater than the original separation by a factor γ.

2. The main issue in this problem is that it depends on exactly how we choose to accelerate an extended object. If we accelerate a stick by pushing on the back end (or by pulling on the front end), then its length will remain essentially the same in its own frame, so it will become shorter in the original frame. (Assume that the acceleration is small enough so that the required forces don't rip the stick apart.) But if we arrange for each end (or perhaps a number of points on the stick) to speed up in such a way that they always move at the same speed with respect to the original frame, then the stick will necessarily be torn apart.

3. This problem gives the key to the classic problem of the relativistic wagon wheel, which can be stated as follows. (You may want to cover up the following paragraph, so that you can solve the problem on your own.) A wheel is spun faster and faster, until the points on the rim move at a relativistic speed. In the lab frame, the circumference is length contracted, but the spokes aren't (because they always lie perpendicular to the direction of motion). So if the rim has length $2\pi r$ in the lab frame, then it has length $2\pi\gamma r$ in the wheel frame. Therefore, in the wheel frame, the ratio of the circumference to the diameter is larger than π. So the question is: Is this really true? And if so, how does the circumference become longer in the wheel frame?

   I'm putting this sentence here just in case you happened to see the first sentence of this paragraph when you were trying to cover it up and solve the problem on

[5] This also follows from the Equivalence Principle and the general-relativistic time-dilation effect (discussed in Chapter 5), which states that high clocks in a gravitational field run fast. Since A and B are accelerating, they may be considered (by the Equivalence Principle) to be in a gravitational field, with B higher in the field. And since high clocks run fast, A sees B's clock running fast (and B sees A's clock running slow).

your own. The answer is that the circumference is indeed longer in the wheel frame. If we imagine little rocket engines placed around the rim, and if we have them all accelerate with the same proper acceleration, then from the above result, the separation between the engines will gradually increase, thereby increasing the length of the circumference. Assuming that the material in the rim can't stretch indefinitely, the rim will eventually break between each engine. So in the rotating wheel frame, the ratio of the circumference to the diameter is larger than π. In other words, space is curved in the wheel frame. ♣

2.14. **Successive Lorentz transformations**

It isn't necessary to use matrices to solve this problem, of course. But things look nicer if we do. The desired composite L.T. is obtained by multiplying the matrices for the individual L.T.'s. So the composite matrix M is

$$\begin{aligned} M &= \begin{pmatrix} \cosh\phi_2 & \sinh\phi_2 \\ \sinh\phi_2 & \cosh\phi_2 \end{pmatrix} \begin{pmatrix} \cosh\phi_1 & \sinh\phi_1 \\ \sinh\phi_1 & \cosh\phi_1 \end{pmatrix} \\ &= \begin{pmatrix} \cosh\phi_1\cosh\phi_2 + \sinh\phi_1\sinh\phi_2 & \sinh\phi_1\cosh\phi_2 + \cosh\phi_1\sinh\phi_2 \\ \cosh\phi_1\sinh\phi_2 + \sinh\phi_1\cosh\phi_2 & \sinh\phi_1\sinh\phi_2 + \cosh\phi_1\cosh\phi_2 \end{pmatrix} \\ &= \begin{pmatrix} \cosh(\phi_1+\phi_2) & \sinh(\phi_1+\phi_2) \\ \sinh(\phi_1+\phi_2) & \cosh(\phi_1+\phi_2) \end{pmatrix}, \end{aligned} \tag{2.106}$$

where we have used the sum formulas for sinh and cosh, which can be proved by writing sinh and cosh in terms of the $e^{\pm\phi}$ exponentials. The result in Eq. (2.106) is an L.T. with $v = \tanh(\phi_1 + \phi_2)$, as desired. Except for a few minus signs, this proof is the same as the one for successive rotations in a plane.

2.15. **Accelerator's time**

Eq. (2.53) gives the speed of the spaceship as a function of the spaceship's time (which we are denoting by t' here):

$$\beta(t') \equiv \frac{v(t')}{c} = \tanh(at'/c). \tag{2.107}$$

The person in the lab frame sees the spaceship's clock run slow by the factor $1/\gamma = \sqrt{1-\beta^2}$, which means that $dt = dt'/\sqrt{1-\beta^2}$; dt is larger than dt'. So we have

$$t = \int_0^t dt = \int_0^{t'} \frac{dt'}{\sqrt{1-\beta(t')^2}} = \int_0^{t'} \cosh(at'/c)\,dt', \tag{2.108}$$

where we have used $1 - \tanh^2 x = (\cosh^2 x - \sinh^2 x)/\cosh^2 x = 1/\cosh^2 x$. The integral of cosh is sinh, so we end up with

$$t = \frac{c}{a}\sinh\left(\frac{at'}{c}\right). \tag{2.109}$$

This is the desired relation between t and t'. For small a or t' (more precisely, for $at'/c \ll 1$), we can use $\sinh z \approx z$ (which you can verify by using the definition of sinh, along with $e^z \approx 1 + z$). This yields $t \approx t'$, as it should. For large times, we have $\sinh z \approx e^z/2$, so

$$t \approx \frac{c}{2a}e^{at'/c} \qquad \text{or} \qquad t' = \frac{c}{a}\ln\left(\frac{2at}{c}\right). \tag{2.110}$$

The first expression here tells us that the lab frame will see the astronaut read all of "Moby Dick," but it will take an exponentially long time (not that it doesn't already!).

Chapter 3

Dynamics

In the first two chapters, we dealt only with abstract particles flying through space and time. We didn't concern ourselves with the nature of the particles, how they got to be moving, or what would happen if various particles interacted. In this chapter we will deal with these issues. That is, we will discuss masses, forces, energy, momentum, etc. The two main results in this chapter are that the momentum and energy of a particle are given by

$$\boxed{\mathbf{p} = \gamma m \mathbf{v}} \quad \text{and} \quad \boxed{E = \gamma m c^2} \tag{3.1}$$

where $\gamma \equiv 1/\sqrt{1 - v^2/c^2}$ and m is the mass of the particle.[1] When $v \ll c$, we have $\gamma \approx 1$, so the expression for $\mathbf{p}$ reduces to $\mathbf{p} = m\mathbf{v}$, as it should for a nonrelativistic particle. And when $v = 0$, the expression for E reduces to the well-known relation $E = mc^2$. For a history of the mass-energy relation, see Fadner (1988).

The outline of this chapter is as follows. In Section 3.1 we give some justification for the expressions for $\mathbf{p}$ and E in Eq. (3.1), and we show that they are consistent with each other. In Section 3.2 we show how p and E transform from one frame to another. We find that the transformations take the same form as the Lorentz transformations for x and t. Section 3.3 covers collisions and decays. As in nonrelativistic physics, the main ingredients are conservation of energy and momentum. It's just that these quantities take different forms in relativistic physics. Section 3.4 gives a brief discussion of the units for energy that particle physicists use. In Section 3.5 we derive the relativistic expression for force, which we will find is *not* equal to ma. Section 3.6 covers relativistic rocket motion, where a rocket is propelled by firing photons out the back. Section 3.7 deals with "relativistic strings," which are massless spring-like objects, but with a force that is independent of the stretching distance.

If you haven't studied Newtonian (nonrelativistic) dynamics before, Appendix E gives a brief review of the concepts needed for this chapter.

3.1 Energy and momentum

In this section, we'll give some theoretical justification for the $\mathbf{p}$ and E in Eq. (3.1). The reasoning here should convince you of their validity. An alternative, and perhaps more convincing, motivation comes from the 4-vector formalism in Chapter 4. In the

[1]There are various ways that people use the word "mass" in relativity. In particular, some people talk about "rest mass" and "relativistic mass." But we won't use these terms. We'll use "mass" to refer only to what some people call "rest mass." See the discussion on page 129 for more on this.

end, however, the real justification for Eq. (3.1) comes from experiments. And indeed, experiments in high-energy accelerators are continually supporting the validity of these expressions. More precisely, they are providing evidence that the momentum and energy in Eq. (3.1) are *conserved* in any type of collision (that is, their values are the same before and after the collision). We therefore conclude, with reasonable certainty, that the above expressions for momentum and energy are correct. But actual experiments aside, let's consider a few thought experiments that motivate the **p** and E in Eq. (3.1).

3.1.1 Momentum

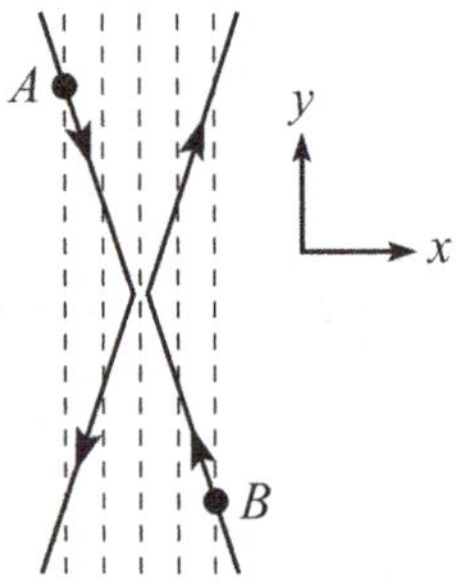

Figure 3.1

Consider the following setup. In the lab frame, identical particles A and B move as shown in Fig. 3.1. They move with equal and opposite small velocities in the x direction, and with equal and opposite large velocities in the y direction. The paths are arranged so that the particles glance off each other and reverse their v_x velocity components. Equivalently, you can imagine that the middle dashed line in Fig. 3.1 is a brick wall that the particles bounce off of. Assume that A and B have clocks that tick every time they cross one of the equally spaced vertical dashed lines.

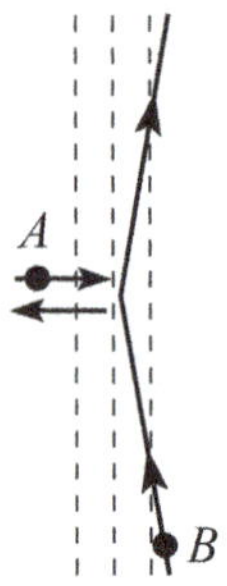

Figure 3.2

Consider now the reference frame that moves in the negative y direction with the same v_y as A. In this frame, the setup is shown in Fig. 3.2. As in the lab frame, the collision reverses the v_x velocity component of each particle. (This is intuitively clear, but you can also show that it follows quickly from the transverse velocity-addition formula. The only before/after difference in the formula is a sign in the numerator.) Therefore, the magnitudes of the x momenta of the two particles must be the same. This is true because if, say, A's $|p_x|$ were larger than B's $|p_x|$, then the total p_x would point to the right before the collision, but to the left after the collision. Since momentum is something we want to be conserved, this cannot be the case. So the p_x magnitudes must be equal.

However, the x *speeds* of the two particles are *not* the same in this new frame. To see why, note that A is essentially at rest in this frame (because the original x speeds were assumed to be small), while B is moving with a very large speed v, mainly in the y direction. Therefore, B's clock is running slower than A's, by a factor essentially equal to $1/\gamma \equiv \sqrt{1 - v^2/c^2}$. And since A's and B's clocks each tick whenever they cross a vertical line (this is a frame-independent fact), B must be crossing the lines at a slower rate than A. In other words, B must be moving *slower* in the x direction, by a factor $1/\gamma$. Therefore, the Newtonian expression $p_x = mv_x$ cannot be the correct one for the x momentum, because B's $|p_x|$ would be smaller than A's (by a factor $1/\gamma$), due to their different x speeds. However, the γ factor in

$$p_x = \gamma m v_x \equiv \frac{mv_x}{\sqrt{1 - v^2/c^2}} \tag{3.2}$$

takes care of this problem, because $\gamma \approx 1$ for A, and $\gamma = 1/\sqrt{1 - v^2/c^2}$ for B, which cancels the effect of B's smaller $|v_x|$. The $|p_x|$ values are then equal, as desired.

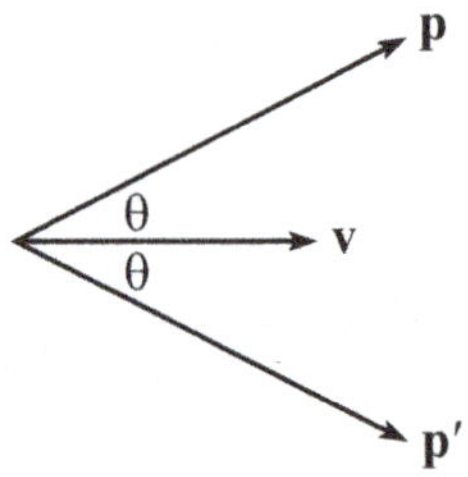

Figure 3.3

To obtain the three-dimensional form of **p**, we can use the fact that the vector **p** must point in the same direction as the vector **v**, because any other direction for **p** would violate rotational invariance. If someone claimed that **p** pointed in the direction shown in Fig. 3.3, then he would be hard-pressed to explain why it didn't instead point along the direction **p′** shown (or anywhere else along the cone on which **p** and **p′** lie, with **v** as the axis). In short, the direction of **v** is the only preferred direction in space, as far as a uniformly moving particle is concerned. Therefore, the momentum must take the form of $\mathbf{p} = A\mathbf{v}$, for some A. This implies that $p_x = Av_x$, whereupon Eq. (3.2) tells us that $A = \gamma m \equiv m/\sqrt{1 - v^2/c^2}$. Hence,

$$\mathbf{p} = \gamma m\mathbf{v} \equiv \frac{m\mathbf{v}}{\sqrt{1 - v^2/c^2}}, \tag{3.3}$$

in agreement with Eq. (3.1). Note that all of the components of $\mathbf{p}$ have the same denominator, which involves the entire speed, $v^2 = v_x^2 + v_y^2 + v_z^2$. The denominator of p_x is *not* $\sqrt{1 - v_x^2/c^2}$, etc. Different denominators would lead to a $\mathbf{p}$ that doesn't point along $\mathbf{v}$.

The above setup is only one specific type of collision among an infinite number of possible types. What we've shown with this setup is that the only possible vector of the form $f(v)m\mathbf{v}$ (where f is some function) that has a chance of being conserved in all collisions is $\gamma m\mathbf{v}$ (or some constant multiple of this). We haven't proved that it actually *is* conserved in all collisions. This is where the gathering of data from experiments comes in. But we've shown above that it would be a waste of time to consider the vector $\gamma^5 m\mathbf{v}$, for example. Any power of γ other than 1 would lead to the $|p_x|$ values of the particles not being equal in Fig. 3.2, since B's x speed is smaller by a factor $1/\gamma$.

3.1.2 Energy

Having given some justification for the momentum expression, $\mathbf{p} = \gamma m\mathbf{v}$, let's now try to justify the energy expression,

$$E = \gamma mc^2. \tag{3.4}$$

More precisely, we'll show that the form of the momentum in Eq. (3.3) implies that γmc^2 is conserved in interactions (or at least in the specific interaction we'll look at below). There are various ways to do this. Perhaps the best way is to use the 4-vector formalism in Chapter 4. But for now we'll just study one simple scenario that should at least make it believable that $E = \gamma mc^2$ is conserved in general.

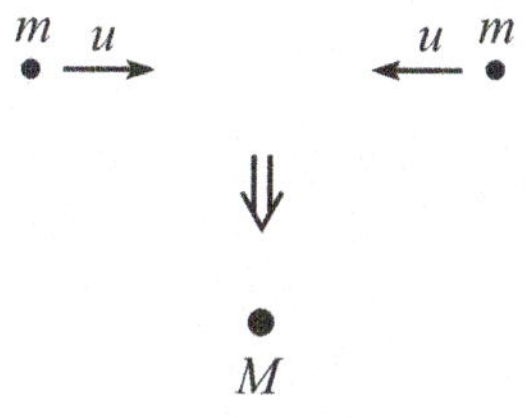

Figure 3.4

Consider the following setup. Two identical particles with mass m head toward each other, both with speed u, as shown in Fig. 3.4. They stick together and form a particle with mass M. M is at rest, due to the symmetry of the setup. At present, we can't assume anything about the size of M, but we'll find below that it is *not* equal to the naive value of $2m$.

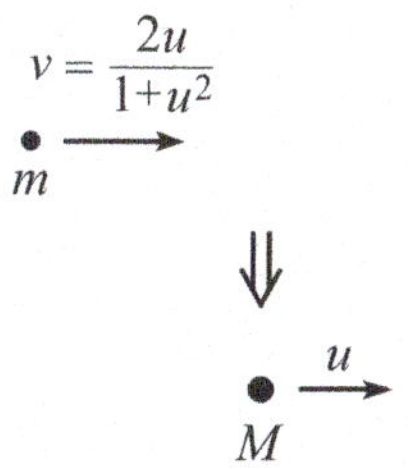

Figure 3.5

This setup is a fairly uninteresting one with regard to momentum, because conservation of momentum just gives $0 = 0$. So let's instead consider a less trivial view of the setup, with respect to the frame moving to the left at speed u. This view is shown in Fig. 3.5. The right mass is at rest, the left mass moves to the right at speed $v = 2u/(1 + u^2)$ from the velocity-addition formula,[2] and the final mass M moves to the right at speed u (because M is at rest in the original frame). Note that the γ factor associated with the speed v is

$$\gamma_v \equiv \frac{1}{\sqrt{1 - v^2}} = \frac{1}{\sqrt{1 - \left(\dfrac{2u}{1 + u^2}\right)^2}} = \frac{1 + u^2}{1 - u^2}. \tag{3.5}$$

Conservation of momentum for the collision then gives

$$\begin{aligned} \gamma_v mv + 0 = \gamma_u Mu \quad &\Longrightarrow \quad m\left(\frac{1 + u^2}{1 - u^2}\right)\left(\frac{2u}{1 + u^2}\right) = \frac{Mu}{\sqrt{1 - u^2}} \\ &\Longrightarrow \quad M = \frac{2m}{\sqrt{1 - u^2}} \equiv 2\gamma_u m. \end{aligned} \tag{3.6}$$

[2]As we've done many times in Chapters 1 and 2, we'll drop the c's, or equivalently set $c = 1$, lest the calculation get too messy. We'll discuss the issue of setting $c = 1$ in more detail at the end of this section.

Conservation of momentum therefore tells us that M is *not* equal to $2m$. But if u is very small, then $\gamma_u \approx 1$, so M is approximately equal to $2m$, as we know from everyday experience.

Using the value of M from Eq. (3.6), let's now check that our candidate for energy, $E = \gamma mc^2$, is conserved in the collision. There is no freedom left in any of the parameters, so γmc^2 is either conserved or it isn't. In the original frame where M is at rest, both of the initial masses m have speed u. So E is conserved if

$$2(\gamma_u mc^2) = \gamma_0 Mc^2 \quad \Longleftrightarrow \quad 2\left(\frac{1}{\sqrt{1-u^2}}\right)m = 1 \cdot \left(\frac{2m}{\sqrt{1-u^2}}\right), \tag{3.7}$$

which is indeed true. Let's also check that E is conserved in the frame moving to the left at speed u. In this frame the right mass is at rest, the left mass moves with speed v, and M moves with speed u. So E is conserved if

$$\begin{aligned} \gamma_v mc^2 + \gamma_0 mc^2 &= \gamma_u Mc^2 \\ \Longleftrightarrow \quad \left(\frac{1+u^2}{1-u^2}\right)m + 1 \cdot m &= \left(\frac{1}{\sqrt{1-u^2}}\right)\left(\frac{2m}{\sqrt{1-u^2}}\right) \\ \Longleftrightarrow \quad \frac{2m}{1-u^2} &= \frac{2m}{1-u^2}, \end{aligned} \tag{3.8}$$

which is again true. This example should convince you that γmc^2 is at least a believable expression for the energy of a particle. But just as in the case of momentum, we haven't proved that γmc^2 actually *is* conserved in all collisions. This is the duty of experiments. But we've shown that it would be a waste of time to consider the quantity $\gamma^4 mc^2$, for example (assuming that we've already accepted the γmv form of the momentum). Any power of γ other than 1 would ruin the equalities in Eqs. (3.7) and (3.8).

One thing that certainly needs to be true is that if E and p are conserved in one reference frame, then they are also conserved in any other frame. We'll demonstrate this rigorously at the end of Section 3.2. After all, a conservation law shouldn't depend on what frame you're in.

A useful relation that follows from Eq. (3.1) is

$$\boxed{\frac{\mathbf{p}}{E} = \frac{\mathbf{v}}{c^2}} \tag{3.9}$$

Given p and E, this is the quickest way to obtain v.

Remarks:

1. In the above discussions, technically the goal wasn't to justify Eq. (3.1). Those two equations by themselves are devoid of any meaning. All they do is define the letters $\mathbf{p}$ and E. Our goal was to make a meaningful physical statement, not just a definition.

   As mentioned above, the meaningful physical statement that we want to make is that the quantities $\gamma m\mathbf{v}$ and γmc^2 are *conserved* (that is, unchanged) in an interaction among particles. (This is what we tried to justify above.) This fact then makes these quantities worthy of special attention, because conserved quantities are very helpful in understanding what is happening in a given setup. And anything worthy of special attention certainly deserves a label, so we may then attach the names "momentum" and "energy" to $\gamma m\mathbf{v}$ and γmc^2. Any other names would work just as well, of course, but we choose these because in the limit of small speeds, $\gamma m\mathbf{v}$ and γmc^2 (as we will soon show) reduce to some other nicely conserved quantities, which someone already tagged with the labels "momentum" and "energy" long ago.

2. As we have noted, the fact of the matter is that we can't *prove* that $\gamma m\mathbf{v}$ and γmc^2 are conserved. In Newtonian physics, conservation of $\mathbf{p} \equiv m\mathbf{v}$ is basically postulated by Newton's third law, and we're not going to do any better than that here. All we can hope to do as physicists is to provide some motivation for considering $\gamma m\mathbf{v}$ and γmc^2, then show that it is consistent for $\gamma m\mathbf{v}$ and γmc^2 to be conserved during an interaction, and then gather a large amount of experimental evidence, all of which is consistent with $\gamma m\mathbf{v}$ and γmc^2 being conserved. As far as the experimental evidence goes, suffice it to say that high-energy accelerators, cosmological observations, and many other forums are continually providing evidence for everything that we think is true about relativistic dynamics. If the theory isn't correct, then we know that it must be the limiting case of a more correct theory, just as the $m\mathbf{v}$ expression for momentum is the limiting case of $\gamma m\mathbf{v}$. (And in fact the relativistic theory *isn't* correct, because it doesn't incorporate quantum mechanics. The more complete theory that includes both special relativity and quantum mechanics is quantum field theory.) But all this experimental induction has to count for something . . .

   "To three, five, and seven, assign
   A name," the prof said, "We'll define."
   But he botched the instruction
   With woeful induction
   And told us the next prime was nine.

3. Conservation of energy in relativistic mechanics is actually a simpler concept than in nonrelativistic mechanics, because $E = \gamma mc^2$ is conserved, period. We don't have to worry about the generation of internal energy (heat or internal potential energy), which ruins conservation of the nonrelativistic "mechanical" energy, $E = mv^2/2$. In relativistic mechanics, the internal energy is simply built into the total energy. In the above example, the two m's collide and generate internal energy (heat) in the resulting mass M. This internal energy shows up as an increase in mass, which makes M larger than $2m$. In the original lab frame, the energy that corresponds to the increase in mass comes from the initial kinetic energy of the two m's.

4. Problem 3.1 gives an alternative derivation of the energy and momentum expressions in Eq. (3.1). The derivation uses additional facts, namely that the energy and momentum of a photon are given by $E = hf$ and $p = hf/c$, where f is the frequency of the light wave, and h is Planck's constant. ♣

Nonrelativistic limit

Since γmc^2 is conserved, any multiple of γmc^2 is also conserved, of course. Why did we pick γmc^2 to label as "E" instead of, say, $5\gamma mc^4$? Because γmc^2 reduces properly (in a certain sense, as we will show) to the standard kinetic energy $mv^2/2$ from Newtonian (nonrelativistic) dynamics. To see why, consider the approximate form that γmc^2 takes in the Newtonian limit, that is, in the limit $v \ll c$. Using the Taylor-series expansion $1/\sqrt{1-x} \approx 1 + x/2 + 3x^2/8$, we can write

$$
\begin{aligned}
E \equiv \gamma mc^2 &= \frac{mc^2}{\sqrt{1 - v^2/c^2}} \\
&= mc^2\left(1 + \frac{v^2}{2c^2} + \frac{3v^4}{8c^4} + \cdots\right) \\
&= mc^2 + \frac{1}{2}mv^2 + \cdots . \qquad (3.10)
\end{aligned}
$$

The dots represent higher-order terms in v^2/c^2, which may be neglected if $v \ll c$. The $mv^2/2$ term in Eq. (3.10) is what we're aiming for, so things look promising. But we have an extra (and very large) mc^2 term, too. How exactly does the relativistic energy

γmc^2 reduce properly to the nonrelativistic kinetic energy $mv^2/2$ if we have an extra mc^2 floating around? Let's consider the two basic kinds of Newtonian collisions.

In an *elastic* collision in Newtonian physics, no heat is generated, so mass is conserved. That is, the mass m of each particle is the same before and after the collision. Imagine writing down all the γmc^2 terms before the collision on one side of the equation, and all the γmc^2 terms after the collision on the other side. Then replace every γmc^2 term with the approximate $mc^2 + mv^2/2$ expression from Eq. (3.10). Even though the huge mc^2 terms from Eq. (3.10) dominate the $mv^2/2$ terms, the mc^2 terms all cancel out because the m's don't change during the collision. So we're left the with statement that the sum of the initial $mv^2/2$ terms equals the sum of the final $mv^2/2$ terms. In other words, conservation of $E \equiv \gamma mc^2$ reduces to the familiar conservation of Newtonian kinetic energy, $mv^2/2$, for elastic collisions in the limit of slow speeds. This is an example of the *correspondence principle*, which says that relativistic formulas must reduce to the familiar nonrelativistic ones in the nonrelativistic limit. Likewise, we picked $\mathbf{p} \equiv \gamma m\mathbf{v}$ for the momentum instead of, say, $6\gamma mc^3\mathbf{v}$, because the former reduces to the familiar Newtonian momentum, $m\mathbf{v}$, in the limit of slow speeds (where $\gamma \approx 1$).

Whether abstract, profound, or just mystic,
Or boring, or somewhat simplistic,
A theory must lead
To results that we need
In limits, nonrelativistic.

In an *inelastic* collision, heat is generated, and this heat shows up as mass. Now, in an everyday-type inelastic collision, the masses change by an *extremely* small amount. But when forming the product mc^2, the largeness of the c^2 factor leads to a nontrivial change in mc^2. So in the conservation-of-energy equation we imagined writing down above, the mc^2 terms from before the collision now *don't* cancel out with the mc^2 terms after the collision. The after-collision terms are larger. We therefore conclude that the Newtonian kinetic energy, $mv^2/2$, is *not* conserved in an inelastic collision (where heat is generated). This is consistent with what we know very well from standard Newtonian mechanics.

Whenever we use the term "energy" in relativistic physics, we mean the total energy, γmc^2. If we use the term "kinetic energy," we mean a particle's excess energy over the energy it has when it is motionless. In other words, the kinetic energy is $\gamma mc^2 - mc^2$. Kinetic energy is *not* necessarily conserved in a collision, because it differs from γmc^2 (which *is* conserved) by the $-mc^2$ term. And mass is not necessarily conserved, as we saw in Eq. (3.6). In the center-of-mass frame in the collision we looked at in Section 3.1.2, there was kinetic energy before the collision, but none after. Kinetic energy is a rather artificial concept in relativity. You virtually always want to use the total energy, γmc^2, when solving a problem.

The "Very Important Relation"

Consider (a little out of the blue) the quantity $E^2 - |\mathbf{p}|^2c^2$. There are good theoretical reasons for considering this quantity, as we'll see below in Section 3.2 and in Chapter 4. But for now let's just note that the quantity can be simplified nicely:

$$\begin{aligned} E^2 - |\mathbf{p}|^2c^2 &= \gamma^2 m^2 c^4 - \gamma^2 m^2 |\mathbf{v}|^2 c^2 \\ &= \gamma^2 m^2 c^4 \left(1 - \frac{v^2}{c^2}\right) \\ &= m^2 c^4. \end{aligned} \tag{3.11}$$

This relation is a primary ingredient in solving relativistic collision problems. It replaces the $K = p^2/2m$ relation between kinetic energy and momentum in Newtonian physics. It can be derived in more profound ways, as we'll see in Chapter 4. It's so important that I like to call it the Very Important Relation. Let's put it in a box, with E isolated on the lefthand side:

$$\boxed{E^2 = p^2c^2 + m^2c^4} \qquad \text{("Very Important Relation")} \tag{3.12}$$

Whenever you know two of the three quantities E, p, and m, this equation gives you the third. In the case where $m = 0$ (as for photons), Eq. (3.12) says that

$$\boxed{E = pc} \qquad \text{(for photons).} \tag{3.13}$$

This is the key equation for massless objects. For photons, the equations $\mathbf{p} = \gamma m\mathbf{v}$ and $E = \gamma mc^2$ don't tell us much, because $m = 0$ and $\gamma = \infty$, which makes the product γm undetermined. But $E^2 - |\mathbf{p}|^2c^2 = m^2c^4$ still holds, so we conclude that $E = pc$ when $m = 0$. Alternatively, $E = pc$ follows from Eq. (3.9) with $v = c$. Of course, knowing that the energy and momentum of a photon are related by $E = pc$ doesn't tell us what E and p actually are. To determine these, we need to know the frequency f of the light wave. The energy and momentum of a photon are then given by $E = hf$ and $p = hf/c$, where f is the frequency of the light wave, and h is Planck's constant (which equals $6.63 \cdot 10^{-34}\ \text{m}^2\,\text{kg/s}$). But these expressions belong more in the study of quantum mechanics than relativity.

Note that any massless particle must have $\gamma = \infty$. That is, it must travel at speed c. If this weren't the case, then $E = \gamma mc^2$ would equal zero, in which case the particle wouldn't be much of a particle. We'd have a hard time observing something that has zero energy.

The general size of mc^2

What is the general size of mc^2? If we let $m = 1$ kg, then $mc^2 = (1\ \text{kg})(3 \cdot 10^8\ \text{m/s})^2 \approx 10^{17}$ joules. How big is this? A typical household electric bill might be \$100 per month, or \$1200 per year. At about 12 cents per kilowatt-hour, this translates to 10^4 kilowatt-hours per year, or equivalently 10^7 watt-hours per year. Since there are 3600 seconds in an hour, this converts to $10^7 \cdot 3600 = 3.6 \cdot 10^{10}$ watt-seconds per year, or equivalently $3.6 \cdot 10^{10}$ joules per year (because a watt is a joule per second). We therefore see that if one kilogram were converted completely into usable energy (that is, kinetic energy, which can then be used to drive a turbine), it would be enough to provide electricity to about $10^{17}/(3.6 \cdot 10^{10}) \approx 3$ million homes for a year. That's a lot.

In a nuclear reactor, only a small fraction of the mass is converted into usable energy. Most of the mass remains in the final products, which doesn't help in lighting your home. If a particle were to combine with its antiparticle, then it would be possible for all of the mass energy to be converted into usable energy. But we're a while away from being able to do this productively. However, even a small fraction of the very large quantity $E = mc^2$ can be large, as evidenced by the use of nuclear power and nuclear weapons. Any quantity with a few powers of c is bound to change the face of the world.

Mass

Some treatments of relativity use the term "rest mass," m_0, to refer to the mass of a motionless particle, and "relativistic mass," m_{rel}, to refer to the quantity γm_0 of a moving particle. We won't use this terminology here. The only thing we'll call "mass" is what the above treatments call "rest mass." (For example, the mass of an electron is $9.11 \cdot 10^{-31}$ kg, and the mass of a liter of water is 1 kg, independent of their speed.) And since we'll refer to only one type of mass, there is no need to use the qualifier "rest" or the subscript "0." We'll therefore simply use the notation "m."

Of course, you can *define* the quantity γm to be anything you want. There's nothing wrong with calling it "relativistic mass." But the point is that γm already goes by another name. It's just the energy, up to factors of c. The use of the word "mass" for this quantity, although quite permissible, is certainly not needed.

Furthermore, the word "mass" is used to describe what is on the righthand side of the equation, $E^2 - |\mathbf{p}|^2c^2 = m^2$. The m^2 here is an *invariant*, that is, it is something that is independent of the reference frame. Although E and $\mathbf{p}$ depend on the frame because they involve v, this v dependence cancels out when taking the difference of the squares (with a c^2 in front of the p^2), yielding a frame-independent m^2. If "mass" is to be used in this definite way to describe an invariant, then it doesn't make sense to also use it to describe the quantity γm, which is frame-dependent. The prefixes "rest-" and "relativistic-" are introduced to avoid this problem, but the result of this is just a watering down of the very important invariance quality of "mass."

However, since there is in fact nothing actually wrong with using "relativistic mass," it mainly comes down to personal preference whether or not you use the term. Certainly one motivation for labeling γm as some kind of mass is that the expression for momentum takes the form, $p = \gamma m_0 v \equiv m_{\text{rel}} v$, which mimics the Newtonian expression. And the expression for energy takes the nice form, $E = \gamma m_0 c^2 \equiv m_{\text{rel}} c^2$. But considering that Newtonian physics is only a limiting case of relativistic physics, it is questionable what is gained by molding a theory that is more correct into one that is less correct. At any rate, my view is that these nice formulas don't outweigh the usefulness of having the word "mass" mean a very specific invariant quantity, with the word "energy" referring to the frame-dependent quantity γm (times c^2). Invariant quantities have a certain sacred place in physics, so they should be given a name that doesn't have any non-invariant connotations.

Setting $c = 1$

For the remainder of this book, we will often work in units where $c = 1$ (as we have already done on many occasions). For example, instead of one meter being the unit of distance, we can make $3 \cdot 10^8$ meters equal to one unit. Or, we can keep the meter as is, and make $1/(3 \cdot 10^8)$ of a second be the unit of time. In such units, our various expressions become

$$\mathbf{p} = \gamma m \mathbf{v}, \qquad E = \gamma m, \qquad E^2 = p^2 + m^2, \qquad \frac{\mathbf{p}}{E} = \mathbf{v}. \tag{3.14}$$

Said in another way, you can simply drop all the c's in your calculations (which will generally save you a lot of strife), and then put them back into your final answer to make the units correct. For example, let's say the goal of a certain problem is to find the time of some event. If you drop the c's and your answer comes out to be ℓ, where ℓ is a given length, then you know that the correct answer has to be ℓ/c, because this has units of time. In order for this procedure to work, there must be only one way to put the c's back in at the end (up to trivial equivalences, like $\ell/c = \ell c/c^2$). This is always the case, because if there were two ways, then we would have $c^a = c^b$ for some numbers $a \neq b$. But this is impossible, because c has units.

3.2 Transformations of E and p

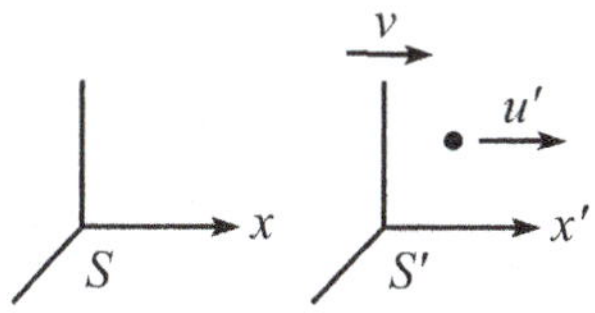

Figure 3.6

Consider the following one-dimensional setup, where all of the motion is in the x direction. A particle with mass m has energy E' and momentum p' in frame S'. Frame S' moves with velocity v with respect to frame S, in the x direction; see Fig. 3.6. What are the particle's E and p in S, in terms of E' and p' (and v)?

Let u' be the particle's velocity in S', which is given by Eq. (3.9) as $u' = c^2p'/E'$. It can also be determined via either $E' = \gamma_{u'}mc^2$ or $p' = \gamma_{u'}mu'$. From the velocity-addition formula, the particle's velocity in S is (dropping the c's)

$$u = \frac{u' + v}{1 + u'v}. \tag{3.15}$$

In finding E and p, this u is technically all we need to know, because a particle's velocity completely determines its energy and momentum (given m). But we'll need to work through some algebra to accomplish our goal of writing E and p in terms of E' and p'. The γ factor associated with the velocity u is (as you can check)

$$\gamma_u = \frac{1}{\sqrt{1 - \left(\dfrac{u' + v}{1 + u'v}\right)^2}} = \frac{1 + u'v}{\sqrt{(1 - u'^2)(1 - v^2)}} \equiv \gamma_{u'}\gamma_v(1 + u'v). \tag{3.16}$$

The energy and momentum in S' are

$$E' = \gamma_{u'}m \qquad \text{and} \qquad p' = \gamma_{u'}mu', \tag{3.17}$$

while the energy and momentum in S are, using Eq. (3.16),

$$\begin{aligned} E &= \gamma_u m = \gamma_{u'}\gamma_v(1 + u'v)m, \\ p &= \gamma_u mu = \gamma_{u'}\gamma_v(1 + u'v)m\left(\frac{u' + v}{1 + u'v}\right) = \gamma_{u'}\gamma_v(u' + v)m. \end{aligned} \tag{3.18}$$

Using the E' and p' from Eq. (3.17), you can verify that E and p can be rewritten as (with $\gamma \equiv \gamma_v$)

$$\begin{aligned} E &= \gamma(E' + vp'), \\ p &= \gamma(p' + vE'). \end{aligned} \tag{3.19}$$

If you want to put the factors of c back in, then the vE' term becomes vE'/c^2, to make the units correct:

$$\boxed{\begin{aligned} E &= \gamma(E' + vp') \\ p &= \gamma(p' + vE'/c^2) \end{aligned}} \tag{3.20}$$

These are the transformations for E and p between two frames. They are easy to remember, because they look just like the Lorentz transformations for the coordinates x and t in Eq. (2.2). We'll therefore also use the name "Lorentz transformations" for the equations in Eq. (3.20). However, the placement of the c's in these equations might mislead you on the correct correspondence between the parameters. To identify the correct correspondence, we can rewrite Eq. (3.20) as (with $\beta \equiv v/c$, and switching the order of the equations)

$$\begin{aligned} pc &= \gamma\left[(p'c) + \beta E'\right], \\ E &= \gamma\left[E' + \beta(p'c)\right]. \end{aligned} \tag{3.21}$$

This is symmetric in pc and E, just as Eq. (2.5) is symmetric in x and ct. And since p and x are the components of vectors, while E and t aren't, the correct correspondence must be $pc \leftrightarrow x$ and $E \leftrightarrow ct$. So pc and E transform like x and ct, respectively. That's why we switched the order and wrote the "p" equation first in Eq. (3.21), so that it would match up with Eq. (2.5). It is no coincidence that momentum and energy transform like space and time, as we will see in Chapter 4.

Let's check that Eq. (3.20) works in a few simple cases. If $u' = 0$ (so that $E' = mc^2$ and $p' = 0$), then $E = \gamma mc^2$ and $p = \gamma mv$, as expected; the object is moving with speed v in S. And if $u' = -v$ (so that $E' = \gamma mc^2$ and $p' = -\gamma mv$), then $E = mc^2$ and $p = 0$, as you can verify; the object is at rest in S.

You can use Eq. (3.20) to show that $E^2 - p^2c^2$ is frame independent:

$$\boxed{E^2 - p^2c^2 = E'^2 - p'^2c^2} \tag{3.22}$$

This follows from the same reasoning that led to $c^2t^2 - x^2 = c^2t'^2 - x'^2$ in Eq. (2.25). The proof there was based on the fact that t and x transform under the Lorentz transformations, so exactly the same proof works here with E and p, due to the $E \leftrightarrow ct$ and $pc \leftrightarrow x$ correspondence we learned from Eq. (3.21). Replacing ct with E, and x with pc, turns $c^2t^2 - x^2$ into $E^2 - p^2c^2$. But by all means work through the algebra analogous to Eq. (2.25) if you have your doubts that $E^2 - p^2c^2$ is invariant. Of course, for one particle, we already know that Eq. (3.22) is true, because both sides are equal to m^2c^4, from Eq. (3.11). Eq. (3.22) also holds for many particles, as we'll see in the discussion of linearity below.

Eq. (3.20) applies to the x component of the momentum. How do the transverse components, p_y and p_z, transform? Just as with the Δy and Δz separations between events, p_y and p_z don't change between frames. The analysis in Chapter 4 makes this clear, so for now we'll simply state that if the relative velocity between the frames is in the x direction, then

$$p_y = p'_y \qquad \text{and} \qquad p_z = p'_z. \tag{3.23}$$

If you really want to show explicitly that the transverse components don't change between frames, or if you're worried that a nonzero speed in the y direction might mess up the relationship between p_x and E that we found in Eq. (3.20), then Exercise 3.26 is for you. But it's a bit tedious, so feel free to settle for the much cleaner reasoning in Chapter 4.

Eq. (3.20) also holds for massless particles like photons, which move with speed c (of course). The above derivation itself isn't valid, because $m = 0$ and $\gamma = \infty$ for photons. But the final result in Eq. (3.20) is still valid. It certainly holds in the $m \to 0$ and $\gamma \to \infty$ limit, so it isn't surprising that it also holds if m and γ are exactly equal to zero and infinity.

Example (Photons and the Doppler effect): A photon traveling in the positive x' direction has energy E' and momentum p' (which equals E'/c) in frame S'. If S' moves with speed v in the positive x direction with respect to frame S, what is the photon's energy E in S? Accepting the fact that a photon's energy is proportional to its frequency (it equals $E = hf$, where f is the frequency of the light wave, and h is Planck's constant), verify that your result is consistent with the Doppler result from Section 2.5.1.

Solution: We'll drop the c's for simplicity. The momentum p' in S' is then simply E'. So Eq. (3.19) gives the energy in S as

$$E = \gamma(E' + vp') = \gamma(E' + vE') = E'\frac{1+v}{\sqrt{1-v^2}} = E'\sqrt{\frac{1+v}{1-v}}. \tag{3.24}$$

We see that the energy (and hence frequency) of the photon is larger in S by the factor $\sqrt{(1+v)/(1-v)}$. This is consistent with Eq. (2.36) with $v \leftrightarrow \beta$, because you could imagine a source at rest in S' emitting a wave with frequency f'. If you are standing off to the right in S, then the source is moving toward you. So the frequency f that you observe is Doppler shifted up to $f = f'\sqrt{(1+v)/(1-v)}$. Equivalently, from the point of view of S', the source is at rest and you are moving leftward. The same Lorentz transformation is applicable, and the same Doppler result holds, independent of who is considered to be moving.

Note that if S' is instead moving in the negative x direction, then the Lorentz transformation is $E = \gamma(E' - vp')$. This quickly leads to a factor of $\sqrt{(1-v)/(1+v)}$ in the energy, and hence also in the frequency. This is the correct Doppler shift for a source moving away from you.

We have found that the frequency changes by a factor of $\sqrt{(1 \pm v)/(1 \mp v)}$ when going form S' to S. Of course, if you actually want to measure the frequency of a photon by observing it, you need to be in front of it (running whichever way you want), so that it will eventually hit you. If the photon is instead, say, on your right and also traveling rightward, then you will never be able to see it, even though its frequency is still perfectly well defined in your frame, taking on the value of $f = f'\sqrt{(1+v)/(1-v)}$.

Linearity

The transformations in Eq. (3.20), which we derived for a single particle, are *linear*. That is, they involve only the first power of E and p. There are no terms like p^2 or E^5, etc. This linearity implies that the equations also hold if E and p represent the total energy and momentum of a *collection* of particles. That is,

$$\begin{aligned} E_{\text{total}} &= \gamma\left(E'_{\text{total}} + vp'_{\text{total}}\right), \\ p_{\text{total}} &= \gamma\left(p'_{\text{total}} + vE'_{\text{total}}/c^2\right). \end{aligned} \tag{3.25}$$

In fact, any (corresponding) linear combinations of the energies and momenta are valid here, in place of the total E and p. For example, we can use the combinations $(E_1 + 3E_2 - 7E_5)$ and $(p_1 + 3p_2 - 7p_5)$ in Eq. (3.25), where the subscripts indicate the particle. You can verify this by simply writing down the equations in Eq. (3.20) for each particle, and then adding up the equations with the appropriate coefficients to generate the desired linear combination of the E's and p's. This consequence of linearity is a very important and useful result, as will become clear below.

Because Eq. (3.20) holds for any (corresponding) linear combinations of E's and p's, the E's and p's in Eq. (3.22) can represent any (corresponding) linear combinations of the E's and p's of the various particles, because the validity of Eq. (3.22) followed from the Lorentz-transformation nature of Eq. (3.20). For example, as we did in going from Eq. (3.20) to Eq. (3.25), we can tack on "total" subscripts in Eq. (3.22):

$$E^2_{\text{total}} - p^2_{\text{total}}c^2 = E'^2_{\text{total}} - p'^2_{\text{total}}c^2. \tag{3.26}$$

For a collection of particles, the invariant quantity $E^2_{\text{total}} - p^2_{\text{total}}c^2$ equals the square of the total energy in the center-of-mass (CM) frame (which reduces to m^2c^4 for one particle), because $p_{\text{total}} = 0$ in the CM frame, by definition. (CM really means "center of momentum" in relativity.)

On page 126 we stated that if E and p are conserved in one reference frame during a collision, then they are conserved in any other frame. (A conservation law shouldn't depend on what frame you're in.) This can be shown as follows. The total change in energy, ΔE, during a collision (which we're trying to show is zero) is a linear combination of the initial and final E's (namely, the sum of the final E's minus the sum of the initial E's). Likewise for Δp. Therefore, since Eq. (3.20) is a linear equation in the E's and p's, it also holds for the ΔE's and Δp's. That is,

$$\Delta E = \gamma(\Delta E' + v\,\Delta p') \qquad \text{and} \qquad \Delta p = \gamma(\Delta p' + (v/c^2)\Delta E'). \tag{3.27}$$

These equations tell us that if the total $\Delta E'$ and $\Delta p'$ in S' are zero, then the total ΔE and Δp in S are also zero, as we wanted to show.

Eq. (3.27) makes it clear that if you accept the fact that $p = \gamma mv$ is conserved in all frames, then you must also accept the fact that $E = \gamma mc^2$ is conserved in all frames (and vice versa). This is true because the second of Eqs. (3.27) says that if Δp and $\Delta p'$ are both zero, then $\Delta E'$ must also be zero. The E and p expressions in Eq. (3.1) have no choice but to go hand in hand.

3.3 Collisions and decays

The strategy for studying relativistic collisions is the same as for nonrelativistic ones (reviewed in Appendix E). You just have to write down all the conservation of energy and momentum equations, and then solve for whatever variables you want to solve for. The conservation principles are the same as they've always been. The only difference is that now the energy and momentum take the new forms in Eq. (3.1). Let's dive right in with an example.

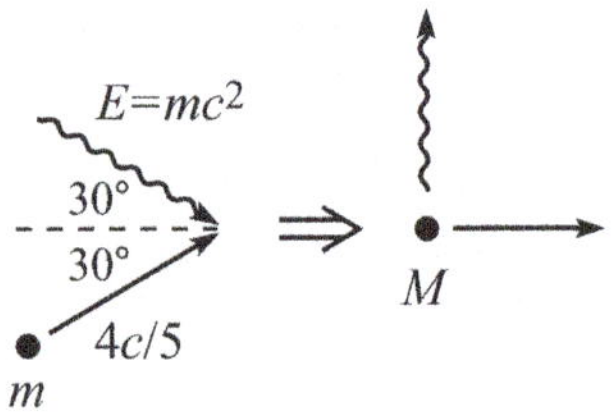

Figure 3.7

Example: A mass m whose speed is $4c/5$, and a photon whose energy is chosen to be mc^2, move at 30° angles with respect to the horizontal, as shown in Fig. 3.7. They collide and produce a photon moving upward (with unknown energy) and a mass M moving rightward. What is M?

Solution: Since $\gamma_{4/5} = 5/3$, the energy and momentum of the mass m are $E_m = \gamma mc^2 = (5/3)mc^2$ and $p_m = \gamma mv = (5/3)m(4c/5) = (4/3)mc$. From Eq. (3.13) the momentum of the photon is $E/c = mc$. Let's drop the c's and put them back in at the end. We then have $E_m = 5m/3$, $p_m = 4m/3$, and $E_{\text{ph}} = p_{\text{ph}} = m$. We can write down three conservation statements:

- Since the final photon has no p_x, all of the final p_x comes from M. So conservation of p_x tells us that the momentum of M is

$$p_M = \frac{4m}{3}\cos 30^\circ + m\cos 30^\circ = \frac{7m}{3}\cdot\frac{\sqrt{3}}{2} = \frac{7m}{2\sqrt{3}}. \tag{3.28}$$

- Since the final M has no p_y, all of the final p_y comes from the photon. So conservation of p_y tells us that the momentum of the final photon is

$$\frac{4m}{3}\sin 30^\circ - m\sin 30^\circ = \frac{m}{3}\cdot\frac{1}{2} = \frac{m}{6}. \tag{3.29}$$

  The final photon's energy is therefore also $m/6$.

- Conservation of energy then gives the energy of M as

$$\frac{5m}{3} + m = E_M + \frac{m}{6} \implies E_M = \frac{5m}{2}. \tag{3.30}$$

Finally, the Very Important Relation in Eq. (3.12) applied to the mass M gives

$$M^2 = E_M^2 - p_M^2 = \left(\frac{5m}{2}\right)^2 - \left(\frac{7m}{2\sqrt{3}}\right)^2 = \left(\frac{25}{4} - \frac{49}{12}\right)m^2 = \frac{13m^2}{6}. \tag{3.31}$$

So $M = \sqrt{13/6}\,m \approx (1.47)m$. The units are correct, so there is no need to put any c's back in.

If you want to also find the velocity of M, the easiest way is to use Eq. (3.9), which gives (without the c's)

$$\frac{p_M}{E_M} = v \implies \frac{(7/2\sqrt{3})m}{5m/2} = v \implies v = \frac{7c}{5\sqrt{3}}, \tag{3.32}$$

where we have added in the c to make the units correct. The same result also follows from $E_M = \gamma M c^2$, because if we plug in the above values of E_M and M, we obtain $\gamma = (5/2)/\sqrt{13/6} = \sqrt{75/26}$, which yields the above v.

In writing down the conservation of energy and momentum equations, it often proves useful to put E and $\mathbf{p}$ together into one four-component vector,

$$\boxed{P \equiv (E, \mathbf{p}) = (E, p_x, p_y, p_z)} \tag{3.33}$$

This is called the *energy-momentum 4-vector*, or the *4-momentum*, for short. We're ignoring the c's, but if you want to include them, the first component is E/c, although some people instead multiply the $\mathbf{p}$ by c; either convention is fine. Our notation will be to use an uppercase P to denote a 4-momentum and a lowercase $\mathbf{p}$ or p to denote a spatial momentum. The components of a 4-momentum are usually indexed from 0 to 3, so that $P_0 \equiv E$, and $(P_1, P_2, P_3) \equiv \mathbf{p}$. For one particle, we have

$$P = (\gamma m, \gamma m v_x, \gamma m v_y, \gamma m v_z). \tag{3.34}$$

The 4-momentum of a collection of particles consists of the total E and total $\mathbf{p}$ of all the particles. There are strong theoretical reasons for considering the 4-momentum, as we'll see in Chapter 4, but for now we'll just view it as a matter of convenience. If nothing else, it helps with the bookkeeping. Conservation of energy and momentum in a collision reduce to the concise statement,

$$P_{\text{before}} = P_{\text{after}}, \tag{3.35}$$

where these are the total 4-momenta of all the particles. Although you never *have* to use 4-momenta when solving collision problems, it often makes things much more organized.

Given two 4-momenta, $A \equiv (A_0, A_1, A_2, A_3)$ and $B \equiv (B_0, B_1, B_2, B_3)$, we define the *inner product* between A and B to be

$$\boxed{A \cdot B \equiv A_0 B_0 - A_1 B_1 - A_2 B_2 - A_3 B_3} \tag{3.36}$$

The inner product of $P \equiv (E, \mathbf{p})$ with itself is then

$$P \cdot P = E^2 - p_x^2 - p_y^2 - p_z^2 = E^2 - p^2. \tag{3.37}$$

The Very Important Relation in Eq. (3.12), namely $E^2 - p^2 = m^2$, which is valid for one particle, may then be written concisely as

$$P \cdot P = m^2 \quad \text{or} \quad \boxed{P^2 = m^2} \tag{3.38}$$

where $P^2 \equiv P \cdot P$. In other words, the square of a particle's 4-momentum equals the square of its mass (or m^2c^4, with the c's). This relation is very useful in collision problems. Note that it is frame-independent, as we saw in Eq. (3.22).

The above inner product is different from the one we're used to for standard vectors in 3-D space. For 4-momenta, we have one positive sign and three negative signs, in contrast with the three positive signs for standard vectors. But we are free to define it however we wish, and we did indeed pick a good definition, because it turns out that our inner product is invariant under Lorentz transformations, just as the usual 3-D inner product is invariant under rotations. For the inner product of a general 4-momentum (which may be any linear combination of 4-momenta of various particles) with itself, this invariance is (in view of Eq. (3.37)) the statement in Eq. (3.22). As we have noted, this relation holds for a collection of particles because the Lorentz transformations in Eq. (3.20) are linear. For the inner product of two different 4-momenta, we'll prove the invariance in Section 4.3.

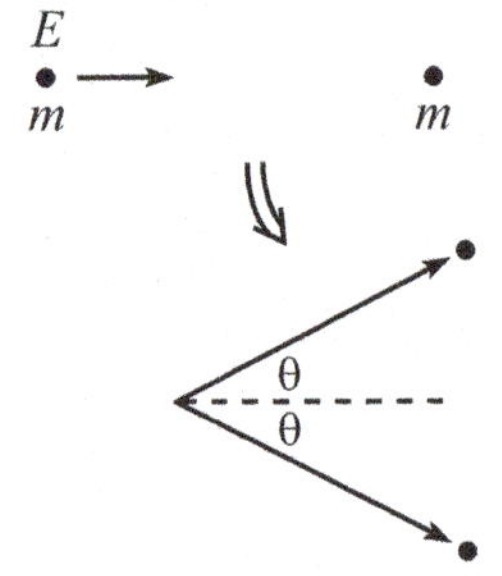

Figure 3.8

Example (Relativistic billiards): A particle with mass m and energy E approaches an identical particle at rest. They collide elastically (which means that the masses don't change) in such a way that they both scatter at an angle θ relative to the incident direction; see Fig. 3.8. What is θ in terms of E and m? What is θ in the nonrelativistic and relativistic limits?

Solution: The first thing we should do is write down the 4-momenta. The two 4-momenta before the collision are

$$P_1 = (E, p, 0, 0), \qquad P_2 = (m, 0, 0, 0), \tag{3.39}$$

where $p = \sqrt{E^2 - m^2}$ from Eq. (3.12) (dropping the c's). The two 4-momenta after the collision are (primes denote "after")

$$P_1' = (E', p'\cos\theta, p'\sin\theta, 0), \qquad P_2' = (E', p'\cos\theta, -p'\sin\theta, 0), \tag{3.40}$$

where $p' = \sqrt{E'^2 - m^2}$. Conservation of energy gives $E + m = 2E' \Longrightarrow E' = (E+m)/2$, and conservation of p_x gives $p + 0 = 2p'\cos\theta \Longrightarrow p'\cos\theta = p/2$. Therefore, the two 4-momenta after the collision are

$$P_{1,2}' = \left(\frac{E+m}{2}, \frac{p}{2}, \pm\frac{p}{2}\tan\theta, 0\right). \tag{3.41}$$

From Eq. (3.38), the squares of these 4-momenta must be m^2. Equivalently, Eq. (3.12) holds. The p^2 in that equation is the square of the magnitude of the momentum, which is $p_x^2 + p_y^2$. We therefore obtain (using $1 + \tan^2\theta = 1/\cos^2\theta$, along with the $p^2 = E^2 - m^2$ relation for the initial moving mass),

$$\begin{aligned}
& m^2 = P_{1,2}'^2 \\
\Longrightarrow\quad & m^2 = E'^2 - (p_x'^2 + p_y'^2) \\
\Longrightarrow\quad & m^2 = \left(\frac{E+m}{2}\right)^2 - \left(\frac{p}{2}\right)^2 (1 + \tan^2\theta) \\
\Longrightarrow\quad & 4m^2 = (E+m)^2 - \frac{(E^2 - m^2)}{\cos^2\theta} \\
\Longrightarrow\quad & \cos^2\theta = \frac{E^2 - m^2}{E^2 + 2Em - 3m^2} = \frac{E+m}{E+3m} \\
\Longrightarrow\quad & \cos\theta = \sqrt{\frac{E + mc^2}{E + 3mc^2}},
\end{aligned} \tag{3.42}$$

where we have put the c's back in to make the units right.

The nonrelativistic limit is $E \approx mc^2$. (It is *not* $E \approx 0$, because even if the incoming mass is moving very slowly, it still has energy mc^2.) This yields $\cos\theta \approx 1/\sqrt{2}$ in Eq. (3.42).

So $\theta \approx 45°$, and the particles scatter with a 90° angle between them. This result is very familiar to pool players. Nonrelativistically, assuming the two masses are equal and the collision is elastic, it can be shown that the angle between the final velocities is 90° even if the two resulting angles aren't the same.

The relativistic limit is $E \gg mc^2$, because this implies $\gamma mc^2 \gg mc^2 \implies \gamma \gg 1$, which means that the initial speed is close to c. This yields $\cos\theta \approx 1$ in Eq. (3.42). So $\theta \approx 0$, and both particles scatter almost directly forward. You can convince yourself that θ should be small by looking at the collision in the CM frame and then shifting back to the lab frame. The transverse speeds decrease when shifting to the lab frame (due to the transverse velocity-addition formula in Eq. (2.22)), which makes the angle smaller than the nonrelativistic angle of 45°.

Note that we never wrote down any v's when working through this example. That is, we never used the relations $E = \gamma mc^2$ and $p = \gamma mv$. For the given task, there was no need to find the velocities; doing so would have essentially involved going in circles. Using $E = \gamma mc^2$ and $p = \gamma mv$ certainly provides a valid way of solving the problem, but in many setups like this one, where velocities aren't given or explicitly asked for, it leads to a very messy solution. A far cleaner method is to use the $E^2 = p^2 + m^2$ relation (along with conservation of E and p, of course).

If γmv yields frustration,
And the similar E, irritation,
Just ditch all the v's,
And use (won't you *please*)
The Very Important Relation!

Let's now look at a decay. Decays are basically the same as collisions. All you have to do is apply conservation of E and p (and probably use $E^2 = p^2 + m^2$).

Example (Decay at an angle): A particle with mass M and energy E decays into two identical particles. In the lab frame, one of them is emitted at a 90° angle, as shown in Fig. 3.9. What are the energies of the created particles? We'll give two solutions. The second one shows how 4-momenta can be used in a clever and time-saving way.

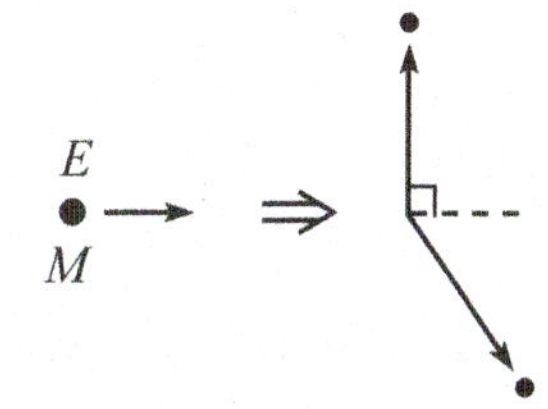

Figure 3.9

First solution: The 4-momentum before the decay is

$$P = (E, p, 0, 0), \tag{3.43}$$

where $p = \sqrt{E^2 - M^2}$. Let m be the common mass of the two created particles, and let the lower particle make an angle θ with the x axis. Then the 4-momenta after the decay take the forms,

$$P_1 = (E_1, 0, p_1, 0), \qquad P_2 = (E_2, p_2\cos\theta, -p_2\sin\theta, 0). \tag{3.44}$$

Conservation of p_x immediately gives $p_2\cos\theta = p$, which then implies that $p_2\sin\theta = p\tan\theta$. The initial p_y is zero, so conservation of p_y says that the final p_y's are equal and opposite. The 4-momenta after the decay are therefore

$$P_1 = (E_1, 0, p\tan\theta, 0), \qquad P_2 = (E_2, p, -p\tan\theta, 0). \tag{3.45}$$

Conservation of energy gives $E = E_1 + E_2$. Using Eq. (3.12) to write E_1 and E_2 in terms of the momenta and masses, we can rewrite the relation $E = E_1 + E_2$ as

$$E = \sqrt{p^2\tan^2\theta + m^2} + \sqrt{p^2(1 + \tan^2\theta) + m^2}. \tag{3.46}$$

Putting the first radical on the lefthand side, squaring, and solving for that radical (which is E_1) gives

$$E_1 = \frac{E^2 - p^2}{2E} = \frac{M^2}{2E}. \tag{3.47}$$

In a similar manner, we find that E_2 equals (using $p^2 = E^2 - M^2$)

$$E_2 = \frac{E^2 + p^2}{2E} = \frac{2E^2 - M^2}{2E}. \tag{3.48}$$

These add up to E, as they should. An alternative way of solving for the energies is to note that if we square each of the expressions for E_1 and E_2 that appear in Eq. (3.46), we find that $E_2^2 - E_1^2 = p^2 \Longrightarrow E_2^2 - E_1^2 = E^2 - M^2$. Dividing this equation by the conservation-of-energy statement, $E_2 + E_1 = E$, gives $E_2 - E_1 = (E^2 - M^2)/E$. Taking the sum or difference of this equation with $E_2 + E_1 = E$ reproduces the above results for E_1 and E_2.

Remark: If you try to determine the values of m and θ, you will find that you are not able to do so. We haven't been given enough information to be able to uniquely determine m and θ. (As long as m is below a certain value, the created particles will be moving fast enough so that we can angle the top one backward to make its x speed in the CM frame (M's frame) be $-v$, where v is the speed of M in the lab frame. The top particle's x speed in the lab frame will then have the given value of zero.) If however, one of the parameters m or θ is given, then we can solve for the other one. The simplest way to do this is to use $p^2 \tan^2\theta + m^2 = E_1^2 \Longrightarrow (E^2 - M^2)\tan^2\theta + m^2 = (M^2/2E)^2$. The maximum possible value of θ occurs when $m = 0$, and the maximum possible value of m occurs when $\theta = 0$.

Given that knowledge of one of m and θ is needed to determine the other, you might find it surprising that even without knowledge of *either* of them, we were still able to determine the final energies. No matter what the (dependent) values of m and θ are, E_1 and E_2 will always take on the definite values we found above. In the end, the reason for this boils down to the fact that the energies are not only related by $E_2 + E_1 = E$, but also by $E_2^2 - E_1^2 = E^2 - M^2$, which is a consequence of the Very Important Relation in Eq. (3.12). ♣

Second solution: With the 4-momenta defined as in Eqs. (3.43) and (3.44), conservation of energy and momentum can be combined into the statement, $P = P_1 + P_2$. Therefore,

$$\begin{aligned} P - P_1 &= P_2 \\ \Longrightarrow \quad (P - P_1)\cdot(P - P_1) &= P_2 \cdot P_2 \\ \Longrightarrow \quad P^2 - 2P\cdot P_1 + P_1^2 &= P_2^2 \\ \Longrightarrow \quad M^2 - 2EE_1 + m^2 &= m^2 \\ \Longrightarrow \quad E_1 &= \frac{M^2}{2E}. \end{aligned} \tag{3.49}$$

And then $E_2 = E - E_1 = (2E^2 - M^2)/2E$. The various (mass)^2 terms in the fourth line above come from Eq. (3.38). And the $P \cdot P_1 = EE_1$ step comes from applying the inner product definition in Eq. (3.36) to the expressions for P and P_1 in Eqs. (3.43) and (3.44); the plethora of zeros leads to the simple EE_1 result.

This solution should convince you that 4-momenta can save you a lot of work. What happened here was that the expression for P_2 was fairly messy, but we arranged things so that it appeared only in the form of P_2^2, which is simply m^2. 4-momenta provide a remarkably organized method for sweeping unwanted garbage under the rug!

3.4 Particle-physics units

A branch of physics that uses relativity as one of its main ingredients is Elementary-Particle Physics, which is the study of the building blocks of matter (electrons, quarks, neutrinos, etc.). However, it is unfortunately the case that most of the elementary particles we want to study don't exist naturally in the world. We therefore have to create them in particle accelerators by colliding other particles together at very high energies. The high speeds involved require the use of relativistic dynamics. Newtonian physics is essentially useless.

What is a typical size of a rest energy, mc^2, of an elementary particle? The mass of a proton (which isn't really elementary; it's made up of quarks, but never mind) is $1.67 \cdot 10^{-27}$ kg. So its rest energy is

$$E_{\text{p}} = m_{\text{p}}c^2 = (1.67 \cdot 10^{-27}\ \text{kg})(3 \cdot 10^8\ \text{m/s})^2 = 1.5 \cdot 10^{-10}\ \text{J}, \qquad (3.50)$$

where J stands for the "joule" unit, which equals $1\ \text{kg}\,\text{m}^2/\text{s}^2$. This $1.5 \cdot 10^{-10}$ number is very small, so a joule probably isn't the best unit to work with. We would get very tired of writing the negative exponents over and over. We could perhaps work with "nanojoules" (10^{-9} joule), but particle physicists like to work instead with the "eV," the *electron-volt*. This is the change in energy of an electron when it passes through a potential difference of one volt. The electron charge is (negative) $e = 1.602 \cdot 10^{-19}$ C (the C is for the "coulomb" unit of charge), and a volt is defined as $1\ \text{V} = 1\ \text{J/C}$. The conversion from eV to joules is therefore[3]

$$1\ \text{eV} = (1.602 \cdot 10^{-19}\ \text{C})(1\ \text{J/C}) = 1.602 \cdot 10^{-19}\ \text{J}. \qquad (3.51)$$

To a couple more significant figures, the proton rest energy in Eq. (3.50) is actually $E_{\text{p}} = 1.503 \cdot 10^{-10}$ J. Dividing this relation by Eq. (3.51), we see that in terms of eV, the rest energy of a proton is $E_{\text{p}} = 938 \cdot 10^6$ eV. We now have the opposite problem of a large positive exponent. But this is easily remedied by the prefix "M," which stands for "mega" or "million." So we finally have a proton rest energy of

$$E_{\text{p}} = 938\ \text{MeV}. \qquad (3.52)$$

An electron has a mass of $9.11 \cdot 10^{-31}$ kg (about 1/2000 of the proton mass), which corresponds to a rest energy of $E_{\text{e}} = 0.511$ MeV. The rest energies of various particles are listed in the table below. The ones preceded by a "$\approx$" are the averages of differently charged particles (all associated with a given name), whose energies differ by a few MeV. These (and the many other) elementary particles have specific properties (spin, charge, etc.), but for the present purposes they need only be thought of as point objects with a definite mass. For higher energies, the prefixes "G" (for "giga," meaning 10^9) and "T" (for "tera," meaning 10^{12}) are used.

[3]This is getting a little picky, but "eV" should technically be written as "eV," because when people write "eV," they actually mean that two things are being multiplied together (in contrast with, for example, the "He" symbol for Helium). The first of these things is the electron charge, which is usually denoted by e.

particle	rest energy (MeV)
electron (e)	0.511
muon (μ)	105.7
tau (τ)	1777
proton (p)	938.3
neutron (n)	939.6
lambda (Λ)	1115.7
sigma (Σ)	≈ 1193
delta (Δ)	≈ 1232
pion (π)	≈ 137
kaon (K)	≈ 496

We now come to a slight abuse of language. When particle physicists talk about masses, they say things like, "The mass of a proton is 938 MeV." This technically doesn't make any sense, because the units are wrong; a mass can't equal an energy. But what they mean is that if you take this energy and divide it by c^2, then you get the mass. It would be a pain to keep saying, "The mass is such-and-such an energy, divided by c^2." For a quick conversion back to kilograms, you can show that

$$1 \text{ MeV}/c^2 = 1.783 \cdot 10^{-30} \text{ kg}. \tag{3.53}$$

So the mass of a proton in kilograms is

$$938 \text{ MeV}/c^2 = (938)(1.783 \cdot 10^{-30} \text{ kg}) = 1.67 \cdot 10^{-27} \text{ kg}, \tag{3.54}$$

as we noted above.

3.5 Force

3.5.1 Force in one dimension

In nonrelativistic physics, Newton's second law is $F = dp/dt$, where $p = mv$. If the mass m is constant, this law reduces to $F = d(mv)/dt \implies F = m\,dv/dt \implies F = ma$. We'll carry the original form of the law over to relativity and continue to write (dealing with just one-dimensional motion for now)

$$\boxed{F = \frac{dp}{dt}} \tag{3.55}$$

However, in relativity we have $p = \gamma mv$, and γ can change with time. This complicates things, and it turns out that F is *not* equal to ma when m is constant. But if dp/dt and ma are different in relativity, why is F equal to dp/dt instead of ma? Why isn't ma the quantity that gets carried over from nonrelativistic physics, instead of dp/dt? Perhaps the best reason arises from the 4-vector formalism in Chapter 4. Another reason is that the F in Eq. (3.55) leads to a familiar work-energy theorem, as we'll see below in Eq. (3.61).

To determine the form that the F in Eq. (3.55) takes in terms of the acceleration $a \equiv dv/dt$, let's first calculate $d\gamma/dt$. All quantities here (a, v, γ, t) are measured with respect to a given inertial frame. (Note that the acceleration a is still defined simply as

dv/dt, just as it is in nonrelativistic physics.) Using the chain rule and dropping the c's, we have

$$\frac{d\gamma}{dt} = \frac{d}{dt}\left((1-v^2)^{-1/2}\right) = -\frac{1}{2}\cdot\frac{-2v}{(1-v^2)^{3/2}}\cdot\frac{dv}{dt} = \gamma^3 va. \tag{3.56}$$

Assuming that m is constant, we therefore have (using the product rule, and with dots denoting time derivatives)

$$F = \frac{d(\gamma m v)}{dt} = m(\dot{\gamma}v + \gamma\dot{v}) = ma\gamma(\gamma^2 v^2 + 1), \tag{3.57}$$

And since $\gamma^2 v^2 + 1 = \gamma^2$ (as you can check), we arrive at

$$\boxed{F = \gamma^3 ma} \tag{3.58}$$

The units are correct here, so there is no need to bring any c's back in. This result doesn't look as nice as $F = ma$, but that's the way it goes. However, F correctly reduces to ma in the limit of small speeds (where $\gamma \approx 1$), as it must.

They *said*, "F is ma, bar none."
What they *meant* wasn't quite as much fun.
It's dp by dt,
Which just happens to be
Good ol' "ma" when γ is 1.

Consider now the quantity dE/dx, where E is the energy, $E = \gamma m$ (dropping the c's). In a manner similar to Eq. (3.56), we obtain

$$\frac{dE}{dx} = \frac{d(\gamma m)}{dx} = m\frac{d}{dx}\left((1-v^2)^{-1/2}\right) = m\frac{v}{(1-v^2)^{3/2}}\cdot\frac{dv}{dx} = \gamma^3 mv\frac{dv}{dx}. \tag{3.59}$$

But $v(dv/dx) = (dx/dt)(dv/dx) = dv/dt = a$. (Yes, it's legal to cancel the dx's here.) Therefore, $dE/dx = \gamma^3 ma$. Combining this with Eq. (3.58) gives

$$\boxed{F = \frac{dE}{dx}} \tag{3.60}$$

Again the units are correct, so there is no need to bring any c's back in. Note that Eqs. (3.55) and (3.60) take exactly the same form as in nonrelativistic physics. The only new thing in relativity is that the expressions for p and E are modified.

The result in Eq. (3.60) suggests another way to motivate the $E = \gamma m$ expression for relativistic energy, assuming that we have accepted the $p = \gamma mv$ expression for relativistic momentum. Define F as we have done, through Eq. (3.55), which leads to Eq. (3.58). Then integrate Eq. (3.58) from x_1 to x_2 to obtain (using $a = v\,dv/dx$)

$$\begin{aligned}\int_{x_1}^{x_2} F\,dx &= \int_{x_1}^{x_2}(\gamma^3 ma)\,dx = \int_{x_1}^{x_2}\left(\gamma^3 mv\frac{dv}{dx}\right)dx = m\int_{v_1}^{v_2}\gamma^3 v\,dv \\ &= m\int_{\gamma_1}^{\gamma_2} d\gamma = \gamma m\Big|_{\gamma_1}^{\gamma_2} \equiv E_2 - E_1 = \Delta E,\end{aligned} \tag{3.61}$$

where we have used the $d\gamma = \gamma^3 v\,dv$ relation (or $d\gamma = \gamma^3 v\,dv/c^2$, with the c's) implicit in either Eq. (3.56) or Eq. (3.59). We see that if we define the energy as $E = \gamma m$, then the work-energy theorem, $\int F\,dx = \Delta E$, holds in relativity just as it does in Newtonian physics. (In 1-D, work is defined as force times distance, or more generally as the integral of force with respect to displacement.) The only difference is that E is now γm instead of $mv^2/2$.[4]

Having shown above in Eq. (3.58) that $F = \gamma^3 ma$, let's now make a list of all the reasons why it is impossible to accelerate a massive object up to speed c. There are (at least) four reasons:

- ENERGY: Since $\gamma = \infty$ when $v = c$, and since $E = \gamma mc^2$, we see that we would need an infinite amount of energy to accelerate a (massive) object up to speed c. Since energy is conserved, this would involve taking an infinite amount of energy away from the rest of the universe, which isn't possible.

- MOMENTUM: Likewise, since $\gamma = \infty$ when $v = c$, and since $p = \gamma mv$, we see that we would need an infinite amount of momentum to accelerate a (massive) object up to speed c. Since momentum is conserved, this would involve taking an infinite amount of momentum away from the rest of the universe, which isn't possible.

- VELOCITY ADDITION: No matter what speed we give to an object relative to the frame it was just in, the velocity-addition formula in Eq. (1.28) tells us that the resulting speed will always be less than c. The only way to obtain a speed of c is for one of the two original speeds to already be equal to c.

- FORCE: Since $a = F/\gamma^3 m$, a given force produces a smaller and smaller acceleration as v approaches c, because γ goes to infinity. The $dv = a\,dt$ increments in speed that the object picks up therefore get smaller and smaller, and the speed can never reach c. This $a \to 0$ reasoning by itself actually isn't completely rigorous, because it isn't enough for a to go to zero as $v \to c$. It needs to go to zero sufficiently fast. The $1/\gamma^3$ factor does in fact go to zero fast enough, and $1/\gamma^2$ would also, but $1/\gamma$ would not. To see where the difference between these cases arises, write a as dv/dt and then separate variables and integrate the $dv/dt = F/\gamma^3 m$ relation to obtain $\int_0^c \gamma^3\, dv = (F/m)t$. You can show that $\int_0^c \gamma^3\, dv$ diverges (whereas $\int_0^c \gamma\, dv$ does not), which implies that it takes an infinite amount of time (which means that it is impossible) to accelerate an object up to speed c.

3.5.2 Force in two dimensions

In two dimensions, the concept of force becomes a little strange. In particular, as we'll see, the acceleration of an object need not point in the same direction as the force. We start with

$$\mathbf{F} = \frac{d\mathbf{p}}{dt}. \tag{3.62}$$

This is a vector equation. Without loss of generality, we'll deal with only two spatial dimensions. We can do this because the momentum vector $\mathbf{p}$ and the force vector $\mathbf{F}$ define a plane, which we can call the x-y plane. We'll define the x axis to point along the $\mathbf{p}$ vector at a given instant. Let us apply a force in an arbitrary direction in the plane. So the force at the given instant takes the general form of $\mathbf{F} = (F_x, F_y)$. Since the force has a y component, the particle will pick up a v_y velocity (which is initially zero, due to the definition of our x axis). A short time later, the particle's momentum is (dropping

[4] Actually, this reasoning suggests only that E is given by γm up to an additive constant. For all we know, E might take the form, $E = \gamma m - m$, which would make the energy of a motionless particle equal to zero. An argument along the lines of the one in Section 3.1.2 is required to show that the additive constant is zero.

the c's)

$$\mathbf{p} = \gamma m \mathbf{v} = \frac{m(v_x, v_y)}{\sqrt{1-v^2}} = \frac{m(v_x, v_y)}{\sqrt{1-v_x^2-v_y^2}}. \tag{3.63}$$

Eq. (3.62) tells us that $\mathbf{F}$ is the derivative of $\mathbf{p}$. To evaluate this derivative, we'll need to use the product rule and the chain rule. Note that $d(v^2)/dt = d(v_x^2 + v_y^2)/dt = 2(v_x\dot{v}_x + v_y\dot{v}_y)$. Evaluating $d\mathbf{p}/dt$ and using the fact that v_y is initially zero, we obtain

$$\begin{aligned}
\mathbf{F} &= \left.\frac{d\mathbf{p}}{dt}\right|_{v_y=0} \\
&= m\frac{d}{dt}\left.\left(v_x\cdot\frac{1}{\sqrt{1-v^2}},\ v_y\cdot\frac{1}{\sqrt{1-v^2}}\right)\right|_{v_y=0} \\
&= m\left.\left(\frac{\dot{v}_x}{\sqrt{1-v^2}} + \frac{v_x(v_x\dot{v}_x+v_y\dot{v}_y)}{(\sqrt{1-v^2})^3},\ \frac{\dot{v}_y}{\sqrt{1-v^2}} + \frac{v_y(v_x\dot{v}_x+v_y\dot{v}_y)}{(\sqrt{1-v^2})^3}\right)\right|_{v_y=0} \\
&= m\left(\frac{\dot{v}_x}{\sqrt{1-v^2}}\left(1+\frac{v^2}{1-v^2}\right),\ \frac{\dot{v}_y}{\sqrt{1-v^2}}\right) \\
&= m\left(\frac{\dot{v}_x}{(\sqrt{1-v^2})^3},\ \frac{\dot{v}_y}{\sqrt{1-v^2}}\right),
\end{aligned} \tag{3.64}$$

where the numerator in the $v^2/(1-v^2)$ term comes from the fact $v_x = v$ at the start. Our result can be rewritten as

$$\boxed{\mathbf{F} = m(\gamma^3 a_x, \gamma a_y)} \tag{3.65}$$

This is *not* proportional to (a_x, a_y). The first component agrees with Eq. (3.58), but the second component has only one factor of γ. The difference comes from the fact that γ has a first-order change if v_x changes, but not if v_y changes, assuming that v_y is initially zero. The particle therefore responds differently to forces in the x and y directions. It is easier to accelerate something in the transverse direction.

3.5.3 Transformation of forces

If a given force acts on a particle, how are the components of the force in the particle's frame S' related to the components of the force in another frame S? To be precise, S' is the instantaneous inertial frame of the particle. Once the force is applied, the particle will accelerate and will therefore no longer be at rest in S'. But for a very small elapsed time, the particle will still essentially be in S'.

Let the relative motion be along the x and x' axes, as shown in Fig. 3.10. As in the above discussion of 2-D forces, we can assume that the force acts in the x-y plane. In frame S, Eq. (3.65) says

$$(F_x, F_y) = m(\gamma^3 a_x, \gamma a_y). \tag{3.66}$$

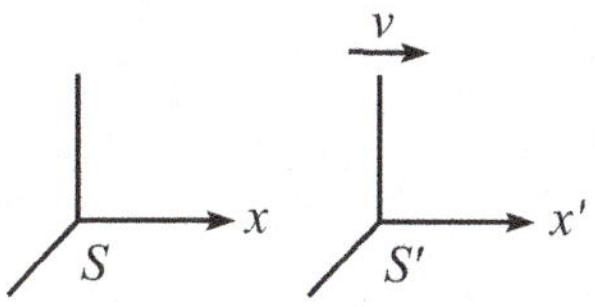

Figure 3.10

And in frame S', the γ factor for the particle equals 1, so Eq. (3.65) reduces to the usual expression,

$$(F'_x, F'_y) = m(a'_x, a'_y). \tag{3.67}$$

Let's now relate these two forces. We'll do this by writing the primed accelerations on the righthand side of Eq. (3.67) in terms of the unprimed accelerations.

First, we claim that $a'_y = \gamma^2 a_y$. This is true because transverse distances are the same in the two frames, but times are shorter in S' by a factor γ. That is, $dt' = dt/\gamma$. We have indeed put the γ in the right place, because the particle is essentially at rest in S', so the usual time-dilation result holds. (See the first remark below for more discussion on this.) Therefore,

$$a'_y \equiv \frac{d^2y'}{dt'^2} = \frac{d^2y}{(dt/\gamma)^2} = \gamma^2\frac{d^2y}{dt^2} \equiv \gamma^2 a_y. \tag{3.68}$$

Second, we claim that $a'_x = \gamma^3 a_x$. In short, this is true because time dilation brings in two factors of γ (as in the a_y case), and length contraction brings in one. In more detail: Let the particle move from one point to another in frame S', as it accelerates from rest in S'. Mark these two points, which are essentially a distance $a'_x(dt')^2/2$ apart, in S'. As S' flies past S, the distance between the two marks is length contracted by a factor γ, as viewed by S. This contracted distance (which is the excess distance the particle travels over what it would have traveled if there were no acceleration) is what S calls $a_x(dt)^2/2$, from standard kinematics in frame S. (Relativity doesn't change the fact that due to the definition of a_x, this kinematics relation is still true for an infinitesimal time dt. What relativity does is affect how a_x is related to F_x.) Therefore, using $dt' = dt/\gamma$, we obtain

$$\frac{1}{2}a_x\,dt^2 = \frac{1}{\gamma}\left(\frac{1}{2}a'_x\,dt'^2\right) \quad\Longrightarrow\quad a'_x = \gamma a_x\left(\frac{dt}{dt'}\right)^2 = \gamma^3 a_x. \tag{3.69}$$

Eq. (3.67) may now be written as

$$(F'_x, F'_y) = m(\gamma^3 a_x, \gamma^2 a_y). \tag{3.70}$$

Comparing Eqs. (3.66) and (3.70) then gives

$$\boxed{F_x = F'_x} \quad\text{and}\quad \boxed{F_y = \frac{F'_y}{\gamma}} \qquad (S' = \text{particle frame}) \tag{3.71}$$

We see that the longitudinal force is the same in the two frames, but the transverse force is larger by a factor of γ in the particle's frame; $F'_y = \gamma F_y$. We will give another derivation of these results in Section 4.5.

Remarks:

1. What if someone comes along and switches the labels of the primed and unprimed frames in Eq. (3.71) and concludes that the transverse force is *smaller* in the particle's frame? That certainly can't be correct, given that Eq. (3.71) is true. But where is the error? The error lies in the fact that we (correctly) used $dt' = dt/\gamma$ above, because this is the relevant expression concerning two events along the particle's worldline. We are interested in two such events, because we want to see how the particle moves. The inverted expression, $dt = dt'/\gamma$, deals with two events located at the same position in S, and therefore has nothing to do with the situation at hand. The particle remains (essentially) at rest in S', but not in any other frame. (Similar reasoning holds for the relation between dx and dx'; the two marks we made above are at rest in S'.) Because we are dealing with a given particle, there is indeed one frame that is special among all possible frames (at least as far as the particle is concerned), namely the particle's instantaneous inertial frame.

2. If you want to compare forces in two frames S and S'', neither of which is the particle's rest frame S', you can do this by using Eq. (3.71) twice and relating each of the forces to the rest-frame forces. It quickly follows that $F''_x = F_x$ and $\gamma'' F''_y = \gamma F_y$, where the γ's are measured relative to the rest frame S'. ♣

Example (Bead on a rod): A spring with a tension has one end attached to the end of a rod, and the other end attached to a bead that is constrained to move along the rod. The rod makes an angle θ' with respect to the x' axis and is fixed at rest in the S' frame; see Fig. 3.11. The bead is released and is pulled along the rod by the spring. Right after the bead is released, what does the situation look like in the frame S of someone moving to the left at speed v? In answering this, draw the directions of:

(a) the rod,

(b) the acceleration of the bead,

(c) the force on the bead.

In frame S, does the rod exert a constraint (normal) force on the bead?

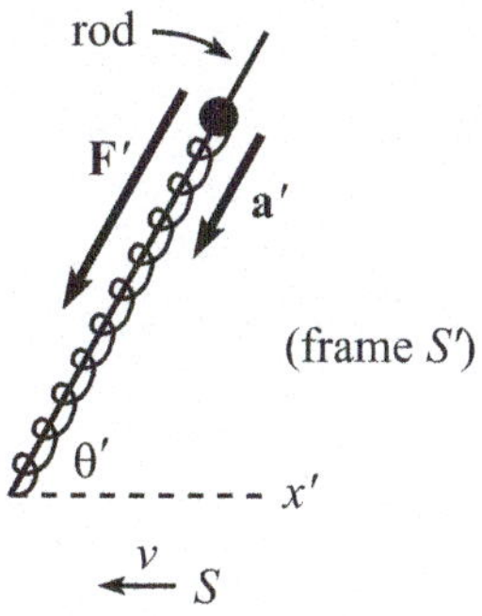

Figure 3.11

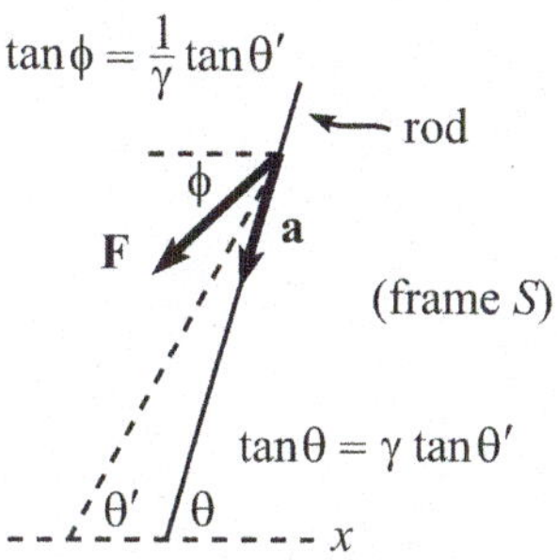

Figure 3.12

Solution: In frame S, the directions of the various items can be found as follows.

(a) The horizontal span of the rod is decreased by a factor γ, due to length contraction, while the vertical span is unchanged. So we have $\tan\theta = \gamma \tan\theta'$, as shown in Fig. 3.12.

(b) The acceleration must point along the rod, because the bead always lies on the rod, and because the rod moves at constant speed in S. Quantitatively, the position of the bead in S takes the form of $(x, y) = (vt - a_x t^2/2,\ -a_y t^2/2)$, by the definition of acceleration. The position relative to the starting point on the rod, which has coordinates $(vt, 0)$, is then $(\Delta x, \Delta y) = (-a_x t^2/2,\ -a_y t^2/2)$. The condition for the bead to stay on the rod is that the ratio of these coordinates equals the slope of the rod in S. Therefore, $a_y/a_x = \tan\theta$, which means that the acceleration points along the rod.

(c) Eq. (3.71) says that in S, the y component of the force on the bead is decreased by a factor γ. The x component is unchanged. So F_y/F_x is smaller than F'_y/F'_x by a factor γ. Therefore, since the force points at an angle θ' in S' (because it points along the rod in S'), the angle in S is given by $\tan\phi = (1/\gamma)\tan\theta'$, as shown in Fig. 3.12. We see that the spring force does *not* point along the spring in S. Strange but true.

As a double-check that the acceleration does indeed point along the rod in S, we can use Eq. (3.65) to write $a_y/a_x = \gamma^2 F_y/F_x$. If we then use Eq. (3.71) to replace F_x with F'_x, and F_y with F'_y/γ, we obtain $a_y/a_x = \gamma F'_y/F'_x = \gamma \tan\theta' = \tan\theta$ (using the result from part (a)). And this is the slope of the rod in S.

The rod does *not* exert a force of constraint. The bead doesn't need to touch the rod in S', so it doesn't need to touch it in S. There is no need to have an extra force to combine with **F** to make the result point along **a**, because **F** simply doesn't have to be collinear with **a** in relativistic physics.

3.6 Rocket motion

Up to this point, we have dealt with situations where the masses of our particles are constant, or where they change abruptly (as in a decay, where the sum of the masses of the products is less than the mass of the initial particle). But in many setups, the mass of an object changes continuously. A rocket is the classic example of this, so we'll use the term "rocket motion" to describe the general class of problems where the mass changes continuously.

The relativistic rocket itself encompasses all of the important ideas, so we'll study that example here. Many more examples are left for the problems. We'll present three solutions to the rocket problem, the last of which is rather slick. In the end, the solutions are all basically the same, but it is instructive to see the various ways of looking at things.

Example (Relativistic rocket): Assume that a rocket propels itself by continually converting mass into photons and firing them out the back. Let m be the rocket's instantaneous mass, and let v be its instantaneous speed with respect to the ground. Show that (dropping the c's)

$$\frac{dm}{m} + \frac{dv}{1 - v^2} = 0. \tag{3.72}$$

If the initial mass is M and the initial speed is zero, integrate Eq. (3.72) to obtain

$$m = M\sqrt{\frac{1 - v}{1 + v}} \quad \Longrightarrow \quad v = \frac{1 - (m/M)^2}{1 + (m/M)^2}. \tag{3.73}$$

First solution: The strategy of this solution will be to use conservation of momentum in the ground frame. Consider the effect of a small mass being converted into photons. The mass of the rocket goes from m to $m + dm$ (where dm is negative). So in the frame of the rocket, photons with total energy $E_r = -dm$ (which is positive) are fired out the back (we're dropping the c's). In the frame of the rocket, these photons have momentum $p_r = dm$ (which is negative), since a photon has $|p| = E/c \to E$.

With v being the instantaneous speed of the rocket with respect to the ground, the momentum of the photons in the ground frame, p_g, can be found via the Lorentz transformation in Eq. (3.19),

$$p_g = \gamma(p_r + vE_r) = \gamma(dm + v(-dm)) = \gamma(1 - v)\,dm. \tag{3.74}$$

This is still negative, of course, since dm is negative.

Remark: A common error is to say that the converted mass $-dm$ takes the form of photons with energy $-dm$ (and hence momentum dm) in the ground frame. This is incorrect because, although the photons have energy $-dm$ in the rocket frame, they are redshifted (due to the Doppler effect) in the ground frame. From Eq. (2.36) we see that the frequency (and hence the energy, because $E = hf$ for a photon) of the photons decreases by a factor of $\sqrt{(1 - v)/(1 + v)}$ when going from the rocket frame to the ground frame. (The negative sign is in the numerator because the photons are fired backward.) This factor equals the $\gamma(1 - v)$ factor in Eq. (3.74). That said, although the Doppler effect explains what is going on physically, you technically don't need to know anything about that. It suffices to simply use the Lorentz transformation in Eq. (3.74). ♣

We can now use conservation of momentum in the ground frame, during the small time in which the small mass $-dm$ is converted into photons. The initial momentum of the rocket equals the final momentum of the rocket plus the momentum of the photons. That is,

$$(\gamma mv)_{\text{old}} = (\gamma mv)_{\text{new}} + \gamma(1 - v)\,dm \quad \Longrightarrow \quad 0 = \gamma(1 - v)\,dm + d(\gamma mv). \tag{3.75}$$

Using the product rule, the $d(\gamma mv)$ term may be expanded to give (making use of the $d\gamma = \gamma^3 v\,dv$ relation implicit in either Eq. (3.56) or Eq. (3.59))

$$\begin{aligned} d(m\gamma v) &= (dm)\gamma v + m(d\gamma)v + m\gamma(dv) \\ &= \gamma v\,dm + m(\gamma^3 v\,dv)v + m\gamma\,dv \\ &= \gamma v\,dm + m\gamma(\gamma^2 v^2 + 1)\,dv \\ &= \gamma v\,dm + m\gamma^3 dv, \end{aligned} \tag{3.76}$$

where we have used $\gamma^2 v^2 + 1 = \gamma^2$. Therefore, Eq. (3.75) gives

$$\begin{aligned} 0 &= \gamma(1-v)\,dm + \gamma v\,dm + m\gamma^3 dv \\ &= \gamma\,dm + m\gamma^3 dv. \end{aligned} \tag{3.77}$$

Hence,

$$\frac{dm}{m} + \frac{dv}{1-v^2} = 0, \tag{3.78}$$

in agreement with Eq. (3.72). We must now integrate this equation. The initial values of m and v are M and 0, respectively, so we have

$$\int_M^m \frac{dm'}{m'} + \int_0^v \frac{dv'}{1-v'^2} = 0. \tag{3.79}$$

We have put primes on the integration variables so that we don't confuse them with the limits of integration. The dm' integral is a log:

$$\int_M^m \frac{dm'}{m'} = \ln m' \Big|_M^m = \ln m - \ln M = \ln\left(\frac{m}{M}\right). \tag{3.80}$$

We can look up the dv' integral in a table,[5] but let's instead do it from scratch. Writing $1/(1-v'^2)$ as the sum of two fractions yields

$$\begin{aligned} \int_0^v \frac{dv'}{1-v'^2} &= \frac{1}{2}\int_0^v \left(\frac{1}{1+v'} + \frac{1}{1-v'}\right) dv' \\ &= \frac{1}{2}\Big(\ln(1+v') - \ln(1-v')\Big)\Big|_0^v \\ &= \frac{1}{2}\ln\left(\frac{1+v}{1-v}\right) = \ln\sqrt{\frac{1+v}{1-v}}. \end{aligned} \tag{3.81}$$

Eq. (3.79) therefore gives

$$\ln\left(\frac{m}{M}\right) = -\ln\sqrt{\frac{1+v}{1-v}} \quad \Longrightarrow \quad m = M\sqrt{\frac{1-v}{1+v}}, \tag{3.82}$$

in agreement with Eq. (3.73). This result is independent of the rate at which the mass is converted into photons. It is also independent of the frequency of the emitted photons. Only the total mass converted is relevant. Note that Eq. (3.82) quickly tells us that the energy of the rocket, as a function of velocity, is

$$E = \gamma m = \frac{1}{\sqrt{1-v^2}} \cdot M\sqrt{\frac{1-v}{1+v}} = \frac{M}{1+v} \longrightarrow \frac{Mc^2}{1+v/c}. \tag{3.83}$$

(See Exercise 3.44 for another derivation of this.) This result has the interesting property of approaching $Mc^2/2$ as $v \to c$. This means that half of the initial energy remains with the rocket, and half ends up as photons. You might think that the energy of the rocket should go to zero since its mass eventually approaches zero. But this is compensated for by the increasing γ factor, which diverges as $v \to c$. It isn't obvious, though, why the energy should approach the nice clean value of $Mc^2/2$.

Remark: From Eq. (3.74), or from the previous remark, we see that at any given instant, the ratio of the energy of the photons (that were just emitted) in the ground frame to their energy in the rocket frame is $\sqrt{(1-v)/(1+v)}$. This factor is the same as the factor in Eq. (3.82). In other words, the just-emitted photons' energy in the ground frame decreases in exactly the same manner as the mass of the rocket (assuming that the photons are ejected with the same frequency in the rocket frame throughout the process). Therefore, in the ground frame, the ratio of the just-emitted photons' energy to the mass of the rocket doesn't change with time. There must be a nice intuitive explanation for this, but it eludes me. ♣

[5]Tables often list the integral of $1/(1-v^2)$ as $\tanh^{-1} v$. You can show that this is equivalent to the result in Eq. (3.81), by using the definition of $\tanh v$.

Second solution: The strategy of this solution will be to use $F = dp/dt$ in the ground frame. Let τ denote the time in the rocket frame. As in the first solution, if a mass $-dm$ is converted into photons, then in the rocket frame these photons have momentum dm (which is negative). The photons therefore acquire momentum at the rate $dp/d\tau = dm/d\tau$ in the rocket frame. Since force is the rate of change in momentum, we see that a force of $dm/d\tau$ pushes the photons backward, and so an equal and opposite force of $F = -dm/d\tau$ pushes the rocket forward in the rocket frame. (Newton's third law still holds in relativity. Given the second law, the third law is equivalent to the statement of conservation of momentum.) Now switch to the ground frame. We know from Eq. (3.71) that the longitudinal force is the same in both frames, so $F = -dm/d\tau$ is also the force on the rocket in the ground frame. And since $dt = \gamma\, d\tau$, where t is the time on the ground (the photon emissions occur at the same place in the rocket frame, so we have indeed put the time-dilation factor of γ in the right place), we have

$$F = -\frac{dm}{d\tau} = -\frac{dm}{dt/\gamma} = -\gamma\frac{dm}{dt}\,. \tag{3.84}$$

Remark: We can also calculate the force on the rocket by working entirely in the ground frame. Consider a mass $-dm$ that is converted into photons. Before the conversion, this mass is traveling along with the rocket, so it has momentum $\gamma(-dm)v$. After the conversion into photons, it has momentum $\gamma(1-v)\,dm$, from the first solution above. The change in momentum is therefore $\gamma(1-v)\,dm - (-dm)\gamma v = \gamma\, dm$. Since force is the rate of change in momentum, a force of $\gamma\, dm/dt$ pushes the photons backward in the ground frame, and so an equal and opposite force of $F = -\gamma\, dm/dt$ pushes the rocket forward. ♣

Now things get a little tricky. It is tempting to write down $F = dp/dt = d(m\gamma v)/dt = (dm/dt)\gamma v + m\, d(\gamma v)/dt$. This, however, is incorrect, because the dm/dt term isn't relevant here. When the force (the force between the rocket and the emitted photons) is applied to the rocket at an instant when the rocket has mass m, the only thing the force cares about is that the mass of the rocket at the given instant is m. It doesn't care that m is changing.[6] Said in a different way, the momentum associated with the converted mass still exists. It's just that it's not part of the rocket anymore; it's in the photons. Therefore, the correct expression we want is

$$F = m\frac{d(\gamma v)}{dt}\,. \tag{3.85}$$

As in the first solution above, or in Eq. (3.57), we have $d(\gamma v)/dt = \gamma^3\, dv/dt$. Using the F from Eq. (3.84), we arrive at

$$-\gamma\frac{dm}{dt} = m\gamma^3\frac{dv}{dt}\,, \tag{3.86}$$

which is equivalent to Eq. (3.77). The solution proceeds as above.

Third solution: The strategy of this solution will be to use conservation of energy and momentum in the ground frame, in a slick way. Consider a clump of photons fired out the back. The energy and momentum of these photons are equal to each other in magnitude but opposite in sign (up to a factor of c, and with the convention that the photons are fired in the negative direction). By conservation of energy and momentum, the same statement must be true about the changes in energy and momentum of the rocket. That is,

$$d(\gamma m) = -d(\gamma m v) \quad\Longrightarrow\quad d(\gamma m + \gamma m v) = 0. \tag{3.87}$$

Therefore, $\gamma m(1+v)$ is a constant. We are given that $m = M$ when $v = 0$. Hence, the constant equals M. So

$$\gamma m(1+v) = M \quad\Longrightarrow\quad m = M\sqrt{\frac{1-v}{1+v}}\,. \tag{3.88}$$

[6]If you have some baseballs in your glove and you throw them in succession, then the force between a baseball and you pushes you backward. Each successive baseball is an external object that applies a force on you, so when applying Newton's second law on you, $F = ma$ works perfectly fine, where m is your mass. At any given instant, the fact that your mass (plus any baseballs that you may be holding) is decreasing is irrelevant.

Now, *that's* a quick solution, if there ever was one! If you want to produce Eq. (3.72), you can take the dm/dv derivative of the above expression for m.

3.7 Relativistic strings

Consider a "massless" string with a tension that is constant, that is, independent of length. By "massless," we mean that the string has no mass in its unstretched (zero-length) state. Once it is stretched, it will have potential energy in its rest frame, and hence mass, because $m = E/c^2$. We'll call these objects *relativistic strings*, and we'll study them for two reasons. First, these strings, or reasonable approximations thereof, actually do occur in nature. For example, the gluon force that holds quarks together is approximately constant over distance. And second, these strings open the door to a whole new class of setups we can study, such as those in the two examples below. Relativistic strings might seem a bit strange, but any one-dimensional problem involving them basically comes down to the two relations in Eqs. (3.55) and (3.60):

$$F = \frac{dp}{dt} \quad \text{and} \quad F = \frac{dE}{dx}. \tag{3.89}$$

Example 1 (Mass connected to a wall): A mass m is connected to a wall by a relativistic string with tension T. The mass starts next to the wall and has an initial speed v rightward away from it; see Fig. 3.13. How far from the wall does the mass get? How much time does it take to reach this point?

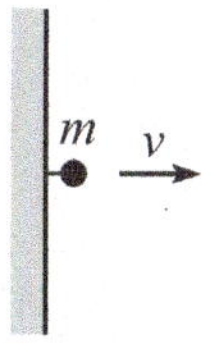

Figure 3.13

Solution: Let ℓ be the maximum distance from the wall. The initial energy of the mass is $E = \gamma m$ (dropping the c's). The final energy at $x = \ell$ is simply m, because the mass is instantaneously at rest there. Integrating $F = dE/dx$ under the assumption that the force is constant gives $F\,\Delta x = \Delta E$. Both F and ΔE are negative in the present setup; $F = -T$ since the tension is directed leftward. So we obtain

$$F\,\Delta x = \Delta E \implies (-T)\ell = m - \gamma m \implies \ell = \frac{m(\gamma - 1)}{T}. \tag{3.90}$$

Let t be the time it takes to reach this point. The initial momentum of the mass is $p = \gamma m v$, and the final momentum is (instantaneously) zero. Integrating $F = dp/dt$ under the assumption that the force is constant gives $F\,\Delta t = \Delta p$. Again, both F and Δp are negative, and we obtain

$$F\,\Delta t = \Delta p \implies (-T)t = 0 - \gamma m v \implies t = \frac{\gamma m v}{T}. \tag{3.91}$$

Note that we *cannot* use $F = ma$ to solve this problem. F does not equal ma. It equals dp/dt (and also dE/dx), which equals $\gamma^3 ma$.

Example 2 (Where the masses meet): A relativistic string with tension T and initial length ℓ connects a mass m and a mass M; see Fig. 3.14. The masses are released from rest. Where do they meet?

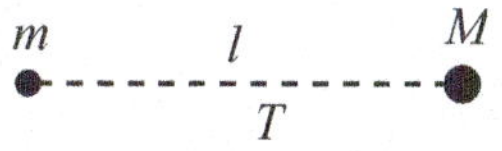

Figure 3.14

Solution: Let the masses meet at a distance x from the initial position of m. At this meeting point, $F = dE/dx \implies F\,\Delta x = \Delta E$ tells us that the energy of m is $m + Tx$ (dropping the c's) and the energy of M is $M + T(\ell - x)$. (Remember that the energies start at mc^2 and Mc^2, and not at zero.) Using $p = \sqrt{E^2 - m^2}$ we see that the magnitudes of the momenta at the meeting point are

$$p_m = \sqrt{(m + Tx)^2 - m^2} \qquad \text{and} \qquad p_M = \sqrt{(M + T(\ell - x))^2 - M^2}. \tag{3.92}$$

But $F = dp/dt$ tells us that these magnitudes must be equal, because the same force T (in magnitude, but opposite in direction) acts on the two masses for the same time. Equating the above p's yields, as you can verify,

$$x = \frac{\ell(M + T\ell/2)}{M + m + T\ell}. \tag{3.93}$$

This result is reassuring, because it is the location of the initial center of mass, with the string being treated (quite correctly) like a stick with length ℓ and mass $T\ell$ (divided by c^2). The above value of x is the weighted average of the positions of the three masses (relative to the initial position of m).

REMARK: Let's check a few limits. In the limit of large T or ℓ (more precisely, in the limit $T\ell \gg Mc^2$ and $T\ell \gg mc^2$), we have $x = \ell/2$. This makes sense, because in this case the masses are negligible and therefore both quickly get moving at essentially speed c, and hence meet in the middle. In the limit of small T or ℓ (more precisely, in the limit $T\ell \ll Mc^2$ and $T\ell \ll mc^2$), we have $x = M\ell/(M + m)$, which is simply the Newtonian result for the center of mass of two objects connected by an everyday-strength massless spring. ♣

3.8 Summary

In this chapter we learned about relativistic dynamics. In particular, we learned:

- The relativistic energy and momentum of a mass m are given by

$$\mathbf{p} = \gamma m\mathbf{v} \qquad \text{and} \qquad E = \gamma mc^2. \tag{3.94}$$

  In the nonrelativistic limit, these expressions reduce properly to $\mathbf{p} = m\mathbf{v}$ and $E = mc^2 + mv^2/2$. The latter is the rest energy mc^2 (which doesn't change in an elastic collision) plus the nonrelativistic kinetic energy $mv^2/2$.

- The "Very Important Relation" is

$$E^2 = p^2c^2 + m^2c^4. \tag{3.95}$$

  Whenever you know two of the three quantities E, p, and m, this equation gives you the third.

- We will often drop the c's in calculations, to keep things from getting messy. They can be put back in at the end, by figuring out where they need to go in order to make the units correct.

- The transformations for E and p are

$$\begin{aligned} E &= \gamma(E' + vp'), \\ p &= \gamma(p' + vE'/c^2). \end{aligned} \tag{3.96}$$

  These are the same Lorentz transformations that apply to t and x. The correct correspondence is $E \leftrightarrow ct$ and $pc \leftrightarrow x$. This implies that $E^2 - p^2c^2$ is invariant, just as $c^2t^2 - x^2$ is.

- As in nonrelativistic collisions, the main ingredients in relativistic collisions are conservation of energy and momentum. It is often helpful to combine E and $\mathbf{p}$ into a four-component vector called the 4-momentum:

$$P \equiv (E, \mathbf{p}) \equiv (E, p_x, p_y, p_z). \tag{3.97}$$

With the inner product defined by Eq. (3.36), we have

$$P^2 \equiv P \cdot P = m^2. \tag{3.98}$$

- A unit of energy that is often used in particle physics is the electron-volt:

$$1 \text{ eV} = 1.602 \cdot 10^{-19} \text{ J}. \tag{3.99}$$

- As in nonrelativistic physics, the force in relativistic physics is $\mathbf{F} = d\mathbf{p}/dt$. For 1-D motion, this leads to

$$F = \gamma^3 ma. \tag{3.100}$$

More generally, for 2-D motion, it leads to

$$\mathbf{F} = m(\gamma^3 a_x, \gamma a_y). \tag{3.101}$$

The force vector therefore doesn't point along the acceleration vector, in general.

- If a force acts on a particle, the longitudinal component of the force is the same in any frame, but the transverse component is smaller (by a factor γ) in any other frame than in the particle's frame:

$$F_x^{\text{other}} = F_x^{\text{particle}} \quad \text{and} \quad F_y^{\text{other}} = \frac{F_y^{\text{particle}}}{\gamma}. \tag{3.102}$$

- If a rocket with initial mass M propels itself by continually converting mass into photons and firing them out the back, the speed v as a function of the mass m at a later time is

$$v = \frac{1 - (m/M)^2}{1 + (m/M)^2}. \tag{3.103}$$

- A relativistic string has a constant tension T and no mass in its unstretched state. Its effect on an object is governed by the two relations, $F\,\Delta t = \Delta p$ and $F\,\Delta x = \Delta E$.

3.9 Problems

Section 3.1: Energy and momentum

3.1. **Deriving E and p** $**$

Accepting the facts that the energy and momentum of a photon are $E = hf$ and $p = hf/c$ (where h is Planck's constant, and f is the frequency of the light wave), derive the relativistic formulas for the energy and momentum of a massive particle, $E = \gamma mc^2$ and $p = \gamma mv$. Do this by considering a mass m that decays into two photons. Look at the decay in the rest frame of the mass, and then in the frame where the mass moves at speed v along the line of the photons' motions. You will need to use the Doppler effect; see Section 2.5. You will also need to use the facts that you must obtain a momentum of mv and a kinetic energy of $mv^2/2$ in the nonrelativistic limit.

Section 3.2: Transformations of E and $\vec{p}$

3.2. **CM velocity** $*$

Given p_{total} and E_{total} for a system of particles moving in 1-D, use the Lorentz transformation to find the velocity of the CM. More precisely, find the velocity of the frame in which the total momentum is zero.

Section 3.3: Collisions and decays

3.3. **Increase in mass** $**$

A particle with large mass M moving at speed v collides with a particle with small mass m at rest. The particles stick together. What is the mass of the resulting object? Assume $m \ll M$, and give your answer to leading order in m.

3.4. **Two-body decay** $**$

A stationary mass M_A decays into masses M_B and M_C. What are the energies of M_B and M_C? What are their momenta?

3.5. **Threshold energy** $**$

A particle with mass m and energy E collides with an identical stationary particle. What is the threshold energy E for a final state containing N particles of mass m? ("Threshold energy" is the minimum energy for which the process can occur.)

3.6. **Maximum energy** $**$

A particle with mass M decays into a number of particles, some of which may be photons. If one of the particles has mass m, and if the sum of the masses of all the other products is μ, what is the maximum possible energy that m can have? *Hint:* Write the conservation of energy and momentum statements as $P_M - P_m = P_\mu$, where P_μ is the total 4-momentum of the other products, and then square this equation. The technique used in Problem 3.5 may be useful.

3.7. **Head-on collision** $***$

A particle with mass M and energy E collides head-on elastically with a stationary particle with mass m. Show that the final energy of the mass M is

$$E' = \frac{2mM^2 + E(m^2 + M^2)}{2Em + m^2 + M^2}. \tag{3.104}$$

Hint: This problem is a bit messy, but you can save yourself a lot of trouble by noting that $E' = E$ must be a root of the quadratic equation you obtain for E'. (Why?) As a reward for trudging through the mess, there are lots of interesting limits you can take.

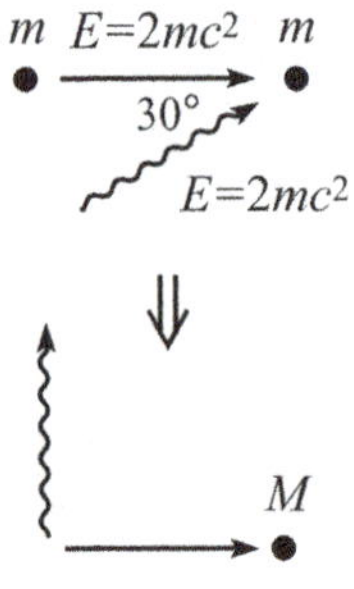

Figure 3.15

3.8. **Colliding photons** $*$

Two photons each have energy E. They collide at an angle θ and create a particle of mass M. What is M?

3.9. **30° collision** $**$

A mass m with total energy $2mc^2$ moves to the right, and a photon also with energy $2mc^2$ moves diagonally at a 30° angle, as shown in Fig. 3.15. They collide with a stationary mass m. If the result of the collision is a photon moving upward and a mass M moving to the right, what is M? What is its velocity?

3.10. **Compton scattering** ∗∗

A photon collides with a stationary electron. If the photon scatters at an angle θ (see Fig. 3.16), show that the resulting wavelength λ' is given in terms of the original wavelength λ by

$$\lambda' = \lambda + \frac{h}{mc}(1 - \cos\theta), \tag{3.105}$$

where m is the mass of the electron and h is Planck's constant. *Note:* The energy of a photon is $E = hf = hc/\lambda$.

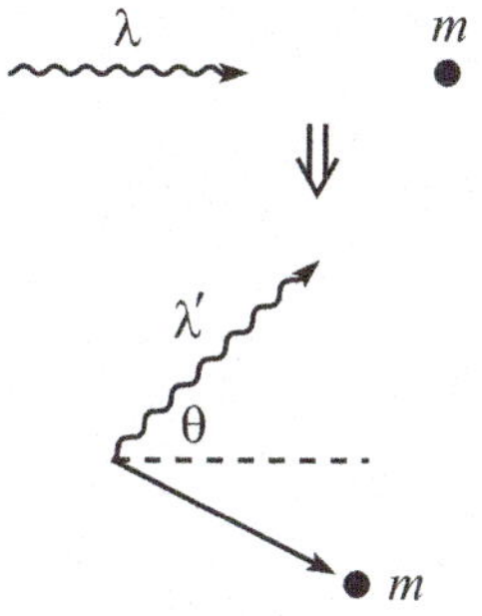

Figure 3.16

Section 3.5: Force

3.11. **Relation between accelerations** ∗∗

In Eq. (3.69) we showed that $a'_x = \gamma^3 a_x$, where the primed frame is the instantaneous inertial frame of a particle. Rederive this relation in the following way. In a small time dt' in S', the particle gains a speed $a'_x\,dt'$, by the definition of acceleration. Find the corresponding increase in speed in S, and then equate the result with $a_x\,dt$. You can calculate this increase by relativistically adding $a'_x\,dt'$ to a given speed v. (Since dt' is assumed to be small, you may drop terms of order dt'^2 and higher in your calculation.)

3.12. **Effective mass of spinning dumbbell** ∗∗

Consider a dumbbell made of two equal masses, m. The dumbbell spins around, with its center pivoted at the end of a stick; see Fig. 3.17. If the speed of the masses is v, then the energy of the system is $2\gamma mc^2$. Treated as a whole, the system is at rest. Therefore, the mass of the system must be $2\gamma m$. (Imagine enclosing it in a box, so that you can't see what's going on inside.) Convince yourself that the system does indeed behave like a mass $M = 2\gamma m$, by pushing on the stick (when the dumbbell is in the "transverse" position shown in the figure) and showing that $F \equiv dp/dt = Ma$.

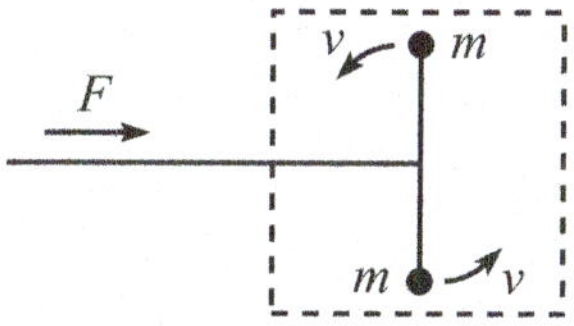

Figure 3.17

3.13. **Relativistic harmonic oscillator** ∗∗

A particle with mass m moves along the x axis, under the influence of the force $F = -m\omega^2 x$. (This is the standard Hooke's-law spring force $-kx$, with $\omega = \sqrt{k/m}$.) The amplitude of the motion (the maximum value of $|x|$) is b. Show that the period (the time of one complete oscillation) is given by

$$T = \frac{4}{c}\int_0^b \frac{\gamma}{\sqrt{\gamma^2 - 1}}\,dx, \quad \text{where } \gamma = 1 + \frac{\omega^2}{2c^2}(b^2 - x^2). \tag{3.106}$$

Section 3.6: Rocket motion

3.14. **Relativistic rocket** ∗∗∗

Consider the relativistic rocket in Section 3.6. Let mass be converted into photons at a rate σ in the (instantaneous inertial) frame of the rocket. Find the time t in the ground frame as a function of v. (Alas, it isn't possible to invert this, to obtain v as a function of t.) You will need to evaluate a slightly tricky integral. Pick your favorite method – pencil, book, or computer.

3.15. **Relativistic dustpan I** ∗∗

A dustpan with initial mass M_0 is given an initial relativistic speed. It gathers up dust with uniform mass density λ per unit length on the floor (as measured in the lab frame). At a general instant when the speed is v, find the rate (as measured in the lab frame) at which the mass of the dustpan-plus-dust-inside system is increasing.

3.16. **Relativistic dustpan II** **

Consider the setup in Problem 3.15. If the initial speed of the dustpan is v_0, find $v(x)$, $v(t)$, and $x(t)$. All quantities here are measured with respect to the lab frame.

3.17. **Relativistic dustpan III** **

Consider the setup in Problem 3.15. Calculate, in both the dustpan frame and the lab frame, the force on the dustpan-plus-dust-inside system (due to the newly acquired dust particles smashing into it) as a function of v, and show that the results are equal.

3.18. **Relativistic cart I** ****

A long cart moves at relativistic speed v. Sand is dropped into the cart (from a stationary funnel) at a rate $dm/dt = \sigma$ in the ground frame. Assume that you stand on the ground next to where the sand falls in, and you push on the cart to keep it moving at constant speed v. What is the force between your feet and the ground? Calculate this force in both the ground frame (your frame) and the cart frame, and show that the results are equal.

3.19. **Relativistic cart II** ****

A long cart moves at relativistic speed v. Sand is dropped into the cart (from a stationary funnel) at a rate $dm/dt = \sigma$ in the ground frame. Assume that you grab the front of the cart and pull on it to keep it moving at constant speed v (while running with it). What force does your hand apply to the cart? Calculate this force in both the ground frame and the cart frame (your frame), and show that the results are equal.

Section 3.7: Relativistic strings

3.20. **Different frames** **

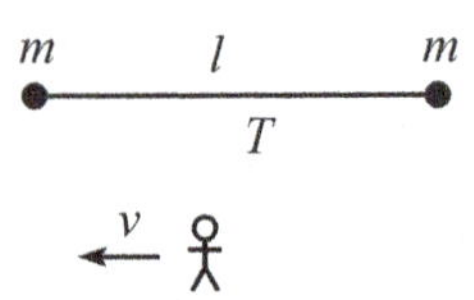

Figure 3.18

(a) Two masses m are connected by a string with initial length ℓ and constant tension T. The masses are released from rest simultaneously, and they collide and stick together. What is the mass M of the resulting blob?

(b) Consider this setup from the point of view of a person moving to the left at speed v; see Fig. 3.18. The energy of the resulting blob must be γMc^2, from part (a). Show that you obtain this same result by calculating the work (force times distance) done on each mass.

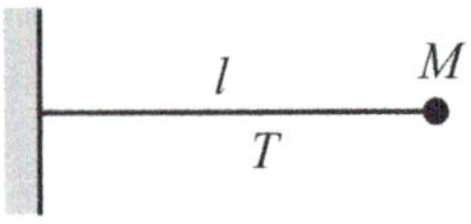

Figure 3.19

3.21. **Splitting mass** **

A massless string with initial length ℓ and constant tension T has one end attached to a wall and the other end attached to a mass M; see Fig. 3.19. The mass is released from rest. Halfway to the wall, the back half of the mass breaks away from the front half (with zero initial relative speed). What is the total time it takes the front half to reach the wall?

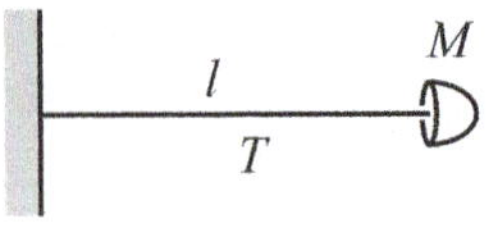

Figure 3.20

3.22. **Relativistic leaky bucket** ***

Let the mass M in Problem 3.21 be replaced by a massless bucket containing an initial mass M of sand; see Fig. 3.20. On the way to the wall, the bucket leaks sand at a rate $dm/dx = M/\ell$, where m denotes the mass at later positions. (The bucket therefore ends up empty when it reaches the wall.) Note that dm and dx are both negative here.

(a) What is the energy of the bucket (the sand inside), as a function of distance from the wall? What is its maximum value? What is the kinetic energy, and its maximum value?

(b) What is the momentum of the bucket (the sand inside), as a function of distance from the wall? Where is it maximum?

3.23. **Relativistic bucket** ***

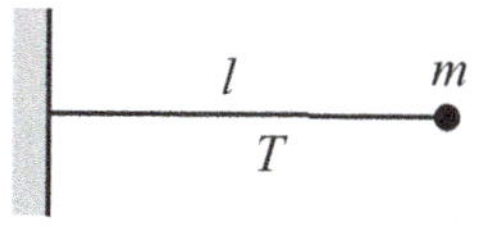

Figure 3.21

(a) A massless string with initial length ℓ and constant tension T has one end attached to a wall and the other end attached to a mass m; see Fig. 3.21. The mass is released from rest. How long does it take to reach the wall?

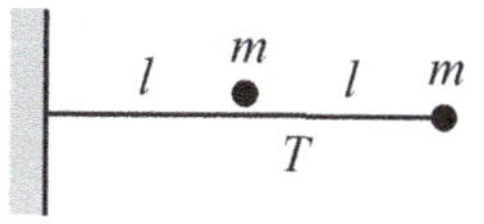

Figure 3.22

(b) Let the string now have initial length 2ℓ, with a mass m at the end. Let another mass m be positioned next to the ℓ mark on the string, but not touching it; see Fig. 3.22. The right mass is released from rest. It heads toward the wall (while the left mass remains motionless) and then sticks to the left mass to make one large blob, which then heads toward the wall.[7] How much time does the entire process take?

Hint: You can solve this in various ways, but one method that generalizes nicely for part (c) is to show that the change in p^2 from the start to a point right before the wall is $\Delta(p^2) = (E_2^2 - E_1^2) + (E_4^2 - E_3^2)$, where the energies of the moving object (that is, the initial m or the resulting blob) are: E_1 at the start, E_2 just before the collision, E_3 just after the collision, and E_4 just before the wall. Note that this method doesn't require knowledge of the mass of the blob (which is *not* $2m$).

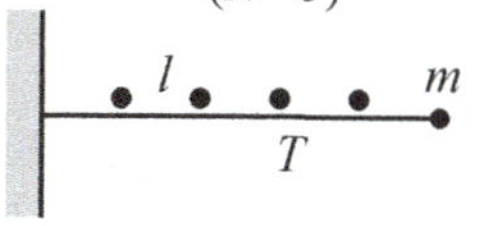

Figure 3.23

(c) Let there now be N masses and a string with initial length $N\ell$. See Fig. 3.23 for the $N = 5$ case. How much time does the entire process take?

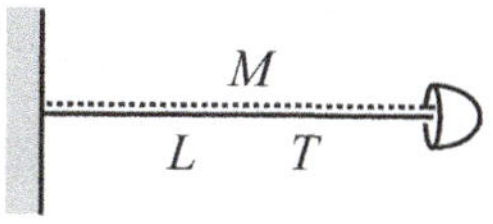

Figure 3.24

(d) Consider now a massless bucket at the end of a string with initial length L and tension T. The bucket gathers up a continuous stream of sand (with total mass M) as it gets pulled to the wall; see Fig. 3.24. How much time does the entire process take? What are the mass and speed of the bucket right before it hits the wall?

3.10 Exercises

Section 3.2: Transformations of E and $\vec{p}$

3.24. **Energy of two masses** *

Two masses m move at speed v, one to the east and one to the west. What is the total energy of the system? Now consider the setup as viewed in a frame moving to the west at speed u. Find the energy of each mass in this frame. Is the total energy larger or smaller than the total energy in the lab frame?

[7]The left mass could actually be attached to the string, and we would still have the same situation. The mass wouldn't move during the first part of the process, because there would be equal tensions T on both sides of it.

3.25. **CM frame** $**$

A mass m travels at speed $3c/5$, and another mass m sits at rest.

(a) Find the energy and momentum of each particle in the lab frame.

(b) Find the speed of the CM of the system, by using a velocity-addition argument, and then again by using the result from Problem 3.2.

(c) Find the energy and momentum of each particle in the CM frame, without using the Lorentz transformations.

(d) Verify that the E's and p's are related by the relevant Lorentz transformations.

(e) Verify that $E^2 - p^2c^2$ for each mass is the same in both frames. Likewise for $E_{\text{total}}^2 - p_{\text{total}}^2 c^2$.

3.26. **Transformations for 2-D motion** $**$

A particle has velocity (u'_x, u'_y) in frame S', which travels with velocity v in the x direction with respect to frame S. Use the velocity-addition formulas in Section 2.2.2 to show that

$$\gamma_u = \gamma_{u'}\gamma_v(1 + u'_x v), \quad \text{where } u = \sqrt{u_x^2 + u_y^2} \text{ and } u' = \sqrt{u_x'^2 + u_y'^2} \tag{3.107}$$

are the speeds in the two frames. Then verify that E and p_x transform according to Eq. (3.20), and also that $p_y = p'_y$.

Section 3.3: Collisions and decays

3.27. **Photon and mass collision** $*$

A photon with energy E collides with a stationary mass m. They combine to form one particle. What is the mass of this particle? What is its speed?

3.28. **A decay** $*$

A stationary mass M decays into a particle and a photon. If the speed of the particle is v, what is its mass? What is the energy of the photon?

3.29. **Two points of view** $*$

In Scenario 1, a mass m moving with speed v collides with a mass $2m$ that is initially at rest. They combine to form a particle with mass M_1. In Scenario 2, a mass $2m$ moving with speed v collides with a mass m that is initially at rest. They combine to form a particle with mass M_2. These two scenarios describe the same setup, just viewed in different reference frames. So it must be the case that $M_1 = M_2$. Verify this explicitly by doing the appropriate calculation in the two different frames.

3.30. **Maximum mass** $**$

A photon and a mass m move directly toward each other. They collide and create a single particle. If the total energy of the system is E, how should it be divided between the photon and the mass m so that the mass of the resulting particle is as large as possible?

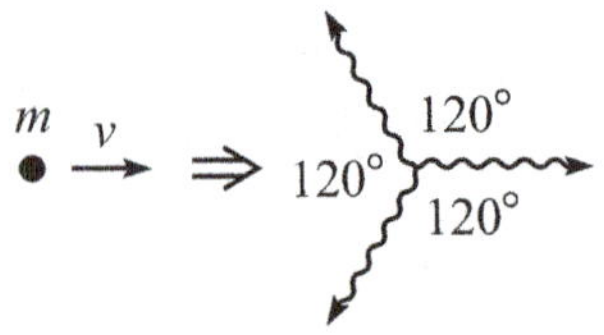

Figure 3.25

3.31. **Three photons** $**$

A mass m moves at speed v. It decays into three photons, one of which travels in the forward direction, and the other two of which move at angles of 120° (in the lab frame) as shown in Fig. 3.25. What are the energies of the three photons?

3.32. Perpendicular photon $**$

A photon with energy E collides with a mass M. The mass M scatters at an angle. If the resulting photon moves perpendicular to the incident photon's direction, as shown in Fig. 3.26, what is its energy, E'?

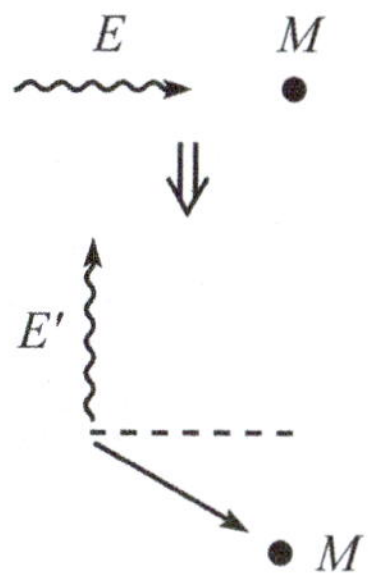

Figure 3.26

3.33. Another perpendicular photon $**$

A mass m moving at speed $4c/5$ collides with another mass m at rest. The collision produces a photon with energy E traveling perpendicular to the original direction, along with a mass M traveling in another direction, as shown in Fig. 3.27. In terms of E and m, what is M? What is the largest value of E (in terms of m) for which this setup is possible?

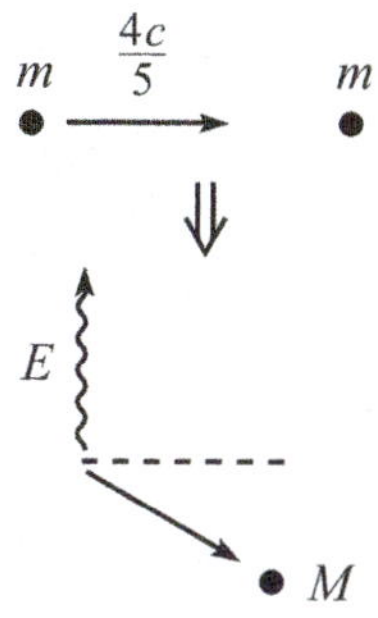

Figure 3.27

3.34. Decay into photons $**$

A mass m moving at speed v decays into two photons. One photon moves perpendicular to the original direction, and the other photon moves at an angle θ, as shown in Fig. 3.28. Show that if $\tan\theta = 1/2$, then $v/c = (\sqrt{5}-1)/2$, which is the inverse of the golden ratio.

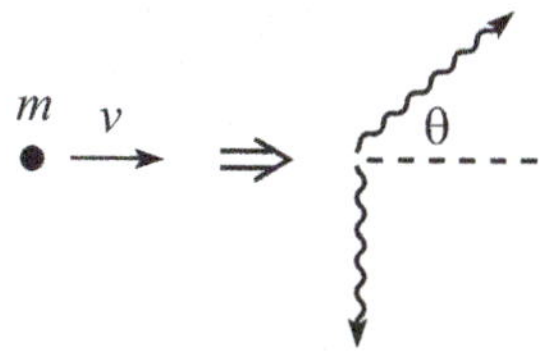

Figure 3.28

3.35. Angled photon $**$

Two photons, each with energy E, move in the directions shown in Fig. 3.29. They collide, and the result of the collision is a photon moving to the right, along with a mass m moving upward.

(a) What is m in terms of E and θ?

(b) What is the largest value of θ for which this scenario is possible?

(c) For the special cases of $\theta = 0$ and $\theta = 90°$, explain why your result for m makes sense.

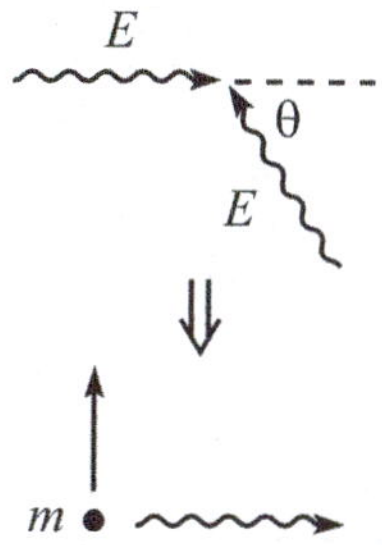

Figure 3.29

3.36. Equal angles $**$

A photon with energy E collides with a stationary mass m. If the mass m and the resulting photon (with unknown energy) scatter at equal angles θ with respect to the initial photon direction, as shown in Fig. 3.30, what is θ in terms of E and m? What is θ in limit $E \ll mc^2$?

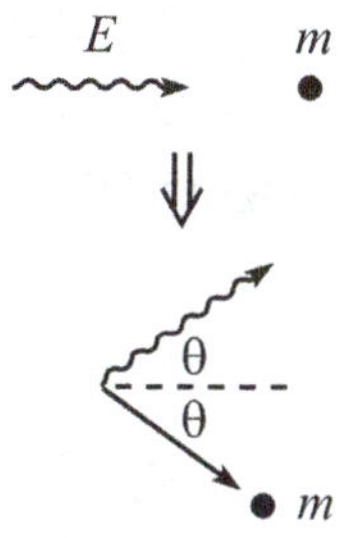

Figure 3.30

Section 3.4: Particle-physics units

3.37. Pion-muon race $*$

A pion and a muon have a 100 m race. If they both have an energy of 10 GeV, by how much distance does the muon win?

3.38. Beta decay $*$

In beta decay, a neutron decays into a proton, an electron, and a neutrino (which is effectively a photon, for the present purposes). The rest energies are $E_n = 939.6$ MeV, $E_p = 938.3$ MeV, $E_e = 0.5$ MeV, and $E_\nu \approx 0$. Use the result from Problem 3.6 to find the maximum energy that the electron can have. Likewise for the neutrino. Explain physically what is going on in the two cases.

3.39. Higgs production $**$

The *Higgs boson* is an elementary particle that was detected at CERN (the European Organization for Nuclear Research) in 2012. Consider a simplified scenario where the Higgs is created by colliding a single proton with a single antiproton. (In reality, two beams of protons collide with each other.) Take the rest energy of a proton (and antiproton) to be 1 GeV. (A "giga electron volt" is about $1.6 \cdot 10^{-10}$ J, but you don't need to know this; just work with GeV.) Assuming that the rest energy of the Higgs is about 125 GeV, how much energy is required (above the rest energies) to produce the Higgs if:

(a) A proton and antiproton have equal and opposite momenta?

(b) A moving proton collides with a stationary antiproton?

Section 3.5: Force

3.40. **Force and a collision** **

Two identical masses m are initially at rest, a distance x apart. A constant force F accelerates one of them toward the other until they collide and stick together. What is the mass of the resulting particle?

3.41. **Pushing on a mass** **

(a) A mass m starts at rest. You push on it with a constant force F. How much time t does it take for the mass to move a distance x? (Both t and x here are measured in the lab frame.)

(b) After a very long time, m's speed will approach c. It turns out that it approaches c sufficiently fast so that after a very long time, m will remain (approximately) a constant distance (as measured in the lab frame) behind a photon that was emitted at $t = 0$ from the starting position of m. What is this distance? (The Taylor series $\sqrt{1+\epsilon} \approx 1 + \epsilon/2$ will be helpful.)

3.42. **Effective mass of a bouncing photon** ***

A photon with energy E bounces back and forth along the x axis inside a massless box. Treated as a whole, the system is at rest. Therefore, the mass of the system must be E/c^2. (This is the same reasoning as in Problem 3.12.) Convince yourself that the system does indeed behave like a mass $M = E/c^2$, by pushing (in the x direction) on the box and showing that $F \equiv dp/dt = Ma$. (This is true in an average sense, at least; a large number of bounces will happen each second. But it suffices to consider one roundtrip of the photon.) *Hints:* You will need to use the facts that the energy and momentum of a photon are $E = hf$ and $p = E/c = hf/c$, where h is Planck's constant and f is the frequency of the light wave. You will need to calculate the change in frequency of a photon bouncing off a moving wall.

3.43. **Momentum paradox** ****

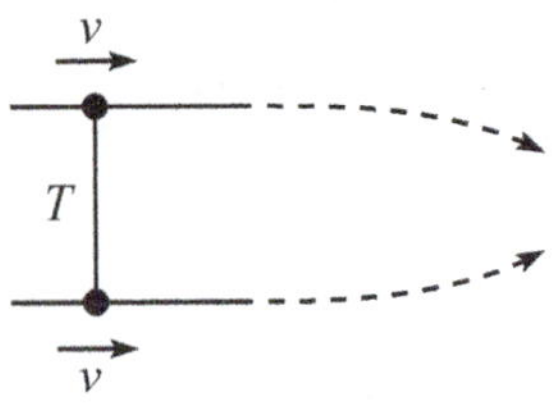

Figure 3.31

Two equal masses are connected by a massless string with tension T. The masses are constrained to move with speed v along parallel lines, as shown in Fig. 3.31. The constraints are then removed, and the masses are drawn together. They collide and make one blob which continues to move to the right. Is the following reasoning correct? If your answer is "no," state what is invalid about whichever of the four sentences is/are invalid.

"The forces on the masses point in the y direction. Therefore, there is no change in the momentum of the masses in the x direction. But the mass of the resulting blob is greater than the sum of the initial masses (because they collide with some relative speed). Therefore, the speed of the resulting blob must be less than v (to keep p_x constant), so the whole apparatus slows down in the x direction."

Section 3.6: Rocket motion

3.44. **Rocket energy** $**$

As we saw in Eq. (3.83) in the first solution to the rocket problem in Section 3.6, the energy of the rocket in the ground frame equals $M/(1+v)$. Derive this result again, by integrating up the amount of energy that the (Doppler shifted) photons have in the ground frame, by the time the rocket's speed is v.

Section 3.7: Relativistic strings

3.45. **Two masses** $**$

A mass m is placed right in front of an identical one. They are connected by a relativistic string with tension T (with initial length zero). The front mass suddenly acquires a speed $3c/5$. How far from the starting point will the masses collide with each other?

3.46. **Relativistic bucket** $**$

One of the results in part (d) of Problem 3.23 was that the bucket moves toward the wall with constant speed $\sqrt{T/(T+\lambda)}$, where $\lambda \equiv M/L$ is the linear mass density of the sand. Derive this again, without using the technique of taking the $N \to \infty$ limit of many masses.

3.11 Solutions

3.1. **Deriving E and p**

Let's derive the energy formula, $E = \gamma mc^2$, first. Let the mass m decay into two photons, and let E_0 be the energy of the mass in its rest frame. Then conservation of energy tells us that each of the resulting photons has energy $E_0/2$ in this frame.

Now look at the decay in a frame where the mass moves at speed v (along the line of the photons' motions). From Eq. (2.36) the frequencies of the photons are Doppler-shifted by the factors $\sqrt{(1+v)/(1-v)}$ and $\sqrt{(1-v)/(1+v)}$ (dropping the c's). Since we are using the fact that the energies of the photons are given by $E = hf$, the energies are shifted by the same Doppler factors, relative to the $E_0/2$ value in the original frame. The total energy of the photons in the frame where the mass moves at speed v is therefore

$$E = \frac{E_0}{2}\sqrt{\frac{1+v}{1-v}} + \frac{E_0}{2}\sqrt{\frac{1-v}{1+v}} = \frac{E_0}{\sqrt{1-v^2}} \equiv \gamma E_0. \tag{3.108}$$

By conservation of energy, this is the energy of the mass m moving at speed v. So we see that a moving mass has an energy that is γ times its rest energy. (If you instead take the mass m to be moving in the direction perpendicular to the line of the photons' motions, then this result also follows quickly from the transverse Doppler effect. Case 1 in Section 2.5.2 is the relevant one.)

We can now use the correspondence principle (which says that relativistic formulas must reduce to the familiar nonrelativistic formulas when $v \ll c$) to find E_0 in terms of m and c. We just found that the difference between the energies of a moving mass and a stationary mass is $\gamma E_0 - E_0$. This must reduce to the familiar kinetic energy $mv^2/2$ in the $v \ll c$ limit. In other words (putting the c's back in, and using the Taylor series $1/\sqrt{1-\epsilon} \approx 1 + \epsilon/2$),

$$\frac{mv^2}{2} \approx \frac{E_0}{\sqrt{1-v^2/c^2}} - E_0 \approx E_0\left(1 + \frac{v^2}{2c^2}\right) - E_0 = \left(\frac{E_0}{c^2}\right)\frac{v^2}{2}. \tag{3.109}$$

Therefore, $E_0/c^2 = m \implies E_0 = mc^2$. So the energy of a moving particle we found in Eq. (3.108) equals $E = \gamma E_0 = \gamma mc^2$, as desired.

We can derive the momentum formula, $p = \gamma mv$, in a similar way. Let the magnitude of the photons' (equal and opposite) momenta in the particle's rest frame be $p_0/2$.[8] We're including a 2 in this definition, so that a 2 won't appear in the final result.

Since the photons' momenta are given by $E = hf/c$, we can use the Doppler-shifted frequencies as we did above to say that the total momentum of the photons in the frame where the mass moves at speed v is

$$p = \frac{p_0}{2}\sqrt{\frac{1+v}{1-v}} - \frac{p_0}{2}\sqrt{\frac{1-v}{1+v}} = \gamma p_0 v. \tag{3.110}$$

Putting the c's back in, we have $p = \gamma p_0 v/c$. By conservation of momentum, this is the momentum of the mass m moving at speed v. Note that since the momentum is a signed quantity, whereas the energy is not, Eq. (3.110) has a relative minus sign between the two terms, whereas Eq. (3.108) does not.

We can now use the correspondence principle to find p_0 in terms of m and c. If $p = \gamma(p_0/c)v$ is to reduce to the familiar $p = mv$ result in the $v \ll c$ limit (where $\gamma \approx 1$), then we must have $p_0 = mc$. Therefore, $p = \gamma mv$, as desired.

Note that the $\gamma \approx 1$ approximation worked fine here with the momentum, whereas we needed the more accurate $\gamma \approx 1 + v^2/2c^2$ approximation in the above calculation of the energy; see Eq. (3.109). This is due to the fact that we were dealing with the difference $\gamma E_0 - E_0$ there, so using $\gamma \approx 1$ would have simply given a result of zero. The $\gamma \approx 1+v^2/2c^2$ approximation yielded the term of order v^2, which was the leading-order term in the result.

3.2. **CM velocity**

Let the CM move with velocity v with respect to the lab frame. Then the Lorentz transformation for the total momentum is $p^{\text{CM}}_{\text{total}} = \gamma_v (p^{\text{lab}}_{\text{total}} - (v/c^2)E^{\text{lab}}_{\text{total}})$. The minus sign comes from the fact that the CM frame sees the lab frame move with velocity $-v$. Using $p^{\text{CM}}_{\text{total}} = 0$, we find

$$\frac{v}{c^2} = \frac{p^{\text{lab}}_{\text{total}}}{E^{\text{lab}}_{\text{total}}}. \tag{3.111}$$

This takes exactly the same form as the $v/c^2 = p/E$ expression in Eq. (3.9) for one particle.

3.3. **Increase in mass**

In the lab frame, conservation of E and p give the energy of the resulting object as $E = \gamma M + m$ (dropping the c's) and the momentum as $p = \gamma Mv + 0$. From Eq. (3.12) the mass M' of the resulting object is therefore

$$M' = \sqrt{E^2 - p^2} = \sqrt{(\gamma M + m)^2 - (\gamma Mv)^2} = \sqrt{M^2 + 2\gamma Mm + m^2}, \tag{3.112}$$

where we have used $\gamma^2(1 - v^2) = 1$. The m^2 term is negligible compared with the other two terms, so we can approximate M' as

$$M' \approx \sqrt{M^2 + 2\gamma Mm} = M\sqrt{1 + \frac{2\gamma m}{M}} \approx M\left(1 + \frac{\gamma m}{M}\right) = M + \gamma m, \tag{3.113}$$

where we have used the Taylor series, $\sqrt{1+\epsilon} \approx 1 + \epsilon/2$. The increase in mass is therefore γ times the mass of the stationary object. This increase is greater than the nonrelativistic answer of m, because heat is generated during the collision, and this heat shows up as mass in the final object. Note that we don't have to worry about putting any c's back in, because the units are already correct.

You might think that since the $2\gamma Mm$ term in Eq. (3.112) is small compared with the M^2 term, we should be able to ignore it along with the m^2 term. But that would produce too coarse an approximation (simply the original mass M). Keeping the $2\gamma Mm$ term yields the leading-order correction to the original M.

---

[8] Since a photon has $p = E/c$, we can invoke the preceding $E_0 = mc^2$ result to quickly conclude that $p_0 = mc$ (because the photons each have energy $E_0/2$). But let's pretend that we haven't found E_0 yet. This will give us an excuse to use the correspondence principle again.

REMARK: The γm result for the increase in mass is clear if we work in the frame where M is initially at rest. In this frame, the mass m comes flying in with energy γm, and then essentially all of this energy shows up as mass in the final object. That is, essentially none of it shows up as overall kinetic energy of the object. This negligible-kinetic-energy result is a general result whenever a small object hits a stationary large object. It follows from the fact that the speed of the large object is proportional to m/M, by momentum conservation (there's a factor of γ if things are relativistic), so the kinetic energy goes like $Mv^2 \propto M(m/M)^2 = m(m/M) \approx 0$, if $M \gg m$. The smallness of v wins out over the largeness of M because v is squared. When a snowball hits a tree, essentially all of the initial energy goes into heat. None of it goes into changing the kinetic energy of the tree/earth. ♣

3.4. **Two-body decay**

M_B and M_C have equal and opposite momenta, so Eq. (3.12) tells us that

$$E_B^2 - M_B^2 = E_C^2 - M_C^2 \quad (= p^2). \tag{3.114}$$

And conservation of energy gives (dropping the c's)

$$E_B + E_C = M_A. \tag{3.115}$$

Eqs. (3.114) and (3.115) are two equations in the two unknowns, E_B and E_C. Solving the equations gives (using the shorthand $a \equiv M_A$, etc.)

$$E_B = \frac{a^2 + b^2 - c^2}{2a} \qquad \text{and} \qquad E_C = \frac{a^2 + c^2 - b^2}{2a}. \tag{3.116}$$

From Eq. (3.114), the magnitude of the (equal and opposite) momentum p of the particles in then

$$\begin{aligned} p &= \sqrt{E_B^2 - b^2} \\ &= \sqrt{\left(\frac{a^2 + b^2 - c^2}{2a}\right)^2 - b^2} \\ &= \frac{1}{2a}\sqrt{a^4 + b^4 + c^4 - 2a^2b^2 - 2a^2c^2 - 2b^2c^2}. \end{aligned} \tag{3.117}$$

REMARK: It turns out that the quantity under the radical in Eq. (3.117) can be factored into

$$(a + b + c)(a + b - c)(a - b + c)(a - b - c). \tag{3.118}$$

This makes it clear that if $a = b + c$ then $p = 0$, because there is no energy left over for the particles to be able to move. Interestingly, the form of p in Eq. (3.117) looks very much like the area of a triangle with sides a, b, c, which is given by Heron's formula as

$$A = \frac{1}{4}\sqrt{2a^2b^2 + 2a^2c^2 + 2b^2c^2 - a^4 - b^4 - c^4}. \tag{3.119}$$

In the given decay, we must have $a \geq b + c$; whereas in a triangle, we must have $a \leq b + c$. ♣

3.5. **Threshold energy**

The initial 4-momenta in the lab frame are

$$(E, p, 0, 0) \qquad \text{and} \qquad (m, 0, 0, 0), \tag{3.120}$$

where $p = \sqrt{E^2 - m^2}$. Therefore, by conservation of energy and momentum, the total 4-momentum of the final particles in the lab frame is given by $(E_{\text{total}}^{\text{lab}}, p_{\text{total}}^{\text{lab}}, 0, 0) = (E + m, p, 0, 0)$.

We claim that in the threshold case, the final particles move together as a blob, with no relative motion. This is true by the following reasoning. From Eq. (3.26), the quantity

$E_{\text{total}}^2 - p_{\text{total}}^2$ is frame independent. So the value it takes is the square of the energy in the CM frame (where $p = 0$). Hence,

$$\begin{aligned}\left(E_{\text{total}}^{\text{lab}}\right)^2 - \left(p_{\text{total}}^{\text{lab}}\right)^2 &= \left(E_{\text{total}}^{\text{CM}}\right)^2 - \left(p_{\text{total}}^{\text{CM}}\right)^2 \\ \Longrightarrow\quad (E+m)^2 - \left(\sqrt{E^2-m^2}\right)^2 &= \left(E_{\text{total}}^{\text{CM}}\right)^2 - 0^2 \\ \Longrightarrow\quad 2Em + 2m^2 &= \left(E_{\text{total}}^{\text{CM}}\right)^2. \end{aligned} \tag{3.121}$$

We see that minimizing E is equivalent to minimizing $E_{\text{total}}^{\text{CM}}$. But $E_{\text{total}}^{\text{CM}}$ is clearly minimized when all of the final particles are at rest in the CM frame, so that there is no kinetic energy added to the rest energy. Therefore, since we have just found that there is no relative motion among the final particles in the CM frame at threshold, there is also no relative motion in any other frame. This means that at threshold, the N masses travel together as a blob in the lab frame, as we claimed above.

Since the particles are at rest in the CM frame at threshold (equivalently, they move as a blob in any other frame), the energy in the CM frame is simply the sum of the rest energies. So $E_{\text{total}}^{\text{CM}} = Nm$, and Eq. (3.121) gives

$$2Em + 2m^2 = (Nm)^2 \quad\Longrightarrow\quad E = \left(\frac{N^2}{2} - 1\right)m. \tag{3.122}$$

This threshold energy E is larger than the naive answer of $(N-1)m$ (obtained by assuming that the final masses have an energy of just $(Nm)c^2$), because in the lab frame the final state has inevitable "wasted" energy in the form of kinetic energy, which is necessitated by conservation of momentum. Note that $E \propto N^2$ for large N.

3.6. **Maximum energy**

With $P_\mu = (E_\mu, \mathbf{p}_\mu)$, Eq. (3.37) tells us that $P_\mu \cdot P_\mu = E_\mu^2 - p_\mu^2$. When applying this relation to the single particles M and m, we obtain $P_M^2 = M^2$ and $P_m^2 = m^2$, as Eq. (3.38) states. The given hint therefore yields

$$(P_M - P_m)^2 = P_\mu^2 \quad\Longrightarrow\quad M^2 + m^2 - 2P_M \cdot P_m = E_\mu^2 - p_\mu^2. \tag{3.123}$$

Since M is initially at rest, its 4-momentum is $P_M = (M, 0, 0, 0)$, so $P_M \cdot P_m = ME_m$. The quantity $E_\mu^2 - p_\mu^2$ is invariant, and it equals the square of the energy in the CM frame (call it $E_{\mu,\text{CM}}$), where the momentum is zero. Eq. (3.123) therefore gives

$$M^2 + m^2 - 2ME_m = E_{\mu,\text{CM}}^2 \quad\Longrightarrow\quad E_m = \frac{M^2 + m^2 - E_{\mu,\text{CM}}^2}{2M}. \tag{3.124}$$

So to maximize E_m, we want to minimize $E_{\mu,\text{CM}}$. But the minimum energy in the CM frame is the sum of the rest energies of all the other products (aside from m). These particles must be at rest in the CM frame, because any nonzero motion would add on kinetic energy. Since the sum of the masses of all the other products is given as μ, the minimum value of $E_{\mu,\text{CM}}$ is just μc^2, or μ without the c's. (Note that if the particles are at rest in the CM frame, they move as a blob in any other frame.) The maximum value of E_m is therefore

$$E_m^{\max} = \frac{M^2 + m^2 - \mu^2}{2M}. \tag{3.125}$$

This is consistent with the E_B and E_C we found in Problem 3.4, where the "other products" were simply one particle.

3.7. **Head-on collision**

The 4-momenta before the collision are

$$P_M = (E, p, 0, 0), \qquad P_m = (m, 0, 0, 0), \tag{3.126}$$

where $p = \sqrt{E^2 - M^2}$. The 4-momenta after the collision are

$$P'_M = (E', p', 0, 0), \qquad P'_m = (\text{we won't need this}), \tag{3.127}$$

where $p' = \sqrt{E'^2 - M^2}$. Conservation of energy and momentum give $P_M + P_m = P'_M + P'_m$. Therefore,

$$P'^2_m = (P_M + P_m - P'_M)^2 \tag{3.128}$$
$$\Longrightarrow \; P'^2_m = P^2_M + P^2_m + P'^2_M + 2P_m \cdot (P_M - P'_M) - 2P_M \cdot P'_M$$
$$\Longrightarrow \; m^2 = M^2 + m^2 + M^2 + 2m(E - E') - 2(EE' - pp')$$
$$\Longrightarrow \; -pp' = M^2 - EE' + m(E - E'). \tag{3.129}$$

Note that because we isolated P'_m on one side of the equation, we didn't need to know anything about it, except that its square equals m^2. Squaring both sides of Eq. (3.129) gives

$$\left(\sqrt{E^2 - M^2}\sqrt{E'^2 - M^2}\right)^2 = \left((M^2 - EE') + m(E - E')\right)^2. \tag{3.130}$$

You can show that this "simplifies" to

$$0 = M^2(E^2 - 2EE' + E'^2) + 2(M^2 - EE')m(E - E') + m^2(E - E')^2. \tag{3.131}$$

As claimed, $E' = E$ is a root of this equation, as it must be, because $E' = E$ and $p' = p$ certainly satisfy conservation of energy and momentum with the initial conditions, by definition. Dividing through by $(E - E')$ yields

$$0 = M^2(E - E') + 2m(M^2 - EE') + m^2(E - E'). \tag{3.132}$$

Solving for E' gives the desired result,

$$E' = \frac{2mM^2 + E(m^2 + M^2)}{2Em + m^2 + M^2}. \tag{3.133}$$

Let's look at a number of limits:

1. If $E \approx M$ (that is, M is barely moving), then $E' \approx M$, because M is still barely moving.
2. If $M = m$, then $E' = M$, because M stops, and m picks up all the energy that M had. (This is true relativistically, just as it is nonrelativistically, as long as the collision is elastic.)
3. If $m \gg E$ ($> M$) (that is, m is a brick wall), then $E' \approx E$, because the heavy mass m picks up essentially no energy. M just bounces back with the same speed it had originally.
4. If ($E >$) $M \gg m$ and $M^2 \gg Em$, then $E' \approx E$, because it's essentially like m isn't there.
5. If ($E >$) $M \gg m$ but $Em \gg M^2$ (which means that E must be huge), then $E' \approx M^2/2m$. This isn't obvious, but it is interesting that it doesn't depend on E. This means that if you throw a large object (M) head-on at a small one (m) fast enough, then independent of exactly how fast you throw it, the large object will always end up with the same energy, $M^2/2m$. And since we're assuming $M^2 \ll Em$, this resulting energy is much less than E. So most of the (very large) initial energy ends up getting transferred to m.
6. If $E \gg m \gg M$, then $E' \approx m/2$. This isn't obvious, but it's similar to the third limit we'll check in the Compton-scattering setup in Problem 3.10. As with the preceding limit, if you throw a small object (M) head-on at a large one (m) fast enough, then independent of exactly how fast you throw it, the small object will always end up with the same energy, $m/2$. And since we're assuming $m \ll E$, this resulting energy is much less than E. So most of the (very large) initial energy ends up getting transferred to m.

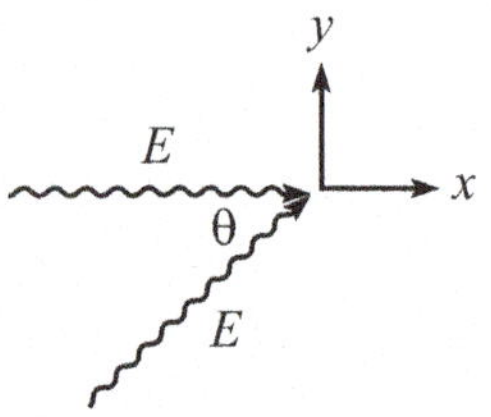

Figure 3.32

3.8. **Colliding photons**

The 4-momenta of the photons are (see Fig. 3.32)

$$P_1 = (E,\ E,\ 0,\ 0) \qquad \text{and} \qquad P_2 = (E,\ E\cos\theta,\ E\sin\theta,\ 0). \tag{3.134}$$

Energy and momentum are conserved, so the 4-momentum of the final particle is $P_M = (2E,\ E + E\cos\theta,\ E\sin\theta,\ 0)$. Therefore,

$$M^2 = P_M \cdot P_M = (2E)^2 - (E + E\cos\theta)^2 - (E\sin\theta)^2. \tag{3.135}$$

Using $\cos^2\theta + \sin^2\theta = 1$, you can quickly show that this simplifies to

$$M = \frac{E\sqrt{2(1-\cos\theta)}}{c^2}, \tag{3.136}$$

where we have added in the c's to make the units right. If $\theta = 180°$ then $M = 2E/c^2$, which is correct; M is at rest, so all of the $2E$ energy takes the form of Mc^2 rest energy. If $\theta = 0°$ then $M = 0$, which is correct; all of the final energy is kinetic (we simply have a photon with twice the energy).

3.9. **30° collision**

The momentum of the initial photon is $E/c = 2mc$, or $2m$ without the c's. We have three facts at our disposal:

- Conservation of p_y: The p_y of the initial photon is $(2m)\sin 30° = m$. The momentum (and hence energy) of the final photon is therefore m.
- Conservation of p_x: The p_x of the initial photon is $(2m)\cos 30° = \sqrt{3}m$. From Eq. (3.12), the p_x of the incoming mass is $\sqrt{E^2 - m^2} = \sqrt{(2m)^2 - m^2} = \sqrt{3}m$. So the total initial p_x is $2\sqrt{3}m$. This is therefore the momentum of the final mass M.
- Conservation of E: The total initial energy is $2m + 2m + m = 5m$. (Remember that the mass sitting at rest still has energy mc^2.) Since the final photon has energy m, this leaves an energy of $4m$ for the mass M.

We therefore know that the mass M has energy $E_M = 4m$ and momentum $p_M = 2\sqrt{3}m$. Eq. (3.12) therefore gives

$$M = \sqrt{E_M^2 - p_M^2} = \sqrt{(4m)^2 - (2\sqrt{3}m)^2} = \sqrt{4m^2} = 2m. \tag{3.137}$$

The units are correct, so there is no need to put any c's back in.

Eq. (3.9) gives the velocity of M as

$$\frac{v}{c^2} = \frac{p}{E} = \frac{2\sqrt{3}mc}{4mc^2} \implies v = \frac{\sqrt{3}c}{2}. \tag{3.138}$$

Alternatively, we have

$$E_M = \gamma M \implies 4m = \gamma(2m) \implies \gamma = 2 \implies v = \frac{\sqrt{3}c}{2}. \tag{3.139}$$

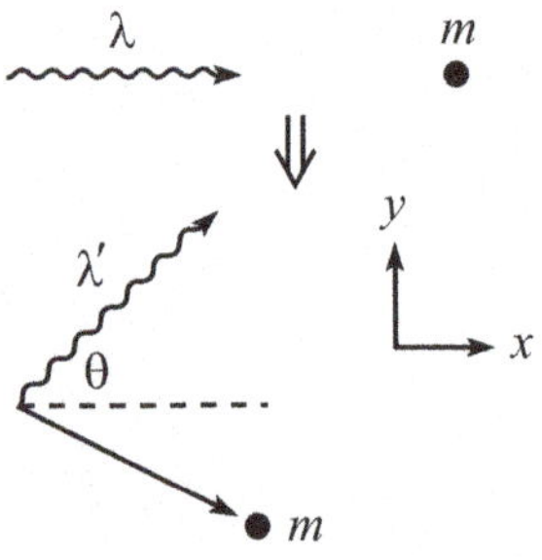

Figure 3.33

3.10. **Compton scattering**

The 4-momenta $(E,\ \mathbf{p}c)$ before the collision are

$$P_\gamma = \left(\frac{hc}{\lambda},\ \frac{hc}{\lambda},\ 0,\ 0\right), \qquad P_m = (mc^2,\ 0,\ 0,\ 0). \tag{3.140}$$

The "γ" subscript here is the symbol commonly used to denote a photon. It has nothing to do with the ubiquitous γ factor in relativity. We'll keep the c's in this calculation, since they don't clutter things up much.

The 4-momenta after the collision are (see Fig. 3.33)

$$P'_\gamma = \left(\frac{hc}{\lambda'},\ \frac{hc}{\lambda'}\cos\theta,\ \frac{hc}{\lambda'}\sin\theta,\ 0\right), \qquad P'_m = (\text{we won't need this}). \tag{3.141}$$

If we wanted to, we could write P'_m in terms of the electron's energy (or momentum) and scattering angle. But we aren't concerned with these quantities, and the nice thing about the following procedure is that we don't need to introduce them. Conservation of energy and momentum give $P_\gamma + P_m = P'_\gamma + P'_m$. Therefore,

$$\begin{aligned}
P'^2_m &= (P_\gamma + P_m - P'_\gamma)^2 \\
\Longrightarrow \quad P'^2_m &= P^2_\gamma + P^2_m + P'^2_\gamma + 2P_m \cdot (P_\gamma - P'_\gamma) - 2P_\gamma \cdot P'_\gamma \\
\Longrightarrow \quad m^2c^4 &= 0 + m^2c^4 + 0 + 2mc^2\left(\frac{hc}{\lambda} - \frac{hc}{\lambda'}\right) - 2\frac{hc}{\lambda}\frac{hc}{\lambda'}(1 - \cos\theta). \qquad (3.142)
\end{aligned}$$

Canceling the m^2c^4 terms and multiplying through by $\lambda\lambda'/(2hmc^3)$ gives the desired result,

$$\lambda' = \lambda + \frac{h}{mc}(1 - \cos\theta). \qquad (3.143)$$

This solution was relatively simple because all the unknown garbage in P'_m disappeared when we squared it.

Let's look at some limits:

1. If $\theta \approx 0$ (not much scattering), then $\lambda' \approx \lambda$, as expected.
2. If $\theta = \pi$ (backward scattering), then $\lambda' = \lambda + 2h/mc$.
3. If $\theta = \pi$ and additionally $\lambda \ll h/mc$ (that is, $mc^2 \ll hc/\lambda = E_\gamma$, so the initial photon's energy is much larger than the electron's rest energy), then $\lambda' \approx 2h/mc$, which means that the final photon's energy is

$$E'_\gamma = \frac{hc}{\lambda'} \approx \frac{hc}{2h/mc} = \frac{1}{2}mc^2. \qquad (3.144)$$

Therefore, the photon bounces back with an essentially fixed E'_γ, which is half of the electron's rest energy, independent of the initial E_γ (as long as E_γ is large enough). This isn't obvious. In fact, it's not even obvious that the photon *can* bounce straight back. You might think that if it has enough energy, it will end up moving forward along with the electron. However, in the CM frame, the photon bounces backward in a head-on collision. So it must bounce backward in every frame, because a photon's direction can't switch when going from one frame to another.

3.11. **Relation between accelerations**

We actually already did this calculation when deriving Eq. (2.52) in our discussion of rapidity in Section 2.6.2. In that discussion, the frame of the spaceship (or object in general) was the unprimed frame, whereas now it is the primed frame. So in the present notation, if the speed of the object with respect to frame S is v at a given time t, then at a time dt later the speed in S is

$$v(t + dt) = \frac{v + a'_x\,dt'}{1 + v \cdot a'_x\,dt'/c^2}. \qquad (3.145)$$

The lefthand side is $v + dv$, by the definition of dv. If we multiply the righthand side by 1 in the form of $(1 - va'_x\,dt'/c^2)/(1 - va'_x\,dt'/c^2)$, and if we ignore the (very small) terms of order dt'^2 in both the numerator and denominator, then Eq. (3.145) becomes

$$v + dv = v + a'_x\,dt' - v^2a'_x\,dt'/c^2. \qquad (3.146)$$

The v's cancel, so we end up with

$$dv = a'_x\left(1 - \frac{v^2}{c^2}\right)dt' \equiv \frac{a'_x\,dt'}{\gamma^2}. \qquad (3.147)$$

But $dv = a_x\,dt$ by the definition of a_x. And $dt' = dt/\gamma$. (The γ is in the right place, because two events located at the object happen at essentially the same place in S'.) So Eq. (3.147) becomes

$$a_x\,dt = \frac{a'_x(dt/\gamma)}{\gamma^2} \quad\Longrightarrow\quad a'_x = \gamma^3 a_x, \qquad (3.148)$$

as desired.

3.12. **Effective mass of spinning dumbbell**

During a short time Δt of application of the force, let the speed of the stick go from zero to ϵ, where $\epsilon \ll v$. Then the masses are now spinning around at speed v in a frame that is moving to the right at speed ϵ. (They are indeed still moving at speed v with respect to the stick, because pushing the stick produces no torque.) The final speeds of the two masses are therefore obtained by relativistically adding or subtracting ϵ from v. We'll assume that Δt is small enough so that the masses are still essentially moving horizontally.

We now need to determine the momenta associated with the speeds obtained from the velocity-addition formula. Conveniently, we've already done the calculation; the result is in Eq. (3.18), with $u' \to \pm\epsilon$. The final momenta of the two masses therefore have magnitudes $\gamma_v\gamma_\epsilon(v \pm \epsilon)m$. But since ϵ is small, we can say that $\gamma_\epsilon \approx 1$, to first order in ϵ. The forward-moving mass then has momentum $\gamma_v(v+\epsilon)m$, and the backward-moving mass has momentum $-\gamma_v(v-\epsilon)m$. The net increase in momentum is therefore $\Delta p = 2\gamma m\epsilon$, where $\gamma \equiv \gamma_v$. Hence (using $a = \epsilon/\Delta t$),

$$F \equiv \frac{\Delta p}{\Delta t} = 2\gamma m \frac{\epsilon}{\Delta t} = 2\gamma m a = Ma, \tag{3.149}$$

where $M = 2\gamma m$, as we wanted to show.

3.13. **Relativistic harmonic oscillator**

With $F = -m\omega^2 x$, the $F = \gamma^3 ma$ relation in Eq. (3.58) gives

$$-m\omega^2 x = \gamma^3 ma \quad\Longrightarrow\quad -\omega^2 x = \gamma^3 \frac{dv}{dt}. \tag{3.150}$$

We must somehow solve this differential equation. A helpful thing to do is multiply both sides by v to obtain $-\omega^2 x\dot{x} = \gamma^3 v\dot{v}$. (This is equivalent to writing the acceleration a as $v\,dv/dx$. See the $v\,dv/dx$ explanation preceding Eq. (3.60).) But from Eq. (3.56), the righthand side of this equation is $d\gamma/dt$. (This doesn't include the c's, so we'll put them in at the end.) Multiplying both sides by dt then gives $-\omega^2 x\,dx = d\gamma$. Integrating this yields $-\omega^2 x^2/2 + C = \gamma$, where C is a constant of integration. We know that $v = 0 \Longrightarrow \gamma = 1$ when $x = b$ (the extreme value of x, where the mass is instantaneously at rest), which tells us that $C = 1 + \omega^2 b^2/2$. So we obtain

$$\gamma = 1 + \frac{\omega^2}{2c^2}(b^2 - x^2), \tag{3.151}$$

where we have put the c's in to make the units right.

Now, $v = dx/dt$ implies $dt = dx/v$. The period is the integral of dt over a complete oscillation, or equivalently four times the integral of dt over a quarter oscillation. The period is therefore

$$T = 4\int_0^b \frac{dx}{v}. \tag{3.152}$$

But $\gamma \equiv 1/\sqrt{1 - v^2/c^2}$ yields $v = (c/\gamma)\sqrt{\gamma^2 - 1}$. Therefore,

$$T = \frac{4}{c}\int_0^b \frac{\gamma}{\sqrt{\gamma^2 - 1}}\,dx. \tag{3.153}$$

Remark: In the limit $\omega b \ll c$ (so that $\gamma \approx 1$, from Eq. (3.151), which means that the speed is always small), we must recover the Newtonian limit. And indeed, if we square the γ in Eq. (3.151) and ignore the term of order $1/c^4$, we obtain $\gamma^2 \approx 1 + (\omega^2/c^2)(b^2 - x^2)$. So Eq. (3.153) becomes

$$T \approx 4\int_0^b \frac{dx}{\omega\sqrt{b^2 - x^2}}, \tag{3.154}$$

where we have set the γ in the numerator equal to 1, to leading order. Eq. (3.154) is the correct nonrelativistic result, because conservation of energy for a nonrelativistic spring gives (with $\omega^2 = k/m$)

$$\frac{1}{2}kb^2 = \frac{1}{2}mv^2 + \frac{1}{2}kx^2 \quad\Longrightarrow\quad \omega^2(b^2 - x^2) = v^2. \tag{3.155}$$

Using this v in the general expression for T in Eq. (3.152) yields Eq. (3.154). ♣

3.14. **Relativistic rocket**

The relation between m and v in Eq. (3.73) is independent of the rate at which mass is converted into photons. The point of this problem is to assume a certain rate, in order to obtain a relation between v and t.

We are given that $dm = -\sigma\, d\tau$, where τ is the time in the rocket frame. Time dilation relates the times in the ground and rocket frames by $dt = \gamma\, d\tau$. (The γ is in the right place, because the mass is always ejected at the same place in the rocket frame, namely right where the rocket is.) So we have $dm = -(\sigma/\gamma)\, dt$ in the ground frame. Taking the differential of the first equation in Eq. (3.73) to obtain another expression for dm, you can show that the result is

$$dm = \frac{-M\, dv}{(1+v)\sqrt{1-v^2}}. \tag{3.156}$$

Alternatively (and more easily), you can just plug the m from Eq. (3.73) into Eq. (3.72). Equating this dm with the above $dm = -(\sigma/\gamma)\, dt$ result, and separating variables and integrating, gives

$$\int_0^t \frac{\sigma\, dt}{M} = \int_0^v \frac{dv}{(1+v)(1-v^2)}. \tag{3.157}$$

(Technically we should put primes on the integration variables, but we won't bother.) We could use a computer to evaluate the dv integral, but let's do it from scratch. Applying the "partial fractions" technique a couple times, we have

$$\begin{aligned}\int \frac{dv}{(1+v)(1-v^2)} &= \int \frac{dv}{(1+v)(1-v)(1+v)} \\ &= \frac{1}{2}\int \left(\frac{1}{1+v} + \frac{1}{1-v}\right)\frac{dv}{1+v} \\ &= \frac{1}{2}\int \frac{dv}{(1+v)^2} + \frac{1}{4}\int \left(\frac{1}{1+v} + \frac{1}{1-v}\right) dv \\ &= -\frac{1}{2(1+v)} + \frac{1}{4}\ln\left(\frac{1+v}{1-v}\right),\end{aligned} \tag{3.158}$$

where we have used the fact that the difference of the logs is the log of the quotient. Equation (3.157) therefore gives (the lower limit, 0, of the v integration produces the "1/2")

$$\frac{\sigma t}{M} = \frac{1}{2} - \frac{1}{2(1+v)} + \frac{1}{4}\ln\left(\frac{1+v}{1-v}\right) = \frac{v}{2(1+v)} + \frac{1}{4}\ln\left(\frac{1+v}{1-v}\right). \tag{3.159}$$

The desired expression for t is then M/σ times the righthand side.

Remark: If $v \ll 1$ (or rather $v \ll c$), the first term in the result in Eq. (3.159) equals $v/2$ (approximately). And the second term also equals $v/2$; this follows from applying the Taylor approximation $\ln(1 \pm x) \approx \pm x$. Eq. (3.159) then becomes $\sigma t/M \approx v$. We see that in the $v \ll c$ limit, v grows linearly with t. The acceleration a in the ground frame (which is v/t) is therefore constant.

Do we recover the correct nonrelativistic $F = ma$ equation for the rocket, in the $v \ll c$ limit? Well, $\sigma t/M \approx v$ can be rewritten as $\sigma \approx M(v/t) = Ma$. And as an exercise you can show that σ is indeed the force acting on the rocket. You just need to explain why $-\sigma$ is the rate of change in momentum of the photons, and then use Newton's third law. (In the $v \ll c$ limit, you don't have to worry about any relativistic differences, such as the Doppler-shift, between the ground and rocket frames.) ♣

3.15. **Relativistic dustpan I**

This problem is essentially the same as Problem 3.3. Let M be the mass of the dustpan-plus-dust-inside system at a given time t in the lab frame. After a small time dt in the lab frame, the system has moved a distance $dx = v\, dt$, so it has effectively collided with an infinitesimal mass $\lambda\,dx = \lambda v\,dt$. The energy of the system therefore increases from γM to $\gamma M + \lambda v\,dt$. Its momentum is still $\gamma M v$, so from Eq. (3.12) its mass is now

$$M + dM = \sqrt{(\gamma M + \lambda v\,dt)^2 - (\gamma M v)^2} \approx \sqrt{M^2 + 2\gamma M \lambda v\,dt}, \tag{3.160}$$

where we have used $\gamma^2(1 - v^2) = 1$ and dropped the second-order dt^2 term. Applying the Taylor series $\sqrt{1+\epsilon} \approx 1 + \epsilon/2$, we can approximate $M + dM$ as

$$M + dM \approx M\sqrt{1 + \frac{2\gamma\lambda v\,dt}{M}} \approx M\left(1 + \frac{\gamma\lambda v\,dt}{M}\right) = M + \gamma\lambda v\,dt. \tag{3.161}$$

Therefore, $dM = \gamma\lambda v\,dt$. The rate (as measured in the lab frame) of increase in the system's mass is therefore $dM/dt = \gamma\lambda v$. As in Problem 3.3, this increase is greater than the nonrelativistic answer of λv, because heat is generated during the collision, and this heat shows up as mass in the final object.

Remarks: As explained in the remark in the solution to Problem 3.3, the $\gamma\lambda v\,dt$ mass increase is clear if we work in the dustpan frame. In this frame, an infinitesimal mass $\lambda v\,dt$ comes flying in with energy $\gamma(\lambda v\,dt)$, and essentially all of this energy shows up as mass in the final object.

What is the rate at which the mass increases, as measured in the *dustpan* frame? If τ is the dustpan time, then $d\tau = dt/\gamma$. (We have indeed put the γ factor in the correct place, because the dust-entering-dustpan events happen at the same location in the dustpan frame.) So $dM/d\tau = dM/(dt/\gamma) = \gamma\,dM/dt = \gamma(\gamma\lambda v) = \gamma^2\lambda v$. (Note that the increase in mass dM itself is invariant, so the only difference between the rates in the two frames comes from the time intervals.) Alternatively, you can think in terms of length contraction. The dustpan sees the dust contracted, so its density is increased to $\gamma\lambda$. And the second γ factor comes from the fact that the dust is moving in the dustpan frame, which puts a γ in the energy. ♣

3.16. **Relativistic dustpan II**

The initial momentum of the dustpan is $\gamma_0 M_0 v_0$. By conservation of momentum, this is the momentum of the dustpan-plus-dust-inside system at any later time. The initial energy is $\gamma_0 M_0$ (dropping the c's). At position x, the dustpan has gathered up a mass λx of dust, so the energy of the dustpan-plus-dust-inside system at a later time is $\gamma_0 M_0 + \lambda x$.

We can quickly find $v(x)$ by using Eq. (3.9), which gives

$$v(x) = \frac{p}{E} = \frac{\gamma_0 M_0 v_0}{\gamma_0 M_0 + \lambda x} = \frac{v_0}{1 + \dfrac{\lambda x}{\gamma_0 M_0}}. \tag{3.162}$$

This equals v_0 when $x = 0$, as it should.

We can find $x(t)$ by identifying the above $v(x)$ with dx/dt and then separating variables and integrating:

$$\frac{dx}{dt} = \frac{v_0}{1 + \dfrac{\lambda x}{\gamma_0 M_0}} \implies \int \left(1 + \frac{\lambda x}{\gamma_0 M_0}\right) dx = \int v_0\,dt$$

$$\implies x + \frac{\lambda x^2}{2\gamma_0 M_0} = v_0 t. \tag{3.163}$$

There is no need to add on a constant of integration, because $x = 0$ when $t = 0$. Solving the above quadratic equation for x yields

$$x(t) = \frac{\gamma_0 M_0}{\lambda}\left(\sqrt{1 + \frac{2\lambda v_0 t}{\gamma_0 M_0}} - 1\right). \tag{3.164}$$

This equals zero when $t = 0$, as it should.

We can obtain $v(t)$ by differentiating $x(t)$ with respect to time. This gives

$$v(t) = \frac{v_0}{\sqrt{1 + \dfrac{2\lambda v_0 t}{\gamma_0 M_0}}}. \tag{3.165}$$

This equals v_0 when $t = 0$, as it should. As a double check, if we equate the expressions for v in Eqs. (3.162) and (3.165) and solve for x, we arrive at Eq. (3.164).

Remarks:

1. For small t, you can use the Taylor series $\sqrt{1+\epsilon} \approx 1 + \epsilon/2$ to quickly show that Eq. (3.164) reduces to $x \approx v_0 t$, as it should. (The dustpan hasn't had any time to slow down.) For large t, x has the interesting property of being proportional to $\sqrt{t}$.
2. In the nonrelativistic limit, the answers are obtained by simply erasing all of the γ_0's in the above results, since $\gamma_0 \approx 1$ if $v_0 \ll c$. If you want to solve the nonrelativistic problem from scratch, then conservation of momentum gives $p = M_0 v_0$ at all times, and conservation of *mass* gives the mass of the dustpan-plus-dust-inside system as $m = M_0 + \lambda x$ at a later time. The nonrelativistic relation $p = mv$ then yields $v = p/m \implies v = M_0 v_0/(M_0 + \lambda x)$. This is simply Eq. (3.162) with the γ_0's erased, so all the steps proceed as above, sans γ_0's. ♣

3.17. **Relativistic dustpan III**

Dustpan frame: Let D denote the dustpan-plus-dust-inside system at a given time, and consider a small bit of dust (call this system B) that enters the dustpan. Assume that D is moving in the positive direction in the lab frame, which means that B is moving in the negative direction in D's frame. In D's frame, the density of the dust is $\gamma\lambda$, due to length contraction. Therefore, in a time $d\tau$, where τ is the time in the dustpan frame, a little B system of dust with length $v\,d\tau$, and hence mass $\gamma\lambda(v\,d\tau)$, crashes into D and loses its negative momentum of $-\gamma(\gamma\lambda v\,d\tau)v = -\gamma^2 v^2 \lambda\,d\tau$. So B's momentum changes by the positive amount $dp = \gamma^2 v^2 \lambda\,d\tau$. The force on B is therefore $F = dp/d\tau = \gamma^2 v^2 \lambda$. The desired force on D is equal and opposite to this, which gives

$$F = -\gamma^2 v^2 \lambda. \tag{3.166}$$

Lab frame: In a time dt, where t is the time in the lab frame, a little system B of dust with length $v\,dt$, and hence mass $\lambda(v\,dt)$, gets picked up by the dustpan. What is the change in momentum of B? It is tempting to say that it is $\gamma(\lambda v\,dt)v$, but this would lead to a force of $-\gamma v^2 \lambda$ on the dustpan, which doesn't agree with the result we found above in the dustpan frame. This would be a problem, because longitudinal forces should be the same in different frames; see Eq. (3.71).

The key point to realize is that the mass of whatever is moving increases at a rate $\gamma\lambda v$, instead of λv; see Problem 3.15. We therefore see that the change in momentum of the additional moving mass is $\gamma(\gamma\lambda v\,dt)v = \gamma^2 v^2 \lambda\,dt$. The original moving system D therefore loses this much momentum, so $dp = -\gamma^2 v^2 \lambda\,dt$. The force on D is therefore $F = dp/dt = -\gamma^2 v^2 \lambda$, in agreement with the result in the dustpan frame.

3.18. **Relativistic cart I**

Ground frame (your frame): Using reasoning similar to that in Problem 3.3 and Problem 3.15, we see that the mass of the cart-plus-sand-inside system increases at a rate $\gamma\sigma$. Therefore, its momentum increases at a rate (using the fact that v is constant)

$$\frac{dp}{dt} = \gamma\left(\frac{dm}{dt}\right)v = \gamma(\gamma\sigma)v = \gamma^2\sigma v. \tag{3.167}$$

Since $F = dp/dt$, this is the force that you exert on the cart. Therefore, it is also the force that the ground exerts on your feet, because the net force on you is zero (since your momentum is constant, and in fact zero).

Cart frame: In the cart frame, you and the sand funnel (and the ground) are moving at speed v leftward (assuming the cart is moving rightward in the ground frame). The sand-entering-cart events happen at the same location in the ground frame, so time dilation says that the sand enters the cart at a slower rate in the cart frame, that is, at a rate σ/γ. (If you are holding a clock next to where the sand comes in, the cart sees the clock run slow.) The sand flies in sideways at speed v, and then eventually comes to rest on the cart, so the magnitude of its momentum decreases at a rate $\gamma(\sigma/\gamma)v = \sigma v$. The signed momentum therefore *increases* at a rate σv, because the momentum increases from a negative value up to zero. Since you are the cause of this change in momentum, σv must be the average force that you exert on the cart.

If this were the only change in momentum in the problem, then we would have a problem, because the force on your feet would be σv in the cart frame, whereas we found above that it is $\gamma^2\sigma v$ in the ground frame. This would contradict the fact that longitudinal forces are the same in different frames. What is the resolution to this apparent paradox? (You should think about this before reading further.)

The resolution is that as you are pushing on the cart, *your mass is decreasing*, because your energy is decreasing. Energy is continually being transferred from you (who are moving) to the cart (which is at rest). Because your mass is decreasing, the same is true for your momentum. This is the missing change in momentum that we need. The quantitative reasoning is as follows.

Go back to the ground frame for a moment. We saw above that the mass of the cart-plus-sand-inside system (call this system "C") increases at a rate $\gamma\sigma$ in the ground frame. So the energy of C increases at a rate $\gamma(\gamma\sigma)$ in the ground frame. The sand provides σ of this energy, so you must provide the remaining $(\gamma^2 - 1)\sigma$ part. Therefore, since you are losing energy at this rate, you must also be losing mass at this rate in the ground frame (because you are at rest there).

Now go back to the cart frame. Due to time dilation, you lose mass at a rate of only $(\gamma^2 - 1)\sigma/\gamma$. This mass goes from moving at speed v (that is, along with you), to speed zero (that is, at rest on the cart). Therefore, the rate of decrease in the magnitude of the momentum of this mass is $\gamma((\gamma^2 - 1)\sigma/\gamma)v = (\gamma^2 - 1)\sigma v$. Adding this result to the σv result we found for the sand, we see that the rate of decrease in the magnitude of the total momentum is $\gamma^2\sigma v$. The force that the ground applies to your feet is the cause of this change in momentum. The ground force is therefore $\gamma^2\sigma v$, in agreement with the above calculation in the ground frame.

Note that the reason why we didn't have to worry about your changing mass when doing the calculation in the ground frame was that your speed there was zero. Your momentum was therefore always zero, independent of what was happening to your mass.

Remark: We just showed that the force that the ground applies to your feet is the same in the two frames, as it should be. But what about the force that you apply to the cart? In the above ground-frame reasoning, we found that the force you exert on the cart is $\gamma^2\sigma v$. But in the cart-frame reasoning, we found that the average force you exert on the cart is σv. These results seem to contradict the fact that the (longitudinal) forces should be the same in the two frames. What's going on here? If you grab the cart with your hand and push, is your hand's force the same in the two frames, or not? You should try to answer this before reading further. It's another 4-star problem in itself.

The answer to the preceding question is that the force from your hand is indeed the same in the two frames, as Eq. (3.71) says must be the case. But then how can the above two forces of $\gamma^2\sigma v$ and σv be consistent? The answer is . . . the loss of simultaneity! Whenever your hand pushes on the cart, it does indeed apply the same force (namely $\gamma^2\sigma v$), independent of the frame. But we escape the paradox because the force acts *for a smaller fraction of the time* in the cart frame. So the *average* force from your hand is smaller in the cart frame. Let's be quantitative about this.

Let's assume that in the ground frame, one of your hands grabs the cart at the origin in the ground frame and pushes it over a distance L for a time L/v. Then it lets go, and simultaneously (in the ground frame) your other hand grabs the cart at the origin and repeats the process. And so on. So at every moment in the ground frame, the same force (which we know is $\gamma^2\sigma v$) is applied.

Now look at things in the cart frame. Due to time dilation, each hand grabs a given point on the cart for a time of only $L/\gamma v$. (You can imagine a clock bolted to the cart where you grab it. The ground frame sees this clock run slow, so less time than L/v elapses on it.) And due to length contraction, each hand grabs the cart at a point that is a proper distance of γL on the cart away from the preceding grab. (This distance is what it length-contracted down to the separation of L between your successive hands in the ground frame.) Also, due to the rear-clock-ahead effect as observed in the ground frame, if one hand lets go of the cart when a clock on the cart at that location says zero, then the next hand grabs the cart when a clock at that location says $(\gamma L)v/c^2$, or γLv without the c's. We have correctly used the proper length γL here.

In the cart frame, the process is therefore the following. A hand applies a force (which is $\gamma^2\sigma v$) for a time $L/\gamma v$. It then lets go, and we wait for a time of γLv while no force is applied; this is the time between letting go and grabbing that we found above. Then the next hand again applies a force for a time $L/\gamma v$, and then we wait for another time of γLv with no force. And so on. The fraction of the time that a force is applied is therefore

$$\frac{L/\gamma v}{\gamma Lv + L/\gamma v} = \frac{1}{\gamma^2 v^2 + 1} = \frac{1}{\gamma^2}. \qquad (3.168)$$

This is the desired factor that explains why the σv average force in the cart frame is $1/\gamma^2$ times the $\gamma^2\sigma v$ force in the ground frame. Your hands apply the same force (whenever they are acting) in the cart frame as in the ground frame (namely $\gamma^2\sigma v$), but for only $1/\gamma^2$ of the time in the cart frame. Due to the nonuniformity of the force in the cart frame, the cart will inevitably undergo a somewhat jerky motion. But on average, the force is σv in the cart frame. ♣

3.19. **Relativistic cart II**

GROUND FRAME: Using reasoning similar to that in Problem 3.3 and Problem 3.15, we see that the mass of the cart-plus-sand-inside system increases at a rate $\gamma\sigma$. Therefore, its momentum increases at a rate $\gamma(\gamma\sigma)v = \gamma^2\sigma v$ (the same as in Problem 3.18). However, this is *not* the force that your hand exerts on the cart. The reason is that your hand is receding from the location where the sand enters the cart, so your hand cannot immediately be aware of the need for additional momentum. No matter how rigid the cart is, it can't transmit information faster than c. In a sense, there is a sort of Doppler effect going on, and your hand needs to be responsible for only a certain fraction of the momentum increase. Let's be quantitative about this.

Consider two grains of sand that enter the cart a time t apart, as measured in the ground frame. What is the difference between the two times that your hand becomes aware that the grains have entered the cart? Assuming maximal rigidity (that is, assuming that signals propagate along the cart at speed c), the relative speed (as measured in the ground frame) of the signals and your hand is $c - v$. The distance between the two successive signals is ct, because both signals start at the same place (at the location of the funnel). They therefore arrive at your hand separated by a time of $ct/(c - v)$. This is larger than t by the factor $c/(c - v)$, which means that the rate (in the ground frame) at which you feel sand entering the cart is $(c - v)/c$ (or $1 - v$, without the c's) times the given σ rate. This is the factor by which we must multiply the naive $\gamma^2\sigma v$ result for the force we found above. The force you apply is therefore

$$F = (1 - v)\gamma^2\sigma v = \frac{\sigma v}{1 + v}. \qquad (3.169)$$

CART FRAME (YOUR FRAME): The sand-entering-cart events happen at the same location in the ground frame, so time dilation says that the sand enters the cart at a slower rate in the cart frame, that is, at a rate σ/γ. The sand flies in at speed v, and then eventually comes to rest on the cart, so the magnitude of its momentum decreases at a rate $\gamma(\sigma/\gamma)v = \sigma v$

(the same as in Problem 3.18). But again, this is *not* the force that your hand exerts on the cart. As above, the sand enters the cart at a location that is receding from your hand, so your hand cannot immediately be aware of the need for additional momentum. Let's be quantitative about this.

Consider two grains of sand that enter the cart a time t apart, as measured in the cart frame (your frame). What is the difference between the two times that your hand becomes aware that the grains have entered the cart? Assuming maximal rigidity (that is, assuming that signals propagate along the cart at speed c), the relative speed (as measured in the cart frame) of the signals and your hand is c, because you are at rest. The distance between the two successive signals is $ct + vt$, because during the time t, the funnel moves a distance vt away from you while the preceding signal travels a distance ct toward you. The signals therefore arrive at your hand separated by a time of $(c + v)t/c$. This is larger than t by the factor $(c + v)/c$, which means that the rate (in your frame) at which you feel sand entering the cart is $c/(c + v)$ (or $1/(1 + v)$, without the c's) times the time-dilated σ/γ rate. This is the factor by which we must multiply the naive σv result for the force we found above. The force you apply is therefore

$$F = \left(\frac{1}{1+v}\right)\sigma v = \frac{\sigma v}{1+v}, \tag{3.170}$$

in agreement with Eq. (3.169).

In a nutshell, the two naive results in the two frames, $\gamma^2\sigma v$ and σv, differ by two factors of γ. But the ratio of the two "Doppler-effect" factors (which arose from the impossibility of absolute rigidity) precisely remedies this discrepancy. The reason why we didn't need to consider this Doppler effect in Problem 3.18 is that there your hand is always right next to the point where the sand enters the cart. Note that the loss-of-simultaneity effect that was relevant in the remark in Problem 3.18 isn't relevant here, because you aren't switching hands in the present setup; one hand continually pulls the cart.

3.20. **Different frames**

(a) Integrating Eq. (3.60) with a constant F gives $F\,\Delta x = \Delta E$. The masses meet in the middle, so they are each pulled a distance $\ell/2$ by the string. Each mass therefore gains an energy of $T(\ell/2)$. Since each mass starts with an energy of m, the total energy of the masses right before they collide is $2(m + T\ell/2)$. By conservation of energy, this is the energy of the resulting mass M. Since M is at rest, this is also its mass:

$$M = 2m + T\ell, \tag{3.171}$$

or $2m + T\ell/c^2$ with the c's.

(b) Let the person's frame be S, and let the original frame be S'. Consider the two events at which the two masses start to move. Let the left mass and right mass start moving at positions x_L and x_R in S. Then the Lorentz transformation $\Delta x = \gamma(\Delta x' + v\,\Delta t')$ tells us that $x_R - x_L = \gamma\ell$, because $\Delta x' = \ell$ and $\Delta t' = 0$ for these events. Alternatively, this follows from length contraction, because if we picture things in the original S' frame, then the $x_R - x_L = \gamma\ell$ length in S is what is length contracted down to ℓ in S'.

Let the masses collide at position x_C in S. Then the gain in energy of the left mass is $T(x_C - x_L)$, and the gain in energy of the right mass is $(-T)(x_C - x_R)$, which is negative if $x_C > x_R$. We have used the fact that the longitudinal force is the same in the two frames, so the masses still feel a tension T in frame S. The gain in the sum of the energies of the two masses is therefore

$$\Delta E = T(x_C - x_L) + (-T)(x_C - x_R) = T(x_R - x_L) = T\gamma\ell. \tag{3.172}$$

The initial sum of the energies is $2\gamma m$, because both masses are initially moving with speed v. So in terms of the M in Eq. (3.171), the final energy is

$$E = 2\gamma m + T\gamma\ell = \gamma M \to \gamma Mc^2, \tag{3.173}$$

as desired.

The initial distance between the masses in frame S (before either is released) has the length-contracted value ℓ/γ, but this length is irrelevant here. It is the distance $\gamma\ell$ that is relevant when calculating the work done by the string in frame S, because this is the distance between the locations where the masses are (unsimultaneously) released.

3.21. **Splitting mass**

We'll calculate the times for the two parts of the process to occur. The energy of the mass right before it splits is (with the subscript b for "before") $E_b = M + T(\ell/2)$, because the tension T acts over a distance of $\ell/2$. So the momentum is $p_b = \sqrt{E_b^2 - M^2} = \sqrt{MT\ell + T^2\ell^2/4}$. Using $F = dp/dt \implies t = \Delta p/T$, the time for the first part of the process is

$$t_1 = \frac{\sqrt{MT\ell + T^2\ell^2/4}}{T} = \sqrt{\frac{M\ell}{T} + \frac{\ell^2}{4}}. \tag{3.174}$$

The momentum of the front half of the mass immediately after it splits is $p_a = p_b/2 = (1/2)\sqrt{MT\ell + T^2\ell^2/4}$. The energy of the front half after the split is $E_b/2$, so the energy at the wall is $E_w = E_b/2 + T(\ell/2) = M/2 + 3T\ell/4$. The momentum at the wall is then $p_w = \sqrt{E_w^2 - (M/2)^2} = (1/2)\sqrt{3MT\ell + 9T^2\ell^2/4}$. The change in momentum during the second part of the process is therefore

$$\Delta p = p_w - p_a = (1/2)\sqrt{3MT\ell + 9T^2\ell^2/4} - (1/2)\sqrt{MT\ell + T^2\ell^2/4}. \tag{3.175}$$

Since $t = \Delta p/T$, the time for the second part is

$$t_2 = \frac{1}{2}\sqrt{\frac{3M\ell}{T} + \frac{9\ell^2}{4}} - \frac{1}{2}\sqrt{\frac{M\ell}{T} + \frac{\ell^2}{4}}. \tag{3.176}$$

The total time is $t_1 + t_2$, which simply changes the minus sign in the preceding expression to a plus sign.

3.22. **Relativistic leaky bucket**

(a) Let the wall be at $x = 0$. Then the initial position is $x = \ell$. Consider a small interval during which the bucket moves from x to $x + dx$ (where dx is negative). The bucket's energy changes by $(-T)\,dx$ due to the work done by the string (this is positive), and it also changes by a fraction dx/x due to the leaking (this is negative). Therefore, $dE = (-T)\,dx + E\,dx/x$, or

$$\frac{dE}{dx} = -T + \frac{E}{x}. \tag{3.177}$$

In solving this differential equation, it is convenient to introduce the variable $y \equiv E/x$. With this definition, we have $E' = (xy)' = xy' + y$, where a prime denotes differentiation with respect to x. Eq. (3.177) then becomes $xy' = -T$, or $dy = -T\,dx/x$. Integration gives $y = -T\ln x + C$, which we can write as $y = -T\ln(x/\ell) + B$, in order to have a dimensionless argument in the log. Since $E = xy$, we therefore have

$$E(x) = Bx - Tx\ln(x/\ell), \tag{3.178}$$

where B is a constant of integration. The reasoning up to this point is valid for *both* the total energy and the kinetic energy, because they both change in the two ways described above. Let's look at each quantity.

TOTAL ENERGY: Eq. (3.178) gives

$$E(x) = M(x/\ell) - Tx\ln(x/\ell), \tag{3.179}$$

where the constant of integration, B, has been chosen to be M/ℓ so that $E(x) = M$ when $x = \ell$. In terms of the fraction $z \equiv x/\ell$, we have $E(z) = Mz - T\ell z\ln z$. Setting $dE/dz = 0$ to find the maximum, and plugging the result back into $E(z)$, gives

$$\ln z_{\max} = \frac{M}{T\ell} - 1 \implies E_{\max} = \frac{T\ell}{e}e^{M/T\ell}, \tag{3.180}$$

as you can verify. The fraction z must satisfy $z \le 1$, so we must have $\ln z \le 0$. Therefore, a solution for z exists only if $M \le T\ell$. If $M \ge T\ell$, then the total energy decreases all the way to the wall.

REMARKS:

1. If $M \ll T\ell$ (really $Mc^2 \ll T\ell$), then E achieves its maximum at $\ln z_{\text{max}} \approx -1 \implies z_{\text{max}} \approx 1/e$, where it has the value $E_{\text{max}} = T\ell/e$. Essentially all of the energy is kinetic in this case of a negligible mass M, so these results must agree with the results below for the case of the kinetic energy, as they in fact will; see Eq. (3.182).
2. If $M \gg T\ell$, then it is clear that E decreases all the way to the wall. The bucket always moves very slowly, because the work done by the tension is inconsequential. So essentially all of the energy comes from the mass, which decreases steadily due to the leaking.
3. If M is slightly less then $T\ell$, then $\ln z_{\text{max}}$ is slightly less than zero, which means that z_{max} is slightly less than 1. So E quickly achieves a maximum of slightly more than M, and then decreases for the rest of the way to the wall. ♣

KINETIC ENERGY: Eq. (3.178) gives

$$K(x) = -Tx\ln(x/\ell), \tag{3.181}$$

where the constant of integration, B, has been chosen to be zero so that $K(x) = 0$ when $x = \ell$. Equivalently, $E - K$ must equal the mass at any position, which is $M(x/\ell)$. (This mass correctly starts out at M and then decreases at a rate M/ℓ as x decreases from ℓ to zero.) In terms of the fraction $z \equiv x/\ell$, we have $K(z) = -T\ell z\ln z$. Setting $dK/dz = 0$ to find the maximum, and plugging the result back into $K(z)$, gives

$$z_{\text{max}} = \frac{1}{e} \implies K_{\text{max}} = \frac{T\ell}{e}, \tag{3.182}$$

which is independent of M. This result must reduce properly in the nonrelativistic limit. But since there's nothing that needs reducing (there aren't any terms that are small compared with others when $v \ll c$), this result must be exactly equal to the nonrelativistic result. It is indeed, because the reasoning that led to Eq. (3.178) is valid not only for both the total energy and the kinetic energy, but also for both the relativistic and nonrelativistic cases.

(b) With $z \equiv x/\ell$, the momentum of the bucket is $p(z) = \sqrt{E^2 - m^2} = \sqrt{E^2 - (Mz)^2}$. Using the E from Eq. (3.179), this gives

$$p(z) = \sqrt{(Mz - T\ell z\ln z)^2 - (Mz)^2} = \sqrt{-2MT\ell z^2\ln z + T^2\ell^2 z^2\ln^2 z}\,. \tag{3.183}$$

Setting the derivative equal to zero yields $T\ell\ln^2 z + (T\ell - 2M)\ln z - M = 0$. Using the quadratic formula, the maximum momentum therefore occurs at

$$\ln z_{\text{max}} = \frac{2M - T\ell - \sqrt{T^2\ell^2 + 4M^2}}{2T\ell}. \tag{3.184}$$

We have ignored the other root, because it gives $\ln z > 0 \implies z > 1$.

REMARKS:

1. If $M \ll T\ell$, then $\ln z_{\text{max}} \approx -1 \implies z_{\text{max}} \approx 1/e$. In this case, the bucket immediately moves with $v \approx c$, so we have $E \approx pc$ from Eq. (3.9). Therefore, E and p should achieve their maxima at the same location. And indeed, we saw in part (a) that E_{max} occurs at $z_{\text{max}} \approx 1/e$ in this limit.
2. If $M \gg T\ell$, then $\ln z_{\text{max}} \approx -1/2 \implies z_{\text{max}} \approx 1/\sqrt{e}$. (The second-order $T^2\ell^2$ term in Eq. (3.184) can be ignored.) In this case, the bucket is nonrelativistic, so this result should agree with the nonrelativistic result. You can verify this by working out the latter from scratch.

3. If $M = T\ell$, then $\ln z_{\text{max}} = (1 - \sqrt{5})/2$, which is the negative of the inverse of the golden ratio. ♣

3.23. **Relativistic bucket**

(a) The energy of the mass right before it hits the wall is $E = m + T\ell$ (the initial energy plus the work done). Therefore, the momentum of the mass right before it hits the wall is $p = \sqrt{E^2 - m^2} = \sqrt{2mT\ell + T^2\ell^2}$. So $F = dp/dt$ gives (using the fact that the tension is constant)

$$\Delta t = \frac{\Delta p}{F} = \frac{\sqrt{2mT\ell + T^2\ell^2}}{T}. \tag{3.185}$$

If $m \ll T\ell$, then $\Delta t \approx \ell$ (or ℓ/c with the c's), which makes sense, because the mass travels at essentially speed c. And if $m \gg T\ell$, then $\Delta t \approx \sqrt{2m\ell/T}$. This is the nonrelativistic limit, and it agrees with the result obtained from the familiar expression, $\ell = at^2/2$, where $a = T/m$ is the acceleration.

(b) In the notation given in the hint, the change in p^2 from the start to just before the collision is $\Delta(p^2) = E_2^2 - E_1^2$. This is true because

$$E_1^2 - m^2 = p_1^2 \qquad \text{and} \qquad E_2^2 - m^2 = p_2^2. \tag{3.186}$$

And since m is the same throughout the first half of the process, the difference of these equations gives $\Delta(E^2) = \Delta(p^2)$. Likewise, the change in p^2 during the second half of the process is $\Delta(p^2) = E_4^2 - E_3^2$, because (with M being the mass of the blob)

$$E_3^2 - M^2 = p_3^2 \qquad \text{and} \qquad E_4^2 - M^2 = p_4^2. \tag{3.187}$$

As an exercise, you can show that M happens to be $\sqrt{4m^2 + 2mT\ell}$. But the nice thing about this method is that we don't need to know this. All we need to know is that M is constant.

The total change in p^2 is the sum of the above two changes. Since p starts at zero, the final p^2 is therefore $(E_2^2 - E_1^2) + (E_4^2 - E_3^2)$. Successive E's are related by either adding on a mass m or doing $T\ell$ of work, so we have $E_1 = m$, $E_2 = m + T\ell$, $E_3 = 2m + T\ell$, and $E_4 = 2m + 2T\ell$. Hence,

$$\begin{aligned} p^2 &= (E_2^2 - E_1^2) + (E_4^2 - E_3^2) \\ &= \left((m + T\ell)^2 - m^2\right) + \left((2m + 2T\ell)^2 - (2m + T\ell)^2\right) \\ &= 6mT\ell + 4T^2\ell^2. \end{aligned} \tag{3.188}$$

The total time is therefore

$$\Delta t = \frac{\Delta p}{F} = \frac{\sqrt{6mT\ell + 4T^2\ell^2}}{T}. \tag{3.189}$$

If $m = 0$ then $\Delta t = 2\ell$ (or $2\ell/c$), as expected. Note that although the tension T acts on two different things (the mass m initially, and then the blob), it is valid to use the total Δp to obtain the total time via $\Delta t = \Delta p/F$. This is true because if we wanted to, we could break up Δp into two parts, and then find the two partial times, and then add them back together to get the total Δt, using the fact that the tension remains the same during the entire process.

(c) The reasoning in part (b) tells us that the final p^2 equals the sum of the $\Delta(E^2)$ terms over the N parts of the process. So we have, using an indexing notation analogous to that in part (b),

$$\begin{aligned} p^2 = \sum_{k=1}^{N} \left(E_{2k}^2 - E_{2k-1}^2\right) &= \sum_{k=1}^{N} \left((km + kT\ell)^2 - (km + (k-1)T\ell)^2\right) \\ &= \sum_{k=1}^{N} \left(2kmT\ell + (2k-1)T^2\ell^2\right) \\ &= N(N+1)mT\ell + N^2T^2\ell^2, \end{aligned} \tag{3.190}$$

where we have used the fact that the sum of the integers from 1 to n equals $n(n+1)/2$. Therefore,

$$\Delta t = \frac{\Delta p}{F} = \frac{\sqrt{N(N+1)mT\ell + N^2T^2\ell^2}}{T}. \tag{3.191}$$

This agrees with the results from parts (a) and (b), for $N = 1$ and 2.

(d) We want to take the $N \to \infty$, $\ell \to 0$, and $m \to 0$ limits, with the restrictions that $N\ell = L$ and $Nm = M$. With $\ell = L/N$ and $m = M/N$, we can write Eq. (3.191) in terms of M and L:

$$\Delta t = \frac{\sqrt{(1+1/N)MTL + T^2L^2}}{T} \quad \longrightarrow \quad \frac{\sqrt{MTL + T^2L^2}}{T}, \tag{3.192}$$

as $N \to \infty$. Note that this Δt is the same as the time it takes for one particle of mass $m = M/2$ to reach the wall, from part (a). This isn't intuitively obvious. As an exercise, you can show from scratch that this $m = M/2$ result holds in the nonrelativistic case.

The mass of the bucket at the wall is

$$\begin{aligned} M_{\rm w} = \sqrt{E_{\rm w}^2 - p_{\rm w}^2} &= \sqrt{(M+TL)^2 - (MTL + T^2L^2)} \\ &= \sqrt{M^2 + MTL}. \end{aligned} \tag{3.193}$$

If $TL \ll M$, then $M_{\rm w} \approx M$, which makes sense; the bucket is moving slowly. If $M \ll TL$, then $M_{\rm w} \approx \sqrt{MTL}$, which means that $M_{\rm w}$ is the geometric mean between the mass of the sand and the work done by the string. This isn't obvious.

The speed of the bucket right before it hits the wall is

$$\begin{aligned} v_{\rm w} = \frac{p_{\rm w}}{E_{\rm w}} &= \frac{\sqrt{MTL + T^2L^2}}{M+TL} \\ &= \sqrt{\frac{TL}{M+TL}} = \sqrt{\frac{T}{\lambda + T}}, \end{aligned} \tag{3.194}$$

where $\lambda \equiv M/L$ is the linear mass density of the sand. We see that $v_{\rm w}$ depends only on T and λ. This means that if we increase L by moving the wall to the left (and adding more sand with the same λ), the speed at the wall will still be $v_{\rm w}$. This implies that the bucket must move toward the wall at the *constant* speed $v_{\rm w}$. The task of Exercise 3.46 is to derive this result without taking the $N \to \infty$ limit of many masses. Note that since this result holds relativistically, it must also hold nonrelativistically (a fun exercise).

Chapter 4

4-vectors

We now come to the subject of *4-vectors*, which we have mentioned a number of times in earlier chapters. Although it is possible to derive everything in special relativity without the use of 4-vectors (and indeed, this is the route that we've taken so far), they are extremely helpful in making calculations simpler and concepts more transparent.

I have chosen to postpone the introduction to 4-vectors until now, in order to make it clear that everything in special relativity can be derived without them. In encountering relativity for the first time, it's nice to know that no "advanced" techniques are required. But now that you've seen everything once, let's go back and derive various things in an easier way.

Although special relativity doesn't require knowledge of 4-vectors, the subject of general relativity definitely requires a firm understanding of *tensors*, which are the generalization of 4-vectors. We won't have time to go deeply into general relativity in Chapter 5, so you'll just have to accept this fact. But suffice it to say that an eventual understanding of general relativity requires a solid foundation in the 4-vectors of special relativity. So let's see what they're all about.

The outline of this chapter is as follows. In Section 4.1 we define 4-vectors, and then we give a number of examples in Section 4.2. Section 4.3 covers the properties of 4-vectors that make them so useful. We look at the energy-momentum 4-vector in Section 4.4, and then the force 4-vector and acceleration 4-vector in Section 4.5. Finally, in Section 4.6 we discuss why physical laws in special relativity must be expressed in 4-vector form, instead of standard 3-vector form.

4.1 Definition of 4-vectors

Definition 4.1 *A 4-tuple,* $A = (A_0, A_1, A_2, A_3)$, *is a "4-vector" if the* A_i *transform between frames in the same way that* $(c\Delta t, \Delta x, \Delta y, \Delta z)$ *do. In other words, A is a 4-vector if it transforms according to the Lorentz transformation,*

$$\begin{aligned} A_0 &= \gamma(A_0' + (v/c)A_1'), \\ A_1 &= \gamma(A_1' + (v/c)A_0'), \\ A_2 &= A_2', \\ A_3 &= A_3'. \end{aligned} \tag{4.1}$$

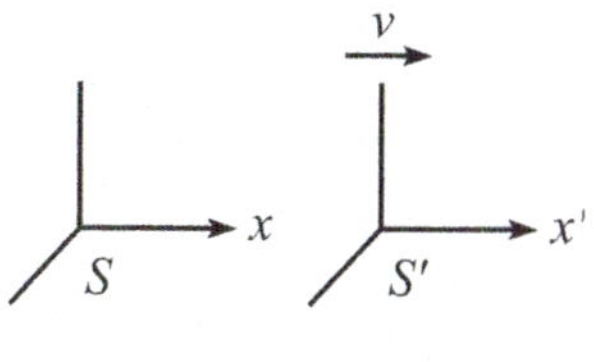

Figure 4.1

The above equations for A_0 and A_1 come from simply replacing ct and x with A_0 and A_1 in the Lorentz transformation in Eq. (2.5). We have assumed, as usual, that the Lorentz transformation is along the x direction; see Fig. 4.1.

By the phrase "transform between frames" in the above definition, we mean the changes that occur to the A_i when x and t transform between frames (that is, when x and t undergo a Lorentz transformation). As always, by x and t here, we really mean Δx and Δt. Due to the types of 4-vectors we'll be working with, we'll replace Δx with dx, etc., in the discussions below. The Lorentz transformation holds for the d's just as it does for the Δ's, because the d's are changes in the coordinates just as the Δ's are. They're simply very small changes.

The A_i may be functions of v (the relative speed between the two frames), the intervals dt, dx, dy, and dz, and also any invariants (that is, frame-independent quantities) such as the mass m or the proper time $d\tau$. We'll look at many examples in Section 4.2.

In addition to the Lorentz-transformation requirement in Eq. (4.1), the last three components of a 4-vector must be a standard vector in 3-D space. That is, they must transform like a usual vector under rotations in 3-D space. So the full definition of a 4-vector is that it must transform like $(c\,dt,\ dx,\ dy,\ dz)$ under Lorentz transformations *and* rotations.

REMARKS:

1. Equations similar to those in Eq. (4.1) must hold, of course, for Lorentz transformations in the y and z directions.
2. We'll use an uppercase italic letter like A to denote a 4-vector. A bold-face letter like $\mathbf{p}$ will denote, as usual, a vector in 3-space.
3. Lest we get tired of writing the c's over and over, we'll work in units where $c = 1$ from now on.
4. The first component of a 4-vector is called the "time" component. The other three are called the "space" components.
5. The components in (dt, dx, dy, dz) are sometimes referred to as (dx_0, dx_1, dx_2, dx_3). Some treatments instead use the indices 1 through 4, with 4 being the time component. We'll use 0 through 3.
6. 4-vectors are the obvious generalization of vectors in regular space. A vector in three dimensions, after all, is something that transforms under a rotation in the same way that (dx, dy, dz) does. We have simply generalized a 3-D rotation to a 4-D Lorentz transformation. ♣

4.2 Examples of 4-vectors

So far, we have only one 4-vector at our disposal, namely (dt, dx, dy, dz) or more generally $(\Delta t, \Delta x, \Delta y, \Delta z)$, which is called the *displacement 4-vector*. (We have previously been referring to this vector as the spacetime separation.) What are some other 4-vectors? Well, $(3\,dt, 3\,dx, 3\,dy, 3\,dz)$ certainly works, as does any other constant multiple of (dt, dx, dy, dz). Indeed, $m(dt, dx, dy, dz)$ is a 4-vector, because the mass m is invariant; it isn't affected by a Lorentz transformation. But how about $A = (dt, 2\,dx, dy, dz)$? What we mean by this 4-tuple is: For any two events, write down the spacetime separation (dt, dx, dy, dz) (or more generally, $(\Delta t, 2\,\Delta x, \Delta y, \Delta z)$) between them, and then multiply the second component by 2 to obtain A. Different frames will have different spacetime separations, and hence different 4-tuples A. Is A a 4-vector? No, because on one hand it must transform (assuming it's a 4-vector) like

$$
\begin{aligned}
A_0 = \gamma(A'_0 + (v/c)A'_1) \quad &\Longrightarrow \quad dt = \gamma\big(dt' + v(2\,dx')\big), \\
A_1 = \gamma(A'_1 + (v/c)A'_0) \quad &\Longrightarrow \quad 2\,dx = \gamma\big((2\,dx') + v\,dt'\big), \\
A_2 = A'_2 \quad &\Longrightarrow \quad dy = dy', \\
A_3 = A'_3 \quad &\Longrightarrow \quad dz = dz',
\end{aligned}
\tag{4.2}
$$

from the definition of a 4-vector. But on the other hand, it must transform like

$$\begin{aligned} dt &= \gamma(dt' + v\,dx'), \\ 2\,dx &= 2 \cdot \gamma(dx' + v\,dt'), \\ dy &= dy', \\ dz &= dz', \end{aligned} \tag{4.3}$$

because this is how dt and the dx_i transform (via the Lorentz transformation). The two preceding sets of equations are inconsistent, so $A = (dt, 2\,dx, dy, dz)$ is not a 4-vector. Note that if we had instead considered the 4-tuple, $A = (dt, dx, 2\,dy, dz)$, then the two preceding equations would have been consistent. But if we had then looked at how A transforms under a Lorentz transformation in the y direction, we would have found that it is not a 4-vector.

The moral of this story is that the above definition of a 4-vector is a nontrivial one because there are two possible ways that a 4-tuple can transform. It can transform according to the 4-vector definition, as in Eq. (4.2). Or it can transform by having each of the A_i transform separately (using how the dx_i, or whatever else the A_i are made of, transform), as in Eq. (4.3). Only for certain special 4-tuples do these two methods give the same result. By definition, we label these special 4-tuples as 4-vectors.

Let's now construct some less trivial examples of 4-vectors. In constructing these, we'll make use of the fact that the proper-time interval, $d\tau \equiv \sqrt{dt^2 - dx^2 - dy^2 - dz^2} \equiv \sqrt{dt^2 - d\mathbf{r}^2}$, is invariant. (See Section 2.3 for discussion of the proper time, relevant to timelike-separated events.)

- **Displacement 4-vector:** For completeness, we'll start by writing down the displacement 4-vector between two events that are infinitesimally close together:

$$dS \equiv (dt, dx, dy, dz). \tag{4.4}$$

If we replace the d's with Δ's, we still have a 4-vector, of course. The separation between the events need not be small.

- **Velocity 4-vector:** We can divide the displacement 4-vector (dt, dx, dy, dz) by $d\tau$, where $d\tau$ is the proper time between the two events that yield dt, dx, dy, and dz. The result is a 4-vector, because $d\tau$ is independent of the frame in which it is measured; it isn't affected by a Lorentz transformation. Using $d\tau = dt/\gamma$, we obtain

$$V \equiv \frac{dS}{d\tau} = \frac{(dt, dx, dy, dz)}{dt/\gamma} = \gamma\left(1, \frac{dx}{dt}, \frac{dy}{dt}, \frac{dz}{dt}\right) = (\gamma, \gamma\mathbf{v}). \tag{4.5}$$

This is known as the *velocity 4-vector*. If we imagine an object traveling at constant velocity between the two events, then in the rest frame of the object we have $\mathbf{v} = 0$, so V reduces to $V = (1, 0, 0, 0)$. For a general v, if we include the c's, the first component in (dt, dx, dy, dz) is really $c\,dt$, so the velocity 4-vector is $V = (\gamma c, \gamma\mathbf{v})$.

Note that we did indeed put the γ in the correct place in the above $d\tau = dt/\gamma$ relation, because the two events happen at the same place in the frame traveling at constant velocity from one event to the other. Everyone else sees a clock in this frame run slow. And it is in this frame that the time is the proper time $d\tau$, by definition.

- **Energy-momentum 4-vector:** If we multiply the velocity 4-vector by the invariant mass m of an object, we obtain another 4-vector,

$$P \equiv mV = (\gamma m, \gamma m\mathbf{v}) = (E, \mathbf{p}). \tag{4.6}$$

This is known as the *energy-momentum 4-vector* (or the *4-momentum* for short), for obvious reasons. In the rest frame of the object, P reduces to $P = (m, 0, 0, 0)$. For a general v, if we include the c's, we have $P = (\gamma mc, \gamma m\mathbf{v}) = (E/c, \mathbf{p})$. Some treatments multiply through by c, so that the 4-momentum is $(E, \mathbf{p}c)$.

We now see how the 4-momentum follows quickly and naturally from the spacetime separation: Just write down the $(dt,\ dx,\ dy,\ dz)$ interval between two nearby events on a particle's worldline, then divide by the proper time (to obtain V) and multiply by the mass (to obtain P). In the end, the E and p that we wrote down (a bit out of the blue) in Eq. (3.1) go hand in hand just as much as t and x do.

- **Acceleration 4-vector:** We can also take the derivative of the velocity 4-vector with respect to τ. The result is a 4-vector, because taking the derivative involves taking the (infinitesimal) difference between two 4-vectors, which results in a 4-vector because Eq. (4.1) is linear (more on this below in Section 4.3). The derivative with respect to τ then involves dividing by the invariant $d\tau$, which again results in a 4-vector. Replacing $d\tau$ with dt via $d\tau = dt/\gamma$, we obtain

$$A \equiv \frac{dV}{d\tau} = \frac{d}{d\tau}(\gamma, \gamma\mathbf{v}) = \gamma\left(\frac{d\gamma}{dt}, \frac{d(\gamma\mathbf{v})}{dt}\right). \tag{4.7}$$

Using $d\gamma/dt = \gamma^3 v\dot{v}$ (see Eq. (3.56)), we have

$$A = (\gamma^4 v\dot{v},\ \gamma^4 v\dot{v}\mathbf{v} + \gamma^2\mathbf{a}), \tag{4.8}$$

where $\mathbf{a} \equiv d\mathbf{v}/dt$. A is known as the *acceleration 4-vector*. In the rest frame of the given object (or, rather, in the instantaneous inertial frame), A reduces to $A = (0, \mathbf{a})$. As we usually do, we'll pick the instantaneous velocity $\mathbf{v}$ to point in the positive x direction. That is, $\mathbf{v} = (v_x, 0, 0)$ with $v_x > 0$. Then $v = v_x$ and $\dot{v} = \dot{v}_x \equiv a_x$. (The acceleration vector $\mathbf{a}$ is free to point in any direction, but you can check that the 0's in $\mathbf{v}$ lead to $\dot{v} = a_x$. See Exercise 4.6.) Eq. (4.8) then becomes

$$\begin{aligned} A &= (\gamma^4 v_x a_x,\ \gamma^4 v_x^2 a_x + \gamma^2 a_x,\ \gamma^2 a_y,\ \gamma^2 a_z) \\ &= (\gamma^4 v_x a_x,\ \gamma^4 a_x,\ \gamma^2 a_y,\ \gamma^2 a_z), \end{aligned} \tag{4.9}$$

where we have used $\gamma^2 v_x^2 + 1 = \gamma^2$. We can keep taking derivatives with respect to τ to create other 4-vectors, but these have little relevance in the real world.

- **Force 4-vector:** We define the *force 4-vector* as the derivative of the momentum 4-vector with respect to τ. Using $d\tau = dt/\gamma$, this gives

$$F \equiv \frac{dP}{d\tau} = \gamma\left(\frac{dE}{dt}, \frac{d\mathbf{p}}{dt}\right) = \gamma\left(\frac{dE}{dt}, \mathbf{f}\right), \tag{4.10}$$

where $\mathbf{f} \equiv d(\gamma m\mathbf{v})/dt$ is the usual 3-force. We'll denote the 3-force by $\mathbf{f}$ instead of $\mathbf{F}$ in this chapter, to avoid confusion with the 4-force, F. Consider the case where m is constant. (The mass m wouldn't be constant if the object were being heated, or if extra mass were being added to it. We won't concern ourselves with such cases here.) F can then be written as

$$F \equiv \frac{dP}{d\tau} \equiv \frac{d(mV)}{d\tau} = m\frac{dV}{d\tau} \equiv mA. \tag{4.11}$$

We therefore still have a nice "F equals mA" physical law, but it's now a 4-vector equation instead of the old 3-vector equation. In terms of the acceleration 4-vector, we can use Eqs. (4.8) and (4.9) to write (if m is constant)

$$\begin{aligned} F = mA &= m(\gamma^4 v\dot{v},\, \gamma^4 v\dot{v}\mathbf{v} + \gamma^2\mathbf{a}) \\ &= m(\gamma^4 v_x a_x,\, \gamma^4 a_x,\, \gamma^2 a_y,\, \gamma^2 a_z). \end{aligned} \tag{4.12}$$

Combining this with Eq. (4.10), we see that the 3-force is

$$\mathbf{f} = m(\gamma^3 a_x,\, \gamma a_y,\, \gamma a_z), \tag{4.13}$$

in agreement with Eq. (3.65). In the rest frame of the object (or, rather, the instantaneous inertial frame), the F in Eq. (4.10) reduces to $F = (0, \mathbf{f})$, because $dE/dt = 0$ when $v = 0$. (This is true because $dE/dt = d(m/\sqrt{1-v^2})/dt$, and this derivative contains a factor of v, which is zero here.) Also, in the rest frame of the object, the mA in Eq. (4.12) reduces to $mA = (0, m\mathbf{a})$. So $F = mA$ reduces to $(0, \mathbf{f}) = (0, m\mathbf{a}) \implies \mathbf{f} = m\mathbf{a}$, which is Newton's second law (when the mass m is constant).

4.3 Properties of 4-vectors

4-vectors have many useful properties. Let's look at a few of them.

- **Linear combinations:** If A and B are 4-vectors, then $C \equiv aA + bB$ is also a 4-vector, for any constants a and b. This is true because the transformations in Eq. (4.1) are linear. This linearity implies that the transformation of, say, the time component is (using the given information that A and B are each 4-vectors)

$$\begin{aligned} C_0 \equiv aA_0 + bB_0 &= a(A_0' + vA_1') + b(B_0' + vB_1') \\ &= (aA_0' + bB_0') + v(aA_1' + bB_1') \\ &\equiv C_0' + vC_1', \end{aligned} \tag{4.14}$$

which is the correct transformation for the time component of a 4-vector. Likewise for the other components. This property holds, of course, just as it does for linear combinations of vectors in 3-space.

- **Inner-product invariance:** Consider two arbitrary 4-vectors, A and B. Define their inner product as we did in Eq. (3.36):

$$A \cdot B \equiv A_0B_0 - A_1B_1 - A_2B_2 - A_3B_3 \equiv A_0B_0 - \mathbf{A} \cdot \mathbf{B}. \tag{4.15}$$

Then $A \cdot B$ is invariant. That is, it is independent of the frame in which it is calculated:

$$\boxed{A \cdot B = A' \cdot B'} \tag{4.16}$$

This can be shown by direct calculation, using the transformations in Eq. (4.1):

$$\begin{aligned}
A \cdot B &\equiv A_0B_0 - A_1B_1 - A_2B_2 - A_3B_3 \\
&= \big(\gamma(A_0' + vA_1')\big)\big(\gamma(B_0' + vB_1')\big) - \big(\gamma(A_1' + vA_0')\big)\big(\gamma(B_1' + vB_0')\big) \\
&\quad - A_2'B_2' - A_3'B_3' \\
&= \gamma^2\big(A_0'B_0' + \cancel{v(A_0'B_1' + A_1'B_0')} + v^2A_1'B_1'\big) \\
&\quad - \gamma^2\big(A_1'B_1' + \cancel{v(A_1'B_0' + A_0'B_1')} + v^2A_0'B_0'\big) - A_2'B_2' - A_3'B_3' \\
&= A_0'B_0' \cdot \gamma^2(1 - v^2) - A_1'B_1' \cdot \gamma^2(1 - v^2) - A_2'B_2' - A_3'B_3' \\
&= A_0'B_0' - A_1'B_1' - A_2'B_2' - A_3'B_3' \\
&\equiv A' \cdot B'.
\end{aligned} \tag{4.17}$$

The importance of this result cannot be overstated. This invariance is analogous to the invariance of the inner product $\mathbf{A} \cdot \mathbf{B}$ under rotations in 3-space.

The above inner product is also invariant under rotations in 3-space, because it involves the standard "dot product" $\mathbf{A} \cdot \mathbf{B}$. This dot product is indeed invariant under rotations, because it can be shown that $\mathbf{A} \cdot \mathbf{B} = |\mathbf{A}||\mathbf{B}|\cos\theta$, where θ is the angle between $\mathbf{A}$ and $\mathbf{B}$. And none of $|\mathbf{A}|$, $|\mathbf{B}|$, or $\cos\theta$ are affected by a rotation.

The minus signs in the above definition of the inner product may seem a little strange. But the goal is to construct a combination of two arbitrary 4-vectors that is invariant under a Lorentz transformation, because such combinations are very useful in seeing how a system behaves. The nature of the Lorentz transformations demands that there be opposite signs in the inner product, so that's the way it is. The invariance of $c^2t^2 - x^2$ in Eq. (2.25) and the invariance of $E^2 - p^2c^2$ in Eq. (3.22) are special cases of Eq. (4.17) when $A = B$.

- **Norm:** As a corollary to the invariance of the inner product, we can look at the inner product of a 4-vector with itself. The norm, $|A|$, is defined to be the square root of this inner product (which is invariant). So we have

$$|A|^2 \equiv A \cdot A \equiv A_0A_0 - A_1A_1 - A_2A_2 - A_3A_3 = A_0^2 - |\mathbf{A}|^2. \tag{4.18}$$

The invariance of the norm $\sqrt{A \cdot A}$ is analogous to the invariance of the norm $\sqrt{\mathbf{A} \cdot \mathbf{A}} \equiv |\mathbf{A}|$ for rotations in 3-space. Special cases of the invariance of the 4-vector norm are the invariance of $c^2t^2 - x^2$ and of $E^2 - p^2c^2$.

- **A theorem:** Here's a nice little theorem:

If a certain one of the components of a 4-vector is zero in every frame, then all four components are zero in every frame.

Proof: If one of the space components (say, A_1) is zero in every frame, then the other space components must also be zero in every frame, because otherwise an appropriate rotation to a new frame would make $A_1 \neq 0$ in that frame. Also, the time component A_0 must be zero in every frame, because otherwise a Lorentz transformation (in the x direction) to a new frame would make $A_1 \neq 0$ in that frame.

If the time component, A_0, is zero in every frame, then the space components must also be zero in every frame, because otherwise a Lorentz transformation (in the appropriate direction) to a new frame would make $A_0 \neq 0$ in that frame. ∎

If someone comes along and says that she has a vector in 3-space that has no x component, no matter how you rotate the axes, then you would certainly say that the vector must be the zero vector. The situation in Lorentzian 4-space is the same, because all of the coordinates get intertwined with each other in the Lorentz (and rotation) transformations.

4.4 Energy, momentum

4.4.1 Norm

Many useful things arise from the fact that the P in Eq. (4.6) is a 4-vector. The invariance of the norm implies that $P \cdot P = E^2 - |\mathbf{p}|^2$ is invariant. If we are dealing with a single particle, we can determine the value of P^2 by conveniently working in the rest frame of the particle, where $\mathbf{p} = 0$ and $E = m$. This gives

$$E^2 - p^2 = m^2, \tag{4.19}$$

or $E^2 - p^2c^2 = m^2c^4$, with the c's. We already knew this, of course, from just writing out $E^2 - p^2 = \gamma^2 m^2 - \gamma^2 m^2 v^2 = m^2$ in Eq. (3.11).

The norm can also be useful for a collection of particles. If a process involves many particles, then we can say that for *any* subset of the particles,

$$\left(\sum E\right)^2 - \left(\sum \mathbf{p}\right)^2 \quad \text{is invariant,} \tag{4.20}$$

because this is the norm of the sum of the energy-momentum 4-vectors of the chosen particles. And the sum is again a 4-vector, due to the linearity of Eq. (4.1). What is the value of the invariant in Eq. (4.20)? The most concise description (which is basically a tautology) is that it is the square of the energy in the CM frame, that is, in the frame in which $\sum \mathbf{p} = 0$. For one particle, this reduces to m^2. Note that the sums are taken before squaring in Eq. (4.20). Squaring before adding would simply give the sum of the squares of the masses.

4.4.2 Transformations of E and p

We already know how the energy and momentum transform (see Section 3.2), but let's produce the transformation again here in a very quick and easy manner. We know that (E, p_x, p_y, p_z) is a 4-vector. So it must transform according to Eq. (4.1). Therefore, for a transformation in the x direction, we have

$$\begin{aligned} E &= \gamma(E' + vp'_x), \\ p_x &= \gamma(p'_x + vE'), \\ p_y &= p'_y, \\ p_z &= p'_z, \end{aligned} \tag{4.21}$$

in agreement with Eq. (3.19). That's all there is to it. The fact that E and $\mathbf{p}$ are part of the same 4-vector provides the following easy way to see that if one of them is conserved (in every frame) in a collision, then the other is also. Consider an interaction among a set of particles, and look at the 4-vector $\Delta P \equiv P_{\text{after}} - P_{\text{before}}$. If E is conserved in every frame, then the time component of ΔP is zero in every frame. But then the theorem in Section 4.3 says that all four components of ΔP are zero in every frame. Therefore, $\mathbf{p}$ is conserved. Likewise for the case where one of the p_i is known to be conserved.

4.5 Force and acceleration

The goal in this section is to determine how forces and accelerations transform between frames, and thereby reproduce the results in Section 3.5.3. We'll restrict the discussion to objects with constant mass, which we'll call "particles." The treatment here can be generalized to cases where the mass changes (for example, the object is being heated, or extra mass is being dumped on it), but we won't concern ourselves with these.

4.5.1 Transformation of forces

Let's first look at the force 4-vector in the instantaneous inertial frame S' of a given particle. Eq. (4.10) becomes

$$F' = \gamma\left(\frac{dE'}{dt}, \mathbf{f}'\right) = (0, \mathbf{f}'). \tag{4.22}$$

The first component here is zero due to the reasoning in the paragraph following Eq. (4.13). Equivalently, Eq. (4.22) follows from using Eq. (4.12) with a speed of zero.

We will now write down two expressions for the 4-force F in another frame S in which the particle moves with velocity v in the x direction. First, since F is a 4-vector, it transforms according to Eq. (4.1). So we have, using Eq. (4.22),

$$\begin{aligned} F_0 &= \gamma(F_0' + vF_1') = \gamma v f_x', \\ F_1 &= \gamma(F_1' + vF_0') = \gamma f_x', \\ F_2 &= F_2' = f_y', \\ F_3 &= F_3' = f_z'. \end{aligned} \tag{4.23}$$

But second, from the definition in Eq. (4.10), we have

$$\begin{aligned} F_0 &= \gamma\, dE/dt, \\ F_1 &= \gamma f_x, \\ F_2 &= \gamma f_y, \\ F_3 &= \gamma f_z. \end{aligned} \tag{4.24}$$

Combining Eqs. (4.23) and (4.24) yields

$$\begin{aligned} dE/dt &= v f_x', \\ f_x &= f_x', \\ f_y &= f_y'/\gamma, \\ f_z &= f_z'/\gamma. \end{aligned} \tag{4.25}$$

(frame S', particle frame)

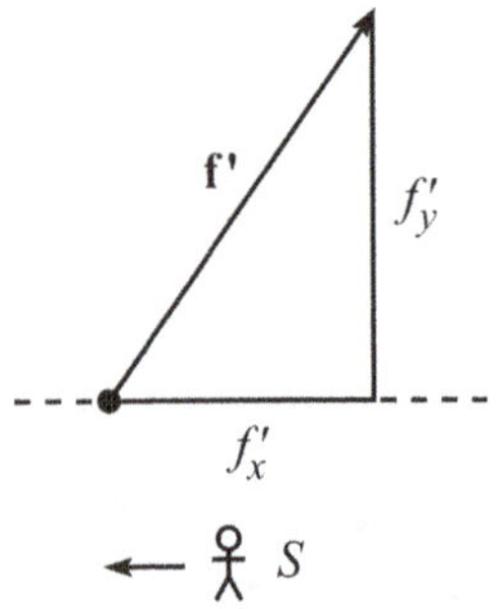

Figure 4.2

(frame S)

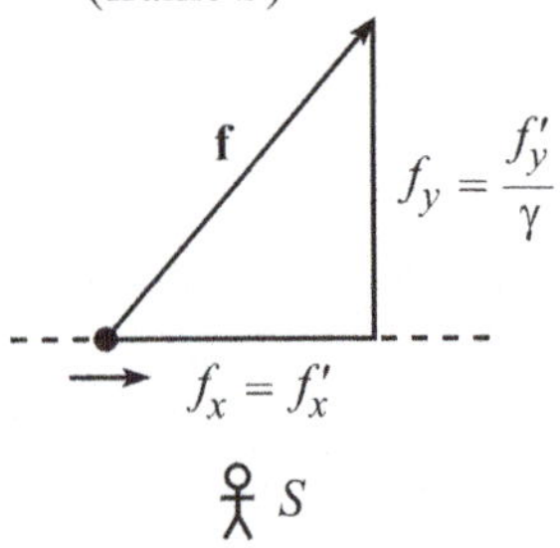

Figure 4.3

We therefore recover the results in Section 3.5.3; see Eq. (3.71). The longitudinal force is the same in both frames, but the transverse forces are larger by a factor of γ in the particle's frame. Hence, f_y/f_x decreases by a factor of γ when going from the particle's frame to any other frame; see Fig. 4.2 and Fig. 4.3. And as a bonus, the F_0 component in Eq. (4.25) tells us (after multiplying through by dt and using $v\,dt = dx$ and $f_x' = f_x$) that $dE = f_x\,dx$, which is the work-energy result in Eq. (3.60).

As noted in the first remark in Section 3.5.3, we can't switch the S and S' frames and write $f_y' = f_y/\gamma$. When talking about the forces on a particle, there is indeed a preferred reference frame, namely the frame S' of the particle. All frames are not equivalent here. When forming all of the 4-vectors in Section 4.2, we explicitly used the $d\tau$, dt, dx, etc., for two events, and it was understood that these two events were located on the particle's worldline.

4.5.2 Transformation of accelerations

The procedure here is similar to the above procedure for the force. Let's first look at the acceleration 4-vector in the instantaneous inertial frame S' of a given particle. Since $v' = 0$ in S', Eq. (4.8) or Eq. (4.9) gives

$$A' = (0, \mathbf{a}'). \tag{4.26}$$

We will now write down two expressions for the 4-acceleration A in another frame S in which the particle moves with velocity v in the x direction. First, since A is a 4-vector, it transforms according to Eq. (4.1). So we have, using Eq. (4.26),

$$\begin{aligned} A_0 &= \gamma(A'_0 + vA'_1) = \gamma v a'_x, \\ A_1 &= \gamma(A'_1 + vA'_0) = \gamma a'_x, \\ A_2 &= A'_2 = a'_y, \\ A_3 &= A'_3 = a'_z. \end{aligned} \tag{4.27}$$

But second, from the expression in Eq. (4.9), we have

$$\begin{aligned} A_0 &= \gamma^4 v a_x, \\ A_1 &= \gamma^4 a_x, \\ A_2 &= \gamma^2 a_y, \\ A_3 &= \gamma^2 a_z. \end{aligned} \tag{4.28}$$

Combining Eqs. (4.27) and (4.28) yields

$$\begin{aligned} a_x &= a'_x/\gamma^3, \\ a_x &= a'_x/\gamma^3, \\ a_y &= a'_y/\gamma^2, \\ a_z &= a'_z/\gamma^2. \end{aligned} \tag{4.29}$$

(The first two equations here are redundant.) We therefore again recover the results in Section 3.5.3; see Eqs. (3.68) and (3.69). So a_y/a_x increases by a factor of $\gamma^3/\gamma^2 = \gamma$ when going from the particle's frame to any other frame; see Fig. 4.4 and Fig. 4.5. This is the opposite of the effect on f_y/f_x.[1] This difference makes it clear that a law of the form $\mathbf{f} = m\mathbf{a}$ wouldn't make any sense. If it were true in one frame, it wouldn't be true in another.

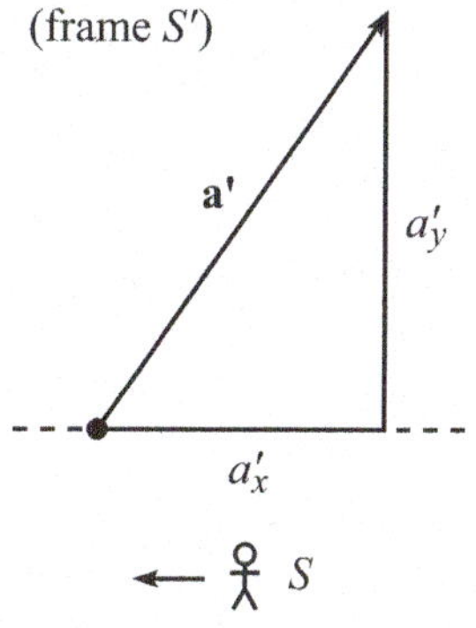

Figure 4.4

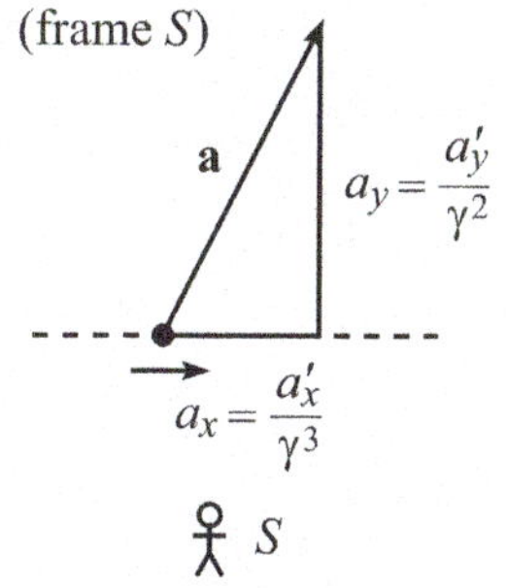

Figure 4.5

Example (Acceleration for circular motion): A particle moves with constant speed v around the circle $x^2 + y^2 = r^2$, $z = 0$, in the lab frame. At the instant the particle crosses the negative y axis (see Fig. 4.6), find the 3-acceleration and 4-acceleration in both the lab frame and the instantaneous inertial frame of the particle (with axes chosen parallel to the lab's axes).

Solution: Let the lab frame be S, and let the particle's instantaneous inertial frame be S' when it crosses the negative y axis. Then S and S' are related by a Lorentz transformation in the x direction. The 3-acceleration in S is simply

$$\mathbf{a} = (0, v^2/r, 0). \tag{4.30}$$

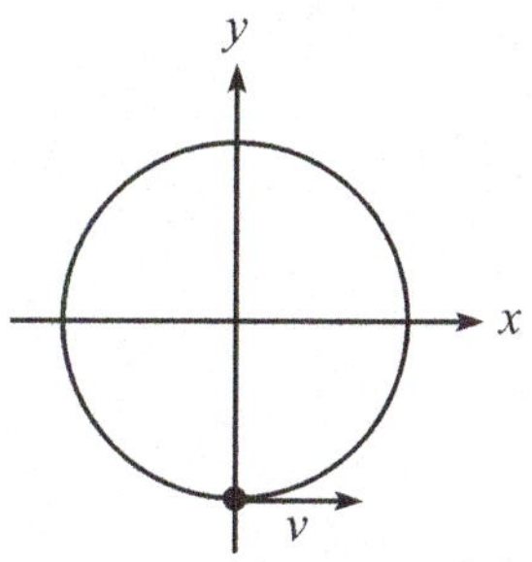

Figure 4.6

[1] In a nutshell, this difference is due to the fact that γ changes with time. When talking about the acceleration 4-vector, there are γ's that we have to differentiate; see Eq. (4.7). This isn't the case with the force 4-vector, because the γ is absorbed into the definition of $\mathbf{p} \equiv \gamma m\mathbf{v}$; see Eq. (4.10). This is what leads to the different powers of γ in Eq. (4.28), in contrast with the identical powers in Eq. (4.24).

There's nothing fancy going on here; the standard nonrelativistic proof of the centripetal acceleration $a = v^2/r$ works just fine again in the relativistic case. Eq. (4.8) or Eq. (4.9) then gives the 4-acceleration in S as

$$A = (0, 0, \gamma^2 v^2/r, 0). \tag{4.31}$$

To find the acceleration vectors in S', we can use the fact S' and S are related by a Lorentz transformation in the x direction. This means that the A_2 component of the 4-acceleration is unchanged. So the 4-acceleration in S' is also

$$A' = A = (0, 0, \gamma^2 v^2/r, 0). \tag{4.32}$$

In the particle's frame, $\mathbf{a}'$ is the space part of A (using Eq. (4.8) or Eq. (4.9), with $v = 0$ and $\gamma = 1$). Therefore, the 3-acceleration in S' is

$$\mathbf{a}' = (0, \gamma^2 v^2/r, 0). \tag{4.33}$$

Note that our results for $\mathbf{a}$ and $\mathbf{a}'$ are consistent with Eq. (4.29). The y acceleration is larger by a factor of γ^2 in the particle's frame . Alternatively, we can arrive at the two factors of γ in $\mathbf{a}'$ by using a simple time-dilation argument. We have (this is just a repeat of Eq. (3.68))

$$a'_y \equiv \frac{d^2y'}{dt'^2} = \frac{d^2y}{(dt/\gamma)^2} = \gamma^2\frac{d^2y}{dt^2} \equiv \gamma^2 a_y = \gamma^2\frac{v^2}{r}, \tag{4.34}$$

where we have used the fact that transverse lengths are the same in the two frames.

4.6 The form of physical laws

The first postulate of special relativity states that all inertial frames are equivalent. Therefore, if a physical law holds in one frame, then it must hold in all frames. Otherwise, it would be possible to differentiate between frames. As noted in the preceding section, the statement "$\mathbf{f} = m\mathbf{a}$" cannot be a physical law. The two sides of this equation transform differently when going from one frame to another, so the statement cannot be true in all frames. If a statement has any chance of being true in all frames, it must involve only 4-vectors. Consider a 4-vector equation (say, "$A = B$") that is true in frame S. Then if we apply to this equation a Lorentz transformation (call it $\mathcal{M}$) from S to another frame S', we have

$$\begin{aligned} A &= B \\ \Longrightarrow \quad \mathcal{M}A &= \mathcal{M}B \\ \Longrightarrow \quad A' &= B'. \end{aligned} \tag{4.35}$$

The law is therefore also true in frame S'. Of course, there are many 4-vector equations that are simply not true in any frame (for example, $F = P$, or $2P = 3P$). Only a small set of such equations (for example, $F = mA$) are true in at least one frame, and hence in all frames.

Physical laws may also take the form of scalar equations, such as $P \cdot P = m^2$. A scalar is by definition a quantity that is frame independent (as we have shown the inner product to be). So if a scalar statement is true in one inertial frame, then it is true in all inertial frames. Physical laws may also be higher-rank "tensor" equations, such as the ones that arise in electromagnetism and general relativity. We won't discuss tensors here, but suffice it to say that they may be thought of as things built up from 4-vectors. Scalars and 4-vectors are special cases of tensors.

All of this is exactly analogous to the situation in 3-D space. In Newtonian mechanics, $\mathbf{f} = m\mathbf{a}$ is a possible law, because both sides are 3-vectors. But $\mathbf{f} = m(2a_x, a_y, a_z)$ is not a possible law, because the righthand side is not a 3-vector; it depends on which axis you label as the x axis. If a particle has acceleration a in the eastward direction, and if you happen to pick this as your x direction, then the force is $2ma$ eastward. But if you happen to pick eastward as your y direction, then the force is ma eastward. It makes no sense for a law to give two different results, depending on your arbitrary choice of axis labels. An example of a frame-independent statement (under rotations) is the claim that a given stick has a length of 2 meters. This is a reasonable statement to make, because it involves the norm, which is a scalar. But if you say that the stick has an x component of 1.7 meters, then this cannot be true in all frames.

God said to his cosmos directors,
"I've added some stringent selectors.
One is the clause
That your physical laws
Shall be written in terms of 4-vectors."

4.7 Summary

In this chapter we learned about 4-vectors. In particular, we learned:

- A 4-tuple, $A = (A_0, A_1, A_2, A_3)$, is a 4-vector if the A_i transform according to the Lorentz transformation,

$$\begin{aligned} A_0 &= \gamma(A_0' + (v/c)A_1'), \\ A_1 &= \gamma(A_1' + (v/c)A_0'), \\ A_2 &= A_2', \\ A_3 &= A_3'. \end{aligned} \tag{4.36}$$

  Additionally, the last three components must transform like a usual vector under rotations in 3-D space.

- Some common 4-vectors are (dropping the c's):

$$\begin{aligned} \text{Displacement:}\quad & dS = (dt, dx, dy, dz). \\ \text{Velocity:}\quad & V = \frac{dS}{d\tau} = (\gamma, \gamma\mathbf{v}) \\ \text{Energy-momentum:}\quad & P = mV = (\gamma m, \gamma m\mathbf{v}) = (E, \mathbf{p}) \\ \text{Acceleration:}\quad & A = \frac{dV}{d\tau} = \frac{d}{d\tau}(\gamma, \gamma\mathbf{v}) = \gamma\left(\frac{d\gamma}{dt}, \frac{d(\gamma\mathbf{v})}{dt}\right) \\ \text{Force:}\quad & F = \frac{dP}{d\tau} = \gamma\left(\frac{dE}{dt}, \frac{d\mathbf{p}}{dt}\right) = \gamma\left(\frac{dE}{dt}, \mathbf{f}\right) \end{aligned} \tag{4.37}$$

- Because the Lorentz transformations are linear, any linear combination of two 4-vectors is again a 4-vector. Also, due to the properties of the Lorentz transformation, the inner product of two vectors,

$$A \cdot B \equiv A_0B_0 - A_1B_1 - A_2B_2 - A_3B_3, \tag{4.38}$$

  is invariant.

- The 4-vector transformations of P, A, and F provide a quick way of reproducing the transformation rules we derived in Chapter 3 for energy/momentum, acceleration, and force.
- Physical laws must be written in terms of 4-vectors (or more generally, tensors). 3-vectors are not permissable, because the laws would take different forms in different frames, in violation of the first postulate of relativity.

4.8 Problems

4.1. **Velocity addition** $**$

In A's frame, B moves to the right with speed u, and C moves to the left with speed v. What is the speed of B with respect to C? In other words, derive the velocity-addition formula (by using 4-vectors).

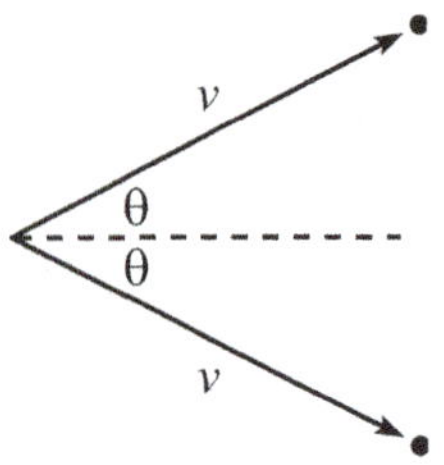

Figure 4.7

4.2. **Relative speed** $**$

In the lab frame, two particles move with speed v along the paths shown in Fig. 4.7. The angle between the trajectories is 2θ. What is the speed of one particle, as viewed by the other?

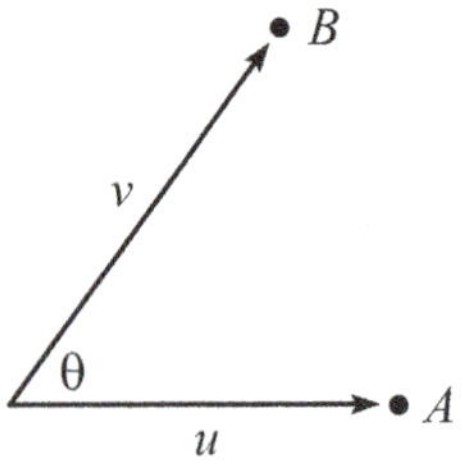

Figure 4.8

4.3. **Another relative speed** $**$

In the lab frame, particles A and B move with speeds u and v along the paths shown in Fig. 4.8. The angle between the trajectories is θ. What is the speed of one particle, as viewed by the other?

4.4. **Acceleration for linear motion** $**$

A spaceship starts at rest with respect to frame S and accelerates with constant proper acceleration a in the x direction. In Section 2.6 we showed that the speed of the spaceship with respect to S is given by $v(t') = \tanh(at')$; see Eq. (2.53). We are using t' instead of t here for the spaceship's proper time, and we have dropped the c's. In frame S, let V be the spaceship's 4-velocity, and let A be its 4-acceleration. In terms of the proper time t',

(a) Find V and A in frame S, by explicitly using $v(t') = \tanh(at')$.

(b) Write down V' and A' in the spaceship's frame, S'.

(c) Verify that V and V' transform like 4-vectors between the two frames. Likewise for A and A'.

4.5. **Linear force** $**$

A particle's velocity v and acceleration $\dot{v}$ (as measured in the lab frame) both point in the x direction. In the spirit of the example in Section 4.5.2, find the 3-force and 4-force in both the lab frame and the instantaneous inertial frame of the particle. Verify that the 3-forces are related according to Eq. (4.25).

4.9 Exercises

4.6. **Acceleration at rest** $*$

Show that the derivative of $v \equiv \sqrt{v_x^2 + v_y^2 + v_z^2}$ equals a_x, independent of how the various v_i's are changing, provided that $v_y = v_z = 0$ at the moment in question.

4.7. **Same speed** $*$

Consider the setup in Problem 4.2. Given v, what should θ be so that the speed of one particle, as viewed by the other, is also v? (The first form of the speed in Eq. (4.44) will make your calculations the simplest.) Do your answers make sense for $v \approx 0$ and $v \approx c$?

4.8. **Doppler effect** $*$

Consider a photon traveling in the positive x direction. Ignoring the y and z components, and setting $c = 1$, the 4-momentum is (p, p). In matrix notation, what are the Lorentz transformations for this vector, to the frames traveling to the left and to the right at speed v? What is the new 4-momentum of the photon in these new frames? Accepting the fact that a photon's energy is proportional to its frequency, verify that your results are consistent with the Doppler results in Section 2.5.1.

4.9. **Three particles** $**$

Three particles head off with equal speeds v, at 120° with respect to each other, as shown in Fig. 4.9. What is the inner product of any two of the 4-velocities, in any frame? Use your result to find the angle θ (see Fig. 4.10) at which two particles travel in the frame of the third.

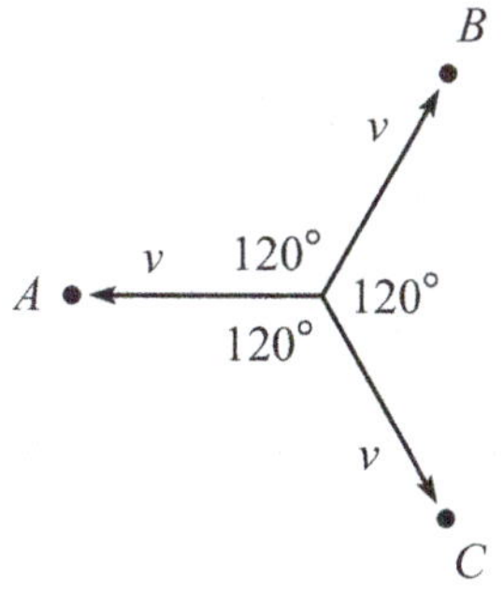

Figure 4.9

4.10. **Linear acceleration** $**$

A particle's velocity v and acceleration $\dot{v}$ (as measured in the lab frame) both point in the x direction. In the spirit of the example in Section 4.5.2, find the 3-acceleration and 4-acceleration in both the lab frame and the instantaneous inertial frame of the particle. Verify that the 3-accelerations are related according to Eq. (4.29).

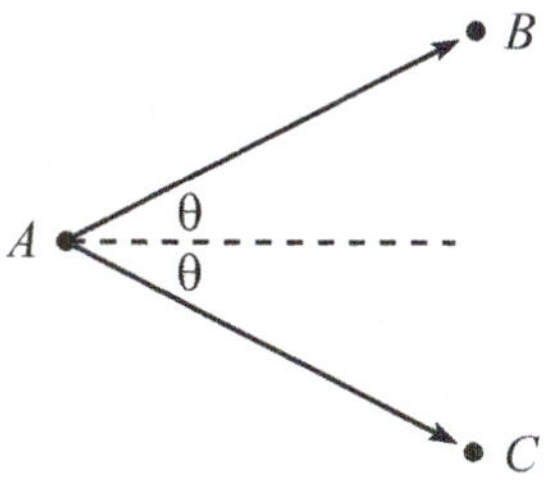

Figure 4.10

4.11. **Circular motion force** $**$

For the setup in the example in Section 4.5.2, find the 3-force and 4-force in both the lab frame and the instantaneous inertial frame of the particle. Verify that the 3-forces are related according to Eq. (4.25). (Solve this from scratch; don't just use the results from the example.)

4.10 Solutions

4.1. **Velocity addition**

Let the desired speed of B with respect to C be w; see the bottom picture in Fig. 4.11. (We've chosen B to be on the right side of C. The relative speed w would be the same if we had chosen B to be on the left.) In A's frame, the 4-velocity of B is $(\gamma_u, \gamma_u u)$, and the 4-velocity of C is $(\gamma_v, -\gamma_v v)$, where we have suppressed the y and z components. In C's frame, the 4-velocity of B is $(\gamma_w, \gamma_w w)$, and the 4-velocity of C is $(1, 0)$. The invariance of the inner product implies that

$$
\begin{aligned}
(\gamma_u, \gamma_u u) \cdot (\gamma_v, -\gamma_v v) &= (\gamma_w, \gamma_w w) \cdot (1, 0) \\
\Longrightarrow \quad \gamma_u \gamma_v (1 + uv) &= \gamma_w \\
\Longrightarrow \quad \frac{1 + uv}{\sqrt{1 - u^2}\sqrt{1 - v^2}} &= \frac{1}{\sqrt{1 - w^2}} . \qquad (4.39)
\end{aligned}
$$

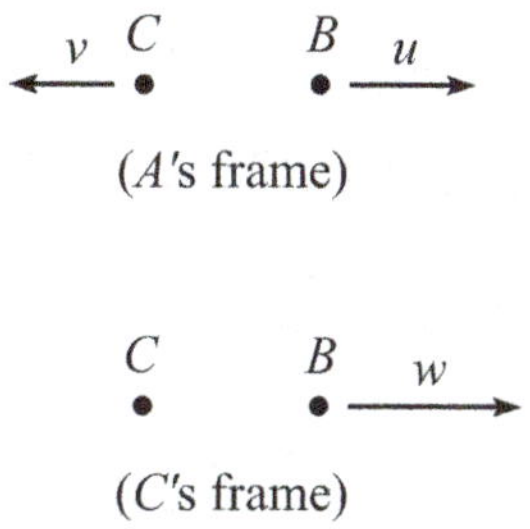

Figure 4.11

Inverting both sides and then squaring and solving for w gives (as you can verify)

$$
w = \frac{u + v}{1 + uv}, \qquad (4.40)
$$

which is the desired velocity-addition formula (without the c's).

4.2. **Relative speed**

In the lab frame, the 4-velocities of the particles are (suppressing the z component)

$$(\gamma_v, \gamma_v v\cos\theta, \gamma_v v\sin\theta) \qquad \text{and} \qquad (\gamma_v, \gamma_v v\cos\theta, -\gamma_v v\sin\theta). \tag{4.41}$$

Let w be the desired speed of one particle as viewed by the other. Then in the frame of one particle, the 4-velocities are (suppressing two spatial components)

$$(\gamma_w, \gamma_w w) \qquad \text{and} \qquad (1, 0), \tag{4.42}$$

where we have rotated the axes so that the relative motion is along the new x axis in this frame. This rotation step isn't necessary, since we know that one particle sees the other one moving vertically. But in a more complicated setup where the direction of the velocity isn't obvious (as in Problem 4.3), the freedom to perform a rotation can be very helpful.

Since the 4-vector inner product is invariant under Lorentz transformations and rotations, we have (using $\cos 2\theta = \cos^2\theta - \sin^2\theta$)

$$\begin{aligned}(\gamma_v, \gamma_v v\cos\theta, \gamma_v v\sin\theta)\cdot(\gamma_v, \gamma_v v\cos\theta, -\gamma_v v\sin\theta) &= (\gamma_w, \gamma_w w)\cdot(1, 0)\\ \Longrightarrow \quad \gamma_v^2(1 - v^2\cos 2\theta) &= \gamma_w.\end{aligned} \tag{4.43}$$

Using the definitions of the γ's, inverting, squaring, and solving for w gives

$$w = \sqrt{1 - \frac{(1-v^2)^2}{(1-v^2\cos 2\theta)^2}} = \frac{\sqrt{2v^2(1-\cos 2\theta) - v^4\sin^2 2\theta}}{1 - v^2\cos 2\theta}. \tag{4.44}$$

If desired, this can be rewritten (using the double-angle formulas) in the form,

$$w = \frac{2v\sin\theta\sqrt{1 - v^2\cos^2\theta}}{1 - v^2\cos 2\theta}. \tag{4.45}$$

See the solution to Problem 2.3 for some limiting cases.

4.3. **Another relative speed**

In the lab frame, the 4-velocities of the particles are (suppressing the z component)

$$V_A = (\gamma_u, \gamma_u u, 0) \qquad \text{and} \qquad V_B = (\gamma_v, \gamma_v v\cos\theta, \gamma_v v\sin\theta). \tag{4.46}$$

Let w be the desired speed of one particle as viewed by the other. Then in the frame of one particle, the 4-velocities are (suppressing two spatial components)

$$(\gamma_w, \gamma_w w) \qquad \text{and} \qquad (1, 0), \tag{4.47}$$

where we have rotated the axes so that the relative motion is along the new x axis in this frame. This rotation greatly simplifies things, because the direction of the velocity of one particle as viewed by the other isn't obvious.

Since the 4-vector inner product is invariant under Lorentz transformations and rotations, we have

$$\begin{aligned}(\gamma_u, \gamma_u u, 0)\cdot(\gamma_v, \gamma_v v\cos\theta, \gamma_v v\sin\theta) &= (\gamma_w, \gamma_w w)\cdot(1, 0)\\ \Longrightarrow \quad \gamma_u\gamma_v(1 - uv\cos\theta) &= \gamma_w.\end{aligned} \tag{4.48}$$

Using the definitions of the γ's, inverting, squaring, and solving for w gives

$$w = \sqrt{1 - \frac{(1-u^2)(1-v^2)}{(1-uv\cos\theta)^2}} = \frac{\sqrt{u^2 + v^2 - 2uv\cos\theta - u^2v^2\sin^2\theta}}{1 - uv\cos\theta}. \tag{4.49}$$

See the solution to Problem 2.4 for some limiting cases.

4.4. **Acceleration for linear motion**

(a) Using $v(t') = \tanh(at')$, along with $\cosh^2(at') - \sinh^2(at') = 1$, we have $\gamma = 1/\sqrt{1-v^2} = \cosh(at')$. Therefore,

$$V = (\gamma, \gamma v) = \big(\cosh(at'), \sinh(at')\big), \tag{4.50}$$

where we have suppressed the two transverse components. Since t' is the spaceship's proper time, Eq. (4.7) tells us that $A = dV/dt'$, so

$$A = \frac{dV}{dt'} = a\big(\sinh(at'), \cosh(at')\big). \tag{4.51}$$

(b) The spaceship is at rest in its instantaneous inertial frame, so with $v = 0$, Eqs. (4.5) and (4.8) give

$$V' = (1, 0) \qquad \text{and} \qquad A' = (0, a). \tag{4.52}$$

Equivalently, these are obtained by setting $t' = 0$ in the results in part (a), because the spaceship hasn't picked up any speed yet at $t' = 0$. And it is always the case that in the spaceship's instantaneous rest frame, it hasn't picked up any speed.

(c) From Eq. (2.7), or equivalently from Eq. (4.1) written in matrix form, the Lorentz-transformation matrix from S' to S is

$$\mathcal{M} = \begin{pmatrix} \gamma & \gamma v \\ \gamma v & \gamma \end{pmatrix} = \begin{pmatrix} \cosh(at') & \sinh(at') \\ \sinh(at') & \cosh(at') \end{pmatrix}. \tag{4.53}$$

We must check that

$$\begin{pmatrix} V_0 \\ V_1 \end{pmatrix} = \mathcal{M}\begin{pmatrix} V'_0 \\ V'_1 \end{pmatrix} \qquad \text{and} \qquad \begin{pmatrix} A_0 \\ A_1 \end{pmatrix} = \mathcal{M}\begin{pmatrix} A'_0 \\ A'_1 \end{pmatrix}. \tag{4.54}$$

Plugging in the results from parts (a) and (b), we quickly see that these relations are true.

4.5. **Linear force**

Let S be the lab frame and S' be the particle's frame. From Eq. (4.13) with $a_x = \dot{v}$, the 3-force in S is

$$\mathbf{f} = m(\gamma^3 \dot{v}, 0, 0). \tag{4.55}$$

And from Eq. (4.12), the 4-force in S is

$$F = m(\gamma^4 v\dot{v}, \gamma^4 \dot{v}, 0, 0). \tag{4.56}$$

The Lorentz transformation (with minus signs because S' sees S move to the left) then gives the 4-force in the particle's frame as

$$\begin{aligned} F'_0 &= \gamma(F_0 - vF_1) = \gamma m\gamma^4(v\dot{v} - v\cdot\dot{v}) = 0, \\ F'_1 &= \gamma(F_1 - vF_0) = \gamma m\gamma^4(\dot{v} - v\cdot v\dot{v}) = m\gamma^3\dot{v}. \end{aligned} \tag{4.57}$$

Therefore, $F' = m(0, \gamma^3\dot{v}, 0, 0)$. And since the velocity in S' is zero, the discussion following Eq. (4.13) tells us that $\mathbf{f}'$ equals the space part of F'. So we have

$$\mathbf{f}' = m(\gamma^3\dot{v}, 0, 0). \tag{4.58}$$

Comparing this with Eq. (4.55) gives $f_x = f'_x$, in agreement with Eq. (4.25).

Chapter 5

General Relativity

This chapter presents an introduction to general relativity (GR).[1] We won't have enough time to get to the heart of the subject, but we'll still be able to get a flavor of it and derive a few interesting GR results. A crucial ingredient of GR is the Equivalence Principle, which says that gravity is equivalent to acceleration. Or in more practical terms, it says that you can't tell the difference. We'll have much to say about this throughout the chapter. Another crucial concept in GR is that of coordinate independence: the laws of physics can't depend on what coordinate system you use. This seemingly innocuous statement has surprisingly far-reaching consequences. However, a discussion of this topic is one of the many things we won't have time for. We would need a whole course on GR to do it justice. But fortunately it's possible to get a sense of the nature of GR without having to master such things. This is the route we'll take in this chapter.

The outline of this chapter is as follows. Section 5.1 introduces the Equivalence Principle, which states that gravity is equivalent to acceleration. Alternatively, it states that "gravitational" mass is equal to (or proportional to) "inertial" mass. In Section 5.2 we use the Equivalence Principle to derive the gravitational time-dilation effect: high clocks in a gravitational field run fast. In Section 5.3 we investigate a reference frame that accelerates in such a way that distances remain fixed in it. A number of interesting properties emerge. Section 5.4 covers the maximal-proper-time principle, which states that a particle under the influence of only gravity has the maximum proper time. In Section 5.5 we revisit the twin paradox, armed with our knowledge of the Equivalence Principle.

5.1 The Equivalence Principle

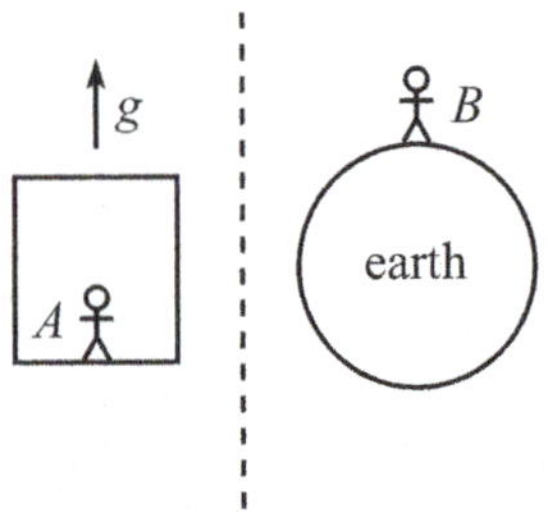

Figure 5.1

Einstein's Equivalence Principle says that it is impossible to locally distinguish gravity from acceleration. This may be stated more precisely in (at least) three ways.

- Let person A be enclosed in a small box, far from any massive objects, that undergoes uniform acceleration (say, g). Let person B stand at rest on the earth; see Fig. 5.1. The Equivalence Principle says that there are no local experiments

[1] A decade after his 1905 paper on the special theory of relativity, Einstein completed his general theory of relativity in 1915, collaborating with Marcel Grossmann during the latter part of this period. David Hilbert also developed many of the final pieces of the theory in parallel with Einstein; see Medicus (1984). For a wonderful account of other historical developments of the theory, see Chandrasekhar (1979).

these two people can perform that will tell them which of the two settings they are in. The physics of each setting is the same.

- Let person A be enclosed in a small box that is in free-fall near a planet. Let person B float freely in space, far away from any massive objects; see Fig. 5.2. The Equivalence Principle says that there are no local experiments these two people can perform that will tell them which of the two settings they are in. The physics of each setting is the same.

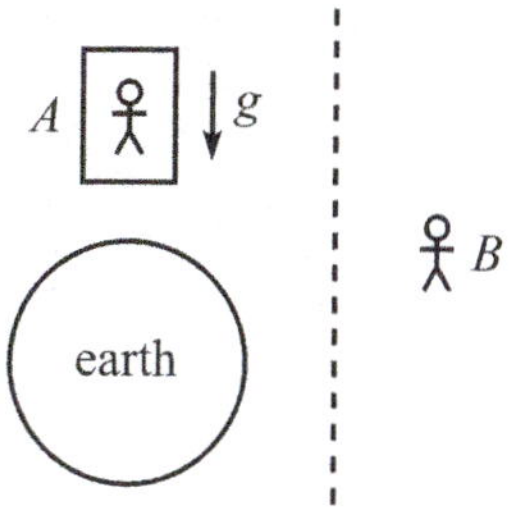

Figure 5.2

- "Gravitational" mass is equal to (or proportional to) "inertial" mass. Gravitational mass is the m_g that appears in Newton's universal law of gravitation, $F = GMm_g/r^2 \equiv m_g g$. Inertial mass is the m_i that appears in Newton's second law, $F = m_i a$. There is no *a priori* reason why these two m's should be the same (or proportional). An object that is dropped on the earth has an acceleration that is given by

$$F = m_i a \implies m_g g = m_i a \implies a = \left(\frac{m_g}{m_i}\right) g. \tag{5.1}$$

For all we know, the ratio m_g/m_i for plutonium might be different from that for copper. But experiments with various materials have detected no difference in the ratios. The Equivalence Principle states that the ratios are equal for any type of mass.

This definition of the Equivalence Principle is equivalent to, say, the second one above for the following reason. A mass that starts at rest near B will stay right next to B as they both float freely in space. But a mass that starts next to A (at rest with respect to A) will stay right next to A if and only if its acceleration is equal to A's, that is, if and only if its m_g/m_i ratio is the same as A's m_g/m_i ratio. If the ratios are different, then the mass will diverge from A, which means that it is possible to distinguish between the two settings, in contradiction to the Equivalence Principle.

The above three statements are all quite believable. Consider the first one, for example. When standing on the earth, you have to keep your legs firm to avoid falling down. And when standing in the accelerating box, you have to keep your legs firm to maintain the same position relative to the floor (that is, to avoid "falling down" with respect to the box). You certainly can't naively tell the difference between the two scenarios. The Equivalence Principle says that it's not just that you're too inept to figure out a way to differentiate between them, but instead that there is no possible local experiment you can perform to tell the difference, no matter how clever you are.

Note the inclusion of the words "small box" and "local" in the first two statements of the Equivalence Principle. Near the surface of the earth, the lines of the gravitational force are not parallel; they converge to the center of the earth. The gravitational force also varies with height, due to the $1/r^2$ factor in the GMm_g/r^2 force law. Therefore, an experiment performed over a nonnegligible distance (for example, dropping two balls next to each other, and watching them converge; or dropping two balls on top of each other and watching them diverge) will have different results from the same experiment in an accelerating box, where the balls maintain a constant separation. The Equivalence Principle says that if your laboratory is small enough, or if the gravitational field is sufficiently uniform, then the two scenarios look essentially the same.

5.2 Time dilation

The Equivalence Principle has a striking consequence with regard to the behavior of clocks in a gravitational field. It implies that higher clocks run faster than lower clocks. If someone puts a watch on top of a tower, and if you stand on the ground, then you will see the watch on the tower tick faster than an identical watch on your wrist. If the watch on the tower is taken down and you compare it with the one on your wrist, it will show more time elapsed.[2] Likewise, someone standing on top of the tower will see a clock on the ground run slow. To be quantitative about this, let's consider the following two scenarios.

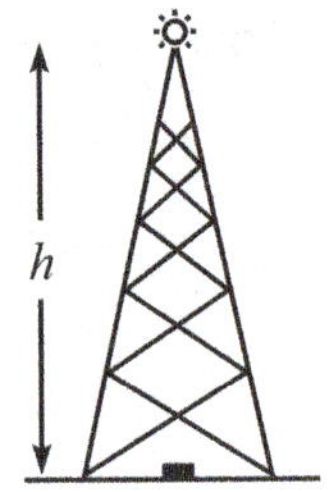

Figure 5.3

- A light source on top of a tower of height h emits flashes at time intervals t_s (as measured by it). A receiver on the ground receives the flashes at time intervals t_r (as measured by it). What is t_r in terms of t_s? See Fig. 5.3.

- A rocket with length h accelerates with acceleration g. A light source at the front end emits flashes at time intervals t_s (as measured by it). A receiver at the back end receives the flashes at time intervals t_r (as measured by it). What is t_r in terms of t_s? See Fig. 5.4.

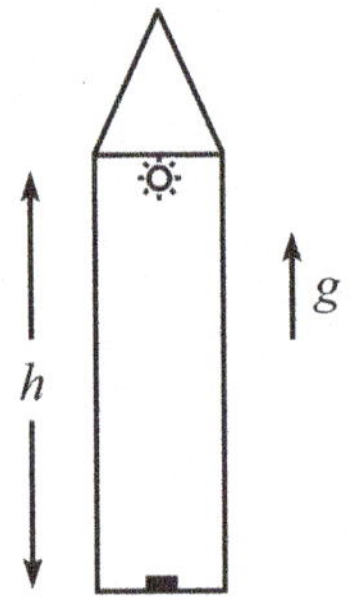

Figure 5.4

The Equivalence Principle tells us that these two scenarios are exactly the same (by the first of the three statements of the Equivalence Principle in Section 5.1), as far as the sources and receivers are concerned. Hence the relation between t_r and t_s must be the same in each. Therefore, to find out what is going on in the first scenario, we will study the second scenario, because we can figure out how that one behaves.

Consider the instantaneous inertial frame S of the rocket at a given moment. In this frame, the rocket is momentarily at rest (at, say, $t = 0$), and then it accelerates out of the frame with acceleration g. The following discussion will be made with respect to the frame S. Consider a series of quick light flashes emitted from the source, starting at $t = 0$. The distance the rocket has traveled out of S at a small time t is $gt^2/2$, which is very small. So if we assume that t_s is extremely small ($t_s \ll t$), then we may say that many flashes are emitted before the rocket moves appreciably. Likewise, the speed of the source, namely gt, is also very small. We may therefore ignore the motion (both displacement and speed) of the rocket, as far as the light source is concerned.

However, the light takes a nonzero time to reach the receiver as it travels the length of the rocket, and by then the receiver will be moving. We therefore *cannot* ignore the motion of the rocket when dealing with the receiver. The time it takes the light to reach the receiver is $t_h = h/c$, at which point the receiver has a speed of $v = gt_h = g(h/c)$.[3] Therefore, by the standard classical Doppler effect, the time between the received flashes is[4]

$$t_r = \frac{t_s}{1 + (v/c)} = \frac{t_s}{1 + (gh/c^2)} . \tag{5.2}$$

[2]Technically, this is true only if the watch on the tower is kept there for a long enough time, because the movement of the watch when it is taken down causes it to run slow, due to the usual special-relativistic time dilation. But the speeding-up effect due to the height can be made arbitrarily large compared with the slowing-down effect due to the motion, by simply keeping the watch on the tower for an arbitrarily long time.

[3]The receiver moves a tiny bit during this time, so the "h" here should really be replaced by a slightly smaller distance. But this yields a negligible second-order effect in the small quantity gh/c^2, as you can show. To sum up, the displacement of the source, the speed of the source, and the displacement of the receiver are all negligible. But the speed of the receiver is quite relevant.

[4]Quick proof of the classical Doppler effect (for a moving receiver): As seen in frame S, when the receiver and a particular flash meet, the next flash is a distance ct_s behind. The receiver and this next flash then travel toward each other at relative speed $c + v$ (as measured by someone in S). The time between receptions is therefore $t_r = ct_s/(c + v)$, which can be rewritten as Eq. (5.2).

The $1 + gh/c^2$ factor here is the same as the factor we derived via a Minkowski diagram in Problem 2.12; see Eq. (2.104). Having found how the times are related, the frequencies $f_r = 1/t_r$ and $f_s = 1/t_s$ are related by

$$f_r = \left(1 + \frac{v}{c}\right) f_s = \left(1 + \frac{gh}{c^2}\right) f_s. \tag{5.3}$$

The flashes therefore hit the receiver at a faster rate (as measured by the receiver) than they are emitted from the source (as measured by the source). By the Equivalence Principle, the relation between the frequencies in Eq. (5.3) also applies to the scenario in Fig. 5.3, with the source on a tower. We therefore conclude that an observer on the ground must see the clock on the tower running fast, by the factor $1 + gh/c^2$. This is true because we can imagine that each emitted flash corresponds to a tick on the clock at the top of the tower. And we just found that an observer on the ground sees these flashes happening at a faster rate than one per second (as measured on a ground clock). This means that the upper clock really *is* running fast, compared with the lower clock.[5] That is,

$$\boxed{\Delta t_h = \left(1 + \frac{gh}{c^2}\right) \Delta t_0} \qquad \text{(gravitational time dilation)} \tag{5.4}$$

The time Δt_0 elapsed on the ground is smaller than the time Δt_h elapsed at height h. If a person on a tower claps her hands and then claps them again $\Delta t_h = 10$ seconds later according to her clock, then a person on the ground will measure a Δt_0 between the two claps (events) that is less than 10 seconds. A twin from Denver will be older than his twin from Boston when they meet up at a family reunion (all other things being equal).

> Greetings! Dear brother from Boulder,
> I hear that you've gotten much older.
> And please tell me why
> My lower left thigh
> Hasn't aged quite as much as my shoulder!

Note that the gh in Eq. (5.4) is the gravitational potential energy, mgh, divided by m.

After a nonzero time has passed, the instantaneous inertial frame S that we used in the above derivation will no longer be of any use to us. But we can always pick a new instantaneous inertial frame of the rocket, so we can repeat the above analysis at any later time. Therefore, the result in Eq. (5.3) holds at all times.

For a constant gravitational field, the result in Eq. (5.4) is technically only an approximation. But it is a very good one, because the inaccuracy involves higher powers of gh/c^2, and this term is very small for any practical purpose. It is easy to see that Eq. (5.4) can't be exactly correct, because the same reasoning that led to it also tells us that a high clock sees a low clock run slow by a factor $(1 - gh/c^2)$. But since a clock can't gain time on itself, this factor must be the inverse of the factor in Eq. (5.4). (This is true because the two clocks are in the same frame. It wouldn't be true for two clocks flying past each other. See Question 47 in Appendix A.) We should therefore have $(1 + gh/c^2)(1 - gh/c^2) = 1$. But since this isn't true, Eq. (5.4) can't be exactly correct. The task of Problem 5.2 is to derive the correct result (for a constant gravitational field).

[5] Unlike the setup in which two people fly past each other (as in the usual special-relativistic twin paradox), we can say here that what an observer *sees* is also what actually *is*. We don't have to worry about "is-ness" depending on the frame, because everyone here is in the same frame. The "turnaround" effect that was present in the twin paradox (see Exercise 1.30) isn't present now. The two clocks here can be slowly moved together without anything exciting or drastic happening to their readings. Since both observers are in the same frame in the tower setup, you might find general-relativistic time dilation even stranger and harder to believe than special-relativistic time dilation.

This gravitational time-dilation effect was first measured by R. Pound and G. Rebka in 1960. They sent gamma rays up a 22 m tower and measured the redshift (that is, the decrease in frequency) at the top. This was a notable feat, considering that they were able to measure a frequency shift of gh/c^2 (which is only a few parts in 10^{15}) to within 10% accuracy. In 1964, R. Pound and J. Snider improved the accuracy to 1%.

REMARK: You might object to the above derivation, because t_r is the time measured by someone in the inertial frame S. And since the receiver is eventually moving with respect to S, we should multiply the f_r in Eq. (5.3) by the usual special-relativistic time-dilation factor, $1/\sqrt{1-(v/c)^2}$ (because the receiver's clocks are running slow relative to S, so the frequency measured by the receiver is greater than that measured in S). However, this is a second-order effect in the small quantity $v/c = gh/c^2$. We already dropped other effects of the same order, so we have no right to keep this one. Of course, if the leading effect in our final answer turned out to be second-order in v/c, then we would know that our answer was garbage. But the leading effect happens to be first order, so we can afford to be careless with the second-order effects. ♣

Example (Airplane's speed): A plane flies at constant height h. What should its speed v be so that an observer on the ground sees the plane's clock tick at the same rate as a ground clock? You may (correctly) work in the approximation where $v \ll c$.

Solution: An observer on the ground sees the plane's clock run slow by the factor $\sqrt{1-v^2/c^2}$ due to SR time dilation. But she also sees it run fast by the factor $(1+gh/c^2)$ due to GR time dilation. We want the product of these two factors to equal 1. Since we're assuming $v \ll c$, we can apply the Taylor series $\sqrt{1-\epsilon} \approx 1-\epsilon/2$ to the first factor. This yields

$$\left(1-\frac{v^2}{2c^2}\right)\left(1+\frac{gh}{c^2}\right)=1 \implies 1-\frac{v^2}{2c^2}+\frac{gh}{c^2}-O\left(\frac{1}{c^4}\right)=1. \tag{5.5}$$

Neglecting the small term of order $1/c^4$ (the O is for "order"), and canceling the 1's, we obtain $v=\sqrt{2gh}$. Interestingly, $\sqrt{2gh}$ is also the answer to a standard question from Newtonian physics: How fast should you throw a ball straight up so that it reaches a height h?

REMARKS:

1. Let's be rigorous about why we can simply multiply the factors of $\sqrt{1-v^2/c^2}$ and $(1+gh/c^2)$, when determining how fast the ground observer sees the plane's clock tick. Consider four clocks: A (the observer) is stationary on the ground, B_1 and B_2 are stationary at height h, and C (the plane) is moving with speed v at height h; see Fig. 5.5. Consider the time separation, as measured by the various people, between the "C passing B_1" and "C passing B_2" events. Due to the GR time dilation, we know that $t_B=(1+gh/c^2)t_A$, where t_B is the (common) time elapsed on B_1 and B_2. And due to the SR time dilation, we know that $t_C=t_B/\gamma$. Therefore,

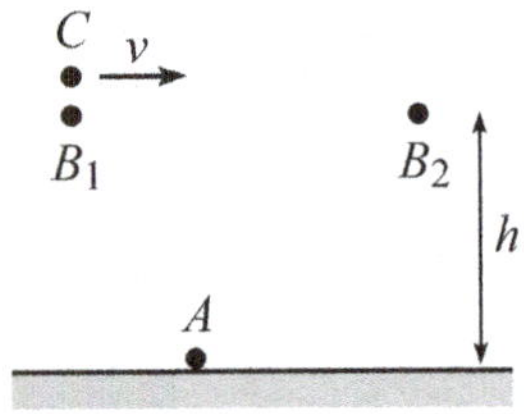

Figure 5.5

$$t_C=\frac{t_B}{\gamma}=\frac{(1+gh/c^2)t_A}{\gamma}=\sqrt{1-\frac{v^2}{c^2}}\left(1+\frac{gh}{c^2}\right)t_A, \tag{5.6}$$

as desired.

2. This is getting a little picky, but there is an ambiguity in this problem, although it is inconsequential at lowest order. When we said that the plane has speed v, we didn't specify whether this is with respect to a fixed object at height h, or with respect to the ground. In the above calculations, we used the first of these two meanings. Given this meaning, if you want to know the speed with respect to the ground, it is $(1+gh/c^2)v$, because the ground sees everything at height h happening faster by a factor $(1+gh/c^2)$. But any ambiguities in the definition of v will yield only tiny corrections of order $(gh/c^2)(v^2/c^2)$.

3. In the above solution, we assumed that the gravitational field is uniform. That is, we assumed that at all points, the field has the same magnitude g and points in the negative y direction. (So the field lines are everywhere parallel.) These assumptions don't actually hold for the earth's field, which drops off with the radius r, and which points radially. (The field lines aren't parallel; they converge toward the center of the earth.) But any errors introduced by these assumptions are negligible. ♣

5.3 Uniformly accelerating frame

Before reading this section, you should think carefully about the "Break or not break" problem in Chapter 2 (Problem 2.13). Don't look at the solution too soon, because chances are you will change your answer after a few more minutes of thought. This is a classic problem, so don't waste it by peeking!

Technically, the uniformly accelerating frame we'll construct here has nothing to do with GR. We won't need to leave the realm of special relativity for the analysis in this section. But the reason we choose to study this special-relativistic setup in detail is that it has many similarities with genuine GR situations, such as black holes.

5.3.1 Uniformly accelerating point particle

In order to understand a uniformly accelerating frame, we first need to understand a uniformly accelerating point particle. By "uniform," we mean that the particle feels a constant force (and hence a constant proper acceleration) in its instantaneous inertial frame. In Section 2.6.2 we briefly discussed the motion of a uniformly accelerating particle. We'll now take a closer look at this motion. Let the particle's instantaneous rest frame be S', and let it start at rest in the inertial frame S. Let its mass be m. We know from Section 3.5.3 that the longitudinal force is the same in the two frames. Therefore, since it is constant in S', it is also constant in S. Call it f. Since $f = dp/dt$, and since f is constant, we see that the momentum in S increases at a constant rate. That is, $p = ft \implies \gamma m v = ft$, where t is the time as measured in S. If we let $g \equiv f/m$ (so g is the proper acceleration felt by the particle), then $ft = \gamma m v$ becomes

$$gt = \gamma v \quad \Longrightarrow \quad (gt)^2 = \frac{v^2}{1 - v^2} \quad \Longrightarrow \quad v = \frac{gt}{\sqrt{1 + (gt)^2}}, \tag{5.7}$$

where we have dropped the c's. As a double-check, this v has the correct behavior for $t \to 0$ (where $v \approx gt$) and for $t \to \infty$ (where $v \to 1$, or $v \to c$ with the c's). If you want to keep the c's, then the $(gt)^2$ in Eq. (5.7) becomes $(gt/c)^2$ to make the units right. As another check on v, the task of Exercise 5.21 is to verify that $f = \gamma^3 ma$ in S.

Having found the speed v in S at time t, the position in S at time t is given by

$$x = \int_0^t v\, dt = \int_0^t \frac{gt\, dt}{\sqrt{1 + (gt)^2}} = \frac{1}{g}\left(\sqrt{1 + (gt)^2} - 1\right). \tag{5.8}$$

For small t, the Taylor series $\sqrt{1 + \epsilon} \approx 1 + \epsilon/2$ turns x into $gt^2/2$. This is the correct nonrelativistic result, valid when t (and hence v) is small.

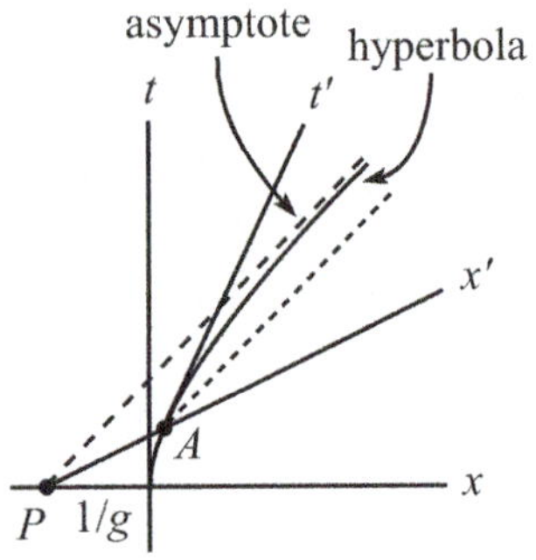

Figure 5.6

For convenience, let P be the point (see Fig. 5.6)

$$(x_P, t_P) = (-1/g, 0). \tag{5.9}$$

Then Eq. (5.8) yields

$$(x - x_P)^2 - t^2 = \frac{1}{g^2}. \tag{5.10}$$

This is the equation for a hyperbola with its center (defined as the intersection of the asymptotes) at point P. This hyperbola is the worldline of the particle. For a large acceleration g, the point P is very close to the particle's starting point. For a small acceleration, it is far away.

Everything has been fairly normal up to this point, but now the fun begins. Consider a point A on the particle's hyperbolic worldline at time t. From Eq. (5.8), A has coordinates

$$(x_A, t_A) = \left(\frac{1}{g}\left(\sqrt{1 + (gt)^2} - 1\right), t\right). \tag{5.11}$$

The slope of the line PA is therefore

$$\frac{t_A - t_P}{x_A - x_P} = \frac{gt}{\sqrt{1 + (gt)^2}}. \tag{5.12}$$

Looking at Eq. (5.7), we see that this slope equals the speed v of the particle at point A. But we know that the speed v (or technically $\beta \equiv v/c$, with the c's) is the slope of the particle's instantaneous x' axis; see Eq. (2.32). Therefore, the line PA (extended in both directions) and the particle's x' axis are the same line, as shown in Fig. 5.6. This holds for any arbitrary time t. So we may say that at any point along the particle's worldline, the line PA is the instantaneous x' axis of the particle. And since the x' axis is a line of simultaneity, we see that no matter where the particle is, the event at P is simultaneous with an event located at the particle, as measured in the instantaneous frame of the particle. In other words, the particle always says that P happens "now." The point P is very much like the event horizon of a black hole. As viewed by our accelerating particle, time seems to stand still at P. And if we went more deeply into GR, we would find that time seems to stand still at the edge of a black hole, as viewed by someone far away.

Here is another strange fact. What is the distance from P to A, as measured in the instantaneous rest frame S' of the particle? Using the v from Eq. (5.7), the γ factor between frames S and S' is

$$\gamma = \frac{1}{\sqrt{1 - v^2}} = \frac{1}{\sqrt{1 - \dfrac{(gt)^2}{1 + (gt)^2}}} = \sqrt{1 + (gt)^2}. \tag{5.13}$$

From Eqs. (5.9) and (5.11), the distance between P and A in frame S is $x_A - x_P = \sqrt{1 + (gt)^2}\,/g$. So the distance between P and A in frame S' is (using the Lorentz transformation $\Delta x = \gamma(\Delta x' + v\,\Delta t')$ with $\Delta t' = 0$)

$$x'_A - x'_P = \frac{1}{\gamma}(x_A - x_P) = \frac{1}{g}. \tag{5.14}$$

(The first equality here also follows from length contraction. P and A are simultaneous in S', so the distance is shorter in S'.) This result has the unexpected property of being independent of t. Therefore, not only do we find that P is always simultaneous with the particle, as measured in the particle's instantaneous rest frame; we also find that P is always the same distance (namely $1/g$, or c^2/g with the c's) away from the particle, in the particle's frame. This is quite strange. The particle accelerates away from point P, but it doesn't get farther away from it, as measured in its own frame.

REMARK: We can give a continuity argument that shows why such a point P must exist. If a point Q is close to you, and if you accelerate away from it, then of course you get farther away from it. Everyday experience is quite valid here. But if Q is sufficiently far away from you, and if you accelerate away from it, then the $at^2/2$ distance you travel can easily be compensated by the decrease in distance due to length contraction, brought about by your newly acquired velocity. (Imagine a stick laid out between you and Q, at rest in the original frame. You will see this stick being length contracted.) The additive (or rather, subtractive) decrease in distance due to length contraction is a given fraction times the distance to Q, so we simply need to pick Q to be sufficiently far away. This effect will then outweigh the $at^2/2$ effect. What this means is that every time you get out of your chair and walk to the door, there are stars very far away behind you that get closer to you (as measured in your instantaneous rest frame) as you walk away from them. By continuity, then, there must exist a point P that remains the same distance from you (in your frame) as you accelerate away from it. ♣

5.3.2 Uniformly accelerating frame

Let's now put a collection of uniformly accelerating particles together to make a uniformly accelerating frame. Our goal will be to create a frame in which the distances between particles (as measured in any particle's instantaneous rest frame) remain constant. Why is this our goal? We know from the "Break or not break" problem in Chapter 2 that if all of the particles accelerate with the same proper acceleration g, then the distances (as measured in a particle's instantaneous rest frame) grow larger. While this is a perfectly possible frame to construct, it is not desirable, for the following reason. Einstein's Equivalence Principle states that an accelerating frame is equivalent to a frame sitting on, say, the earth. We can therefore study the effects of gravity by studying an accelerating frame. But if we want this frame to look anything like the surface of the earth, we certainly can't have distances that change over time. We therefore want to construct a *static* frame, that is, one in which distances do not change (as measured in the frame). This will allow us to say that if we enclose the frame in a windowless box, then for all a person inside knows, he is standing motionless in a static gravitational field (which has a certain definite dependence on position, as we shall see).

Let's figure out how to construct the frame. We'll discuss the acceleration of just two particles here. Others can be added in a similar manner. In the end, the desired frame as a whole is constructed by accelerating each atom in the floor (or rather, wall) of the frame with a specific proper acceleration. From the above discussion in Section 5.3.1, we already have a particle A that is "centered" around the point P. (This will be our shorthand notation for "traveling along a branch of a hyperbola whose center is the point P.") We claim, for reasons that will become clear, that every other particle in the frame should also be "centered" around the same point P.

Consider another particle, B. Let a and b be the initial distances from P to A and B. If both particles are to be centered around P, then their proper accelerations must be, from Eq. (5.9),

$$g_A = \frac{1}{a} \quad \text{and} \quad g_B = \frac{1}{b}. \tag{5.15}$$

Therefore, in order to have all points in the frame be centered around P, we simply have to make their proper accelerations be inversely proportional to their initial distances from P.

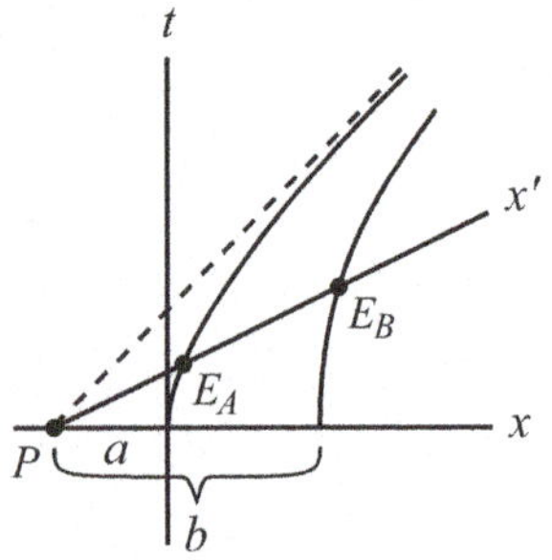

Figure 5.7

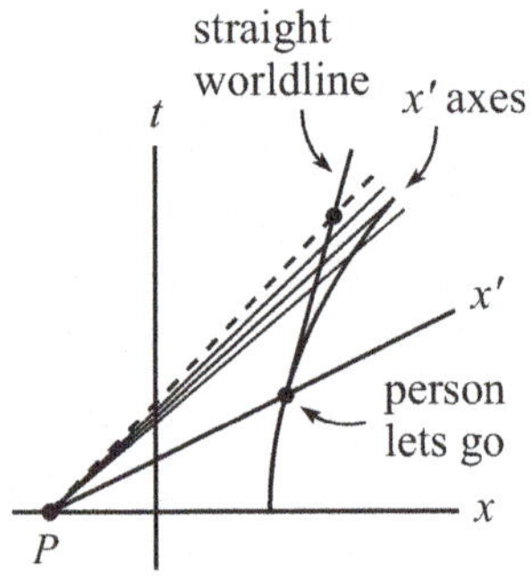

Figure 5.8

Why do we want every particle to be centered around P? Consider two events, E_A and E_B, such that P, E_A, and E_B are collinear in Fig. 5.7. From Section 5.3.1, we know that the line PE_AE_B is the x' axis for both particle A and particle B, at the positions shown. We also know that A is always a distance a from P, and B is always a distance b from P (in their frame). Combining these facts with the fact that A and B measure their distances along the x' axis of the same frame (for the events shown in the figure), we see that both A and B measure the distance between them to be $b - a$. This is independent of time, so A and B measure a constant distance between them. The distance starts off at $b - a$, of course, by construction. But what we just showed is that it remains $b - a$ at all times, as measured in the frame of A and B.

We have therefore constructed our desired static frame. This frame is often called a "Rindler space." If a person walks around in the frame, he will think that he lives in a static world where the acceleration due to gravity takes the form of $g(z) \propto 1/z$, where z is the distance to a certain magical point which is located at the end of the known "universe."

What if a person releases himself from the accelerating frame, so that he forever sails through space at constant speed? His view is that he is falling. He falls past the "magical point" P in a finite proper time, because the hyperbolic worldline of a point infinitesimally close to P is essentially the asymptote of all the hyperbolas, and the person's straight-line worldline (after he lets go) intersects this line; see Fig. 5.8. But his friends who are still in the frame will see him take an infinitely long time to get to P, because the x' axes of points in the frame never quite swing up to the asymptote, even after an infinite amount of time; a few such lines are shown. So the "now" line of any point in the frame never quite passes through the event where the person crosses the asymptote. This is similar to the situation with a black hole. A faraway observer will see it take an infinitely long time for a falling person to reach the "boundary" of a black hole, even though it takes a finite proper time for the person.

Our analysis shows that A and B feel different proper accelerations, because $a \neq b$ in Eq. (5.15). (As mentioned earlier, the word "uniform" in the title of this section refers to the fact that the proper acceleration of each point is constant. It doesn't mean that the accelerations of different points are equal.) There is no way to construct a static frame where all points feel the same proper acceleration, so it is impossible to mimic a constant gravitational field (over a nonzero distance) by using an accelerating frame. The problems and exercises in this chapter offer plenty of opportunity for you to play around with the properties of our uniformly accelerating frame.

5.4 Maximal-proper-time principle

The *maximal-proper-time principle* in general relativity says: Given two events in spacetime, a particle under the influence of only gravity takes the spacetime path between the events that maximizes the proper time. For example, if you throw a ball from given coordinates $(\mathbf{r}_1, t_1)$ to given coordinates $(\mathbf{r}_2, t_2)$, then the ball takes the path that maximizes its proper time. If you instead grab the ball and move it in an arbitrary manner (by applying a force to it) from $(\mathbf{r}_1, t_1)$ to $(\mathbf{r}_2, t_2)$, then its proper time will be smaller than that of the thrown ball. The principle is actually the "*stationary*-proper-time principle," because any type of stationary point (a maximum, minimum, or saddle point) is allowed. But we'll be a little sloppy here and just use the word "maximum," because that's what it will generally turn out to be in the situations we'll consider. However, see Problem 5.11.

The principle is clear for a freely-moving ball in outer space, far from any massive objects. The ball travels at constant speed from one point to another, and we know that this constant-speed motion is the motion with the maximal proper time. This is true because the given ball, A, moving at constant speed sees the clock on any other ball, B, running slow due to the special-relativistic time dilation, if there is a relative speed between them. (We're assuming that B's nonuniform velocity is caused by a non-gravitational force acting on it.) B therefore shows a shorter elapsed time. This argument doesn't work the other way around, because B isn't in an inertial frame and therefore can't use the special-relativistic time-dilation result. B is indeed not in an inertial frame, because B must still satisfy the condition that it moves from the given point $(\mathbf{r}_1, t_1)$ to the given point $(\mathbf{r}_2, t_2)$. There is only one inertial worldline that can do this, and A has already claimed it.

Consistency with Newtonian physics

The most common way of determining the motion of a Newtonian (nonrelativistic) object is to use Newton's second law, $F = ma$ (or really $F = dp/dt$). There is, however, another method, called the *principle of least action.* It can be shown (although we won't do it here; it is outside the scope of this book) that this principle leads to the same results that Newton's second law leads to. (As with the maximal-proper-time principle, the least-action principle is actually the *stationary*-action principle.)

The principle of least action states that given the initial and final coordinates $(\mathbf{r}, t)$ of an object, the path (through space and time) that the object takes between these points is the one that minimizes the *action*, which is defined as the integral $\int (K - U)\,dt$, where K is the object's kinetic energy and U is its potential energy. It is by no means obvious that this principle is equivalent to Newton's second law, but we'll just accept this fact here for the present purposes.

The goal of this section is to show (at least in one specific setup) that the maximal-proper-time principle reduces to the least-action principle in the limit of small velocities. This must be the case if the maximal-proper-time principle has any chance of being valid, because it is an established fact that the least-action principle is valid in the Newtonian (small-velocity) limit. We will demonstrate the equivalence of the two principles in the specific case of a ball thrown vertically on the earth. Assume that the initial and final coordinates are fixed to be (y_1, t_1) and (y_2, t_2).

Before being quantitative, let's get a qualitative idea about what's going on with the ball. There are two competing effects as far as maximizing the proper time goes. On one hand, the ball wants to climb very high, because its clock will run faster there, due to the GR time dilation. But on the other hand, if it climbs very high, it must move very fast to get there (because the total time, $t_2 - t_1$, is fixed), and this will make its clock run slow, due to the SR time dilation. So there is a tradeoff.

Similarly, there are two competing effects as far as minimizing the Newtonian action goes. On one hand, the ball wants to climb very high, because then the action $\int (K - U)\,dt$ will involve a large subtractive $-U = -mgy$ term. But on the other hand, if it climbs very high, it must move very fast to get there (because the total time, $t_2 - t_1$, is fixed), and this will make the positive K term be large. So again there is a tradeoff. Due to the similarities between this paragraph and the preceding one, it isn't much of a surprise that the maximal-proper-time principle leads to the least-action principle, as we will now show.

The proper time is

$$\tau = \int_{\tau_1}^{\tau_2} d\tau. \tag{5.16}$$

Due to the motion of the ball, the usual SR time-dilation effect tells us that the ball's clock runs $\sqrt{1 - v^2/c^2}$ as fast as a clock on the ground (all other things being equal). And due to the height of the ball, the GR time-dilation effect tells us that the ball's clock runs $(1 + gy/c^2)$ as fast as a clock on the ground (all other things being equal). Combining these two effects gives

$$d\tau = \sqrt{1 - \frac{v^2}{c^2}}\left(1 + \frac{gy}{c^2}\right) dt, \tag{5.17}$$

where t is the time measured by an observer standing on the ground. Using the Taylor expansion $\sqrt{1-\epsilon} \approx 1 - \epsilon/2$, and dropping terms of order $1/c^4$ and smaller, we see that we want to maximize

$$\int_{\tau_1}^{\tau_2} d\tau \approx \int_{t_1}^{t_2} \left(1 - \frac{v^2}{2c^2}\right)\left(1 + \frac{gy}{c^2}\right) dt \approx \int_{t_1}^{t_2} \left(1 - \frac{v^2}{2c^2} + \frac{gy}{c^2}\right) dt. \tag{5.18}$$

The "1" term yields a constant (namely $t_2 - t_1$), so we can ignore it. If we multiply the rest of the integral through by the constant factor mc^2 (which doesn't change the maximal nature of the integral) and also by -1 (which changes a maximum to a minimum), we see that *maximizing* the above integral is equivalent to *minimizing* the integral,

$$mc^2 \int_{t_1}^{t_2} \left(\frac{v^2}{2c^2} - \frac{gy}{c^2}\right) dt = \int_{t_1}^{t_2} \left(\frac{mv^2}{2} - mgy\right) dt. \tag{5.19}$$

The integral here is the Newtonian action (the integral of the kinetic energy $K = mv^2/2$ minus the potential energy $U = mgy$). We have therefore demonstrated that the maximal-proper-time principle reduces to the least-action principle in the limit of small velocities, as desired.

For a one-dimensional gravitational problem such as this one, the action is always a (global) minimum, and the proper time is always a (global) maximum, as you can show by considering the second-order change in the action; see Exercise 5.28.

In retrospect, it isn't surprising that the kinetic energy worked out in Eq. (5.19). The factor of $1/2$ comes about in exactly the same way as in the derivation in Eq. (3.10), where we showed that the relativistic form of energy reduces to the familiar Newtonian expression. With regard to the potential energy, the gy in Eq. (5.19) can be traced to the acceleration times a certain time, where this time is proportional to the distance; see the paragraph preceding Eq. (5.2). This yields (when multiplied by mc^2) the usual force-times-distance expression for the potential energy.

Note that although the mgy in Eq. (5.19) has the obvious interpretation of being the gravitational potential energy, our starting point in Eq. (5.16) made no mention of a gravitational force. When working with the maximal-proper-time principle, gravity shouldn't be thought of as a force, but rather as something that affects the proper time.[6]

5.5 Twin paradox revisited

Let's take another look at the standard twin paradox, this time from the perspective of general relativity. We should emphasize that GR is by no means necessary for an understanding of the original formulation of the paradox (the first scenario below). We were able to solve it in Section 1.3.2, after all. The present discussion is given simply to show that the answer to an alternative formulation (the second scenario below) is consistent with what we've learned about GR. Consider the following two twin-paradox setups.

[6]For an interesting article on how the Newtonian action, along with a few ingredients from differential geometry, can lead to general relativity, see Rindler (1994).

- Twin A floats freely in outer space. Twin B flies past A in a spaceship with speed v_0; see Fig. 5.9. At the instant they are next to each other, they both set their clocks to zero. At this same instant, B turns on the reverse thrusters of his spaceship and decelerates with proper deceleration g. B eventually reaches a farthest point from A and then accelerates back toward A, finally passing him with speed v_0 again. When they are next to each other, they compare the readings on their clocks. Which twin is younger?

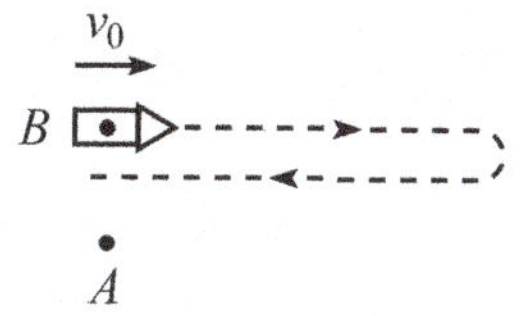

Figure 5.9

- Twin B stands on the earth. Twin A is thrown upward with speed v_0. (Let's say he is fired from a cannon in a hole in the ground; see Fig. 5.10.) At the instant they are next to each other, they both set their clocks to zero. A rises up and then falls back down, finally passing B with speed v_0 again. When they are next to each other, they compare the readings on their clocks. Which twin is younger?

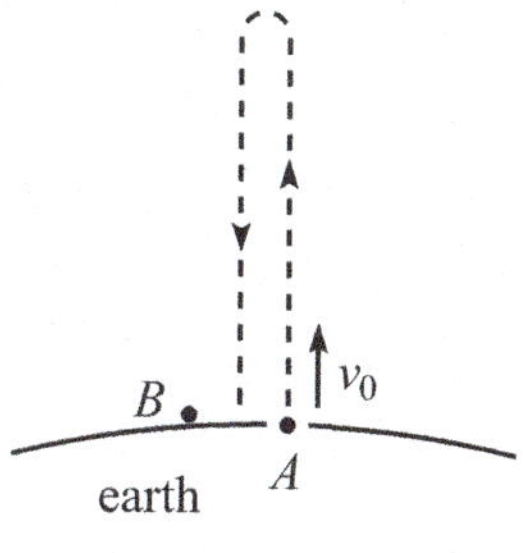

Figure 5.10

The first scenario is easily solved using special relativity. Since A is in an inertial frame, he can apply the results of special relativity. In particular, he sees B's clock running slow, due to the usual special-relativistic time dilation. Therefore, B ends up younger at the end. B cannot use the reverse reasoning, because she is not in an inertial frame.

What about the second scenario? The key point to realize is that the Equivalence Principle says that these two scenarios are exactly the same (ignoring the nonuniformity of the earth's gravitational force), as far as the twins are concerned. Twin B has no way of knowing whether she is in a spaceship accelerating at g or on the surface of the earth. (Rightward in the first scenario corresponds to downward in the second scenario.) And A has no way of knowing whether he is floating freely in outer space or in free-fall in a gravitational field.[7] We therefore conclude that B must be younger in the second scenario, too.

At first glance, this seems incorrect, because in the second scenario, B is sitting motionless while A is the one who is moving. It seems that B should see A's clock running slow, due to the usual special-relativistic time dilation, and hence A should be younger. This reasoning is incorrect because it fails to take into account the gravitational time dilation. A is higher in the gravitational field, and therefore his clock runs faster. It happens that this effect wins out over the special-relativistic time dilation, and A ends up older. You can explicitly show this in Problem 5.14. Note that the fact that A is older is consistent with the maximal-proper-time principle. In both scenarios, A is under the influence of only gravity (zero gravity in the first scenario), whereas B feels a normal force from either the spaceship's floor (or ceiling, the way we've drawn it in Fig. 5.9) in the first scenario or the ground in the second scenario.

The reasoning in this section provides another way to conclude that the Equivalence Principle implies that higher clocks must run faster (in one way or another). Given that we know A is older in the first scenario, the Equivalence Principle implies that A must also be older in the second scenario, which means that there must be some kind of height effect that makes A's clock run fast (fast enough to win out over the special-relativistic time dilation). But it requires some work to show that the factor takes the form of $1 + gh/c^2$.

[7]As mentioned in Section 5.1, this fact is made possible by the equivalence of inertial and gravitational mass. Were it not for this, different parts of A's body would want to accelerate at different rates in the gravitational field in the second scenario. This would certainly clue him in to the fact that he wasn't floating freely in space.

5.6 Summary

In this chapter we learned about the basics of general relativity. In particular, we learned:

- The Equivalence Principle states that gravity is equivalent to acceleration. Alternatively, it states that "gravitational" mass is equal to (or proportional to) "inertial" mass.

- A clock at height h in a gravitational field runs faster than a clock on the ground, according to

$$\Delta t_h = \left(1 + \frac{gh}{c^2}\right)\Delta t_0 \qquad \text{(gravitational time dilation)} \tag{5.20}$$

- The position and velocity of a uniformly accelerating particle are given by

$$x(t) = \frac{1}{g}\left(\sqrt{1+(gt)^2} - 1\right) \qquad \text{and} \qquad v(t) = \frac{gt}{\sqrt{1+(gt)^2}}. \tag{5.21}$$

  The particle says that the event located at the point $P = (x_P, t_P) = (-1/g, 0)$ always happens "now" and is always a distance $1/g$ away, as measured in the particle's instantaneous inertial frame. To build up a frame from accelerating points, in which distances remain fixed, we need the acceleration of each point to be inversely proportional to the distance from the point P.

- The maximal-proper-time principle states that a particle under the influence of only gravity has the maximum proper time. (This is technically the stationary-proper-time principle.) This principle is consistent with the principle of least action (or technically, stationary action).

- We found that a version of the twin paradox involving constant acceleration is consistent (as the Equivalence Principle demands) with a version involving a gravitational field.

5.7 Problems

Section 5.2: Time dilation

5.1. **Clock on a tower** **

A clock starts on the ground and then moves up a tower at constant speed v. It sits on top of the tower for a time T and then descends at constant speed v. If the tower has height h, how long should the clock sit at the top so that when it returns to the ground, it shows the same time as a clock that remained on the ground? (Assume $v \ll c$.)

5.2. **Exact time dilation** **

For a constant gravitational field, the time-dilation result in Eq. (5.4) is only a (very good) approximation. Derive the correct result in two different ways:

(a) If $f(h)$ represents the time-dilation factor as a function of height, explain why f must satisfy $f(h_1+h_2) = f(h_1)f(h_2)$. Find the most general function f with this property, and then demand that it agrees with Eq. (5.4) for small h.

(b) Line up a series of clocks between heights of 0 and h, and look at the successive time-dilation factors between them. *Hint:* Take the log of the product of the factors, and apply the approximation $\ln(1 + x) \approx x$.

5.3. **Varying gravitational field** $**$

When applied to the gravitational field of the earth, the GR time-dilation factor of $1 + gh/c^2$ in Eq. (5.4) holds (up to corrections of order $1/c^4$) only when h is small compared with the radius R of the earth. In the more general case where h isn't small compared with R, a clock at a high radius r_{high} runs faster than a clock at a low radius r_{low} by the factor

$$1 + \frac{GM}{c^2}\left(\frac{1}{r_{\text{low}}} - \frac{1}{r_{\text{high}}}\right), \tag{5.22}$$

where $G = 6.67 \cdot 10^{-11}\,\text{m}^3/(\text{kg s}^2)$ and $M \approx 6 \cdot 10^{24}\,\text{kg}$ (the mass of the earth). Derive this result by lining up a series of clocks and looking at the successive time-dilation factors between them. *Hint:* Take the log of the product of the factors, and apply the approximation $\ln(1 + x) \approx x$. You will need to use the fact that Newton's universal law of gravitation gives the value of g at radius r (outside the earth) as $g_r = GM/r^2$.

5.4. **Global positioning system** $**$

Global Positioning System (GPS) satellites orbit the earth at a radius of about 26,600 km (20,200 km above the earth's surface, which is at a radius of 6400 km). The speed of a satellite is roughly 3900 m/s. (You can show this implies that the time for one orbit is about 12 hours.) Using the result from Problem 5.3, calculate the factor by which a clock on a satellite runs fast or slow relative to a clock on the surface of the earth. Both the SR and GR time-dilation effects are relevant here. After one day, by what fraction of a second will the two clocks differ?

5.5. **Circular motion** $**$

B moves at speed v (with $v \ll c$) in a circle of radius r around A. By what factor does B's clock run slower than A's? Calculate this in three ways. Work in:

(a) The lab frame (A's frame).

(b) The frame whose origin is B and whose axes remain parallel to an inertial set of axes.

(c) The rotating frame that is centered at A and rotates with the same frequency as B.

You will need to use the fact that the centripetal acceleration is given by $a = v^2/r$.

5.6. **More circular motion** $**$

A and B move at speed v (with $v \ll c$) in a circle of radius r, at diametrically opposite points. Each person sees the other person's clock ticking at the same rate as his own. Show this in three ways. Work in:

(a) The lab frame (the inertial frame whose origin is the center of the circle).

(b) The frame whose origin is B and whose axes remain parallel to an inertial set of axes.

(c) The rotating frame that is centered at the origin and rotates with the same frequency as A and B.

You will need to use the fact that the centripetal acceleration is given by $a = v^2/r$.

Section 5.3: Uniformly accelerating frame

5.7. **Time and speed in an accelerating frame** ***

In the spirit of Problem 2.12 ("Acceleration and redshift"), use a Minkowski diagram to solve this problem. A rocket accelerates with proper acceleration g toward a planet. As measured in the instantaneous *inertial* frame of the rocket, the planet is a distance x away and moves at speed v. Everything is in one dimension here. As measured in the *accelerating* frame of the rocket, show that the planet's clock runs at a rate (with $c = 1$),

$$dt_{\mathrm{p}} = dt_{\mathrm{r}}(1 + gx)\sqrt{1 - v^2}, \tag{5.23}$$

and show that the planet's speed is

$$V = (1 + gx)v. \tag{5.24}$$

5.8. **Accelerator's point of view** ***

A rocket starts at rest relative to a planet, a distance ℓ away. It accelerates toward the planet with proper acceleration g. Let τ and t be the readings on the rocket's and planet's clocks, respectively. (The results from Problem 5.7 and Exercise 5.22 will be useful in this problem. See Appendix H for some properties of the hyperbolic trig functions.)

(a) Show that when the astronaut's clock reads τ, he observes (as measured in his instantaneous inertial frame) the rocket-planet distance x to be given by

$$1 + gx = \frac{1 + g\ell}{\cosh(g\tau)}. \tag{5.25}$$

(b) Show that when the astronaut's clock reads τ, he observes the time t on the planet's clock to be given by

$$gt = (1 + g\ell)\tanh(g\tau). \tag{5.26}$$

5.9. **Getting way ahead** ****

A rocket with proper length L accelerates from rest, with proper acceleration g (where $gL \ll c^2$). Clocks are located at the front and back of the rocket. If we look at this setup in the rocket frame, then the GR time-dilation effect tells us that the readings on the front and back clocks are related by $t_{\mathrm{f}} = (1 + gL/c^2)t_{\mathrm{b}}$. Therefore, if we look at things in the ground frame, the readings on the two clocks are related by

$$t_{\mathrm{f}} = t_{\mathrm{b}}\left(1 + \frac{gL}{c^2}\right) - \frac{Lv}{c^2}, \tag{5.27}$$

where the last term comes from the SR rear-clock-ahead result. Derive the above relation by working entirely in the ground frame.

Note: You might find this relation surprising, because it implies that the front clock will eventually be an arbitrarily large time ahead of the back clock, in the ground frame. (The subtractive Lv/c^2 term is bounded by L/c and will therefore eventually become negligible compared with the additive and unbounded $(gL/c^2)t_{\mathrm{b}}$ term.) But both clocks seem to be doing basically the same thing relative to the ground frame, so how can they end up differing by so much? Your task is to find out.

5.10. **Lv/c^2 revisited** **

You stand at rest relative to a rocket that has synchronized clocks at its ends. You and the rocket are then arranged to move with relative speed v. A reasonable question to ask is: As viewed by you at a given instant, what is the difference in readings on the clocks located at the ends of the rocket?

It turns out that this question cannot be answered without further information about how you and the rocket acquired the relative speed v. There are two basic ways this relative speed can come about. The rocket can accelerate while you sit there, or you can accelerate while the rocket sits there. Using the results from Problems 5.8 and 5.9, explain what the answers to the above question are in these two cases.

Section 5.4: Maximal-proper-time principle

5.11. **Circling the earth** ***

Clock A sits at rest on the earth, and clock B circles the earth in an orbit (a freefall orbit, under the influence of only gravity) just above the surface. Both A and B are essentially at the same radius, so the GR time-dilation effect yields no difference in their times. But B is moving relative to A, so A sees B running slow, due to the usual SR time-dilation effect. The orbiting clock, B, therefore shows a *smaller* elapsed proper time on each occasion when it passes A. In other words, the clock under the influence of only gravity (B) does *not* show the maximal proper time, in conflict with what we have been calling the maximal-proper-time principle. Explain.

Section 5.5: Twin paradox revisited

5.12. **Twin paradox** **

A spaceship travels at speed v (with $v \ll c$) to a distant star, a distance ℓ away. Upon reaching the star, the spaceship decelerates and then accelerates back up to speed v in the opposite direction (uniformly, and in a short time compared with the total journey time). When the spaceship returns to the earth, how much younger is the traveler than her twin on the earth? (Ignore the gravity from the earth.) Work in:

(a) The earth frame.

(b) The spaceship frame(s).

5.13. **Twin paradox again** ***

(a) Answer part (b) of the previous problem, except now let the spaceship turn around by moving in a small semicircle while maintaining speed v.

(b) Answer part (b) of the previous problem, except now let the spaceship turn around by moving in an arbitrary manner. The only constraints are that the turnaround is done quickly (compared with the total journey time) and that it is contained in a small region of space (compared with the earth-star distance).

5.14. **Twin paradox times** ***

(a) In the first scenario in Section 5.5, calculate the ratio of B's elapsed time to A's, in terms of v_0 and g. Work in A's frame. Assume $v_0 \ll c$, and drop high-order terms.

(b) Do the same for the second scenario in Section 5.5, but now by working in B's frame. Do this by using the SR and GR time dilations, and then check that your answer agrees (to the accuracy of the calculations) with your answer to part (a), as the Equivalence Principle demands.

5.8 Exercises

Section 5.2: Time dilation

5.15. **Driving on a hill** $*$

You drive up and down a hill of height h at constant speed. The hill is in the shape of an isosceles triangle with altitude h. What should your speed be so that you age the same amount as someone standing at the base of the hill? (Assume $v \ll c$.)

5.16. **Space station** $*$

The space station orbits at a height of about 400 km $= 4 \cdot 10^5$ m, with a speed of about 7500 m/s. What are the numerical values of the SR and GR time-dilation corrections, namely $v^2/2c^2$ and gh/c^2? Which correction is larger, and by what factor?

Note: The $v^2/2c^2$ correction comes from applying the Taylor series $\sqrt{1-x} \approx 1 - x/2$ to the SR time-dilation factor $\sqrt{1 - v^2/c^2}$. The gh/c^2 correction isn't exact here, given the result in Problem 5.3, but it's good enough for the present purpose.

5.17. **Running at the same rate** $**$

Using the result from Problem 5.3, determine the radius at which a satellite (with $v \ll c$) should orbit the earth so that its clock runs at the same rate as a clock on the ground. (Ignore the rotation of the earth.) You will need to use the fact that $F = ma$ gives the satellite's speed v as a function of its radius r as

$$F = ma \quad \Longrightarrow \quad \frac{GMm}{r^2} = m\frac{v^2}{r} \quad \Longrightarrow \quad v = \sqrt{\frac{GM}{r}}. \tag{5.28}$$

5.18. Lv/c^2 **and** gh/c^2 $**$

The SR rear-clock-ahead result, Lv/c^2, looks rather similar to the gh/c^2 term in the GR time-dilation result, Eq. (5.4). For small v, devise a thought experiment that explains how the Lv/c^2 result follows from the gh/c^2 result.

5.19. **Both points of view** $***$

A and B are initially a distance L apart, at rest with respect to each other. At a given time, B accelerates toward A with constant proper acceleration a. Assume $aL \ll c^2$.

(a) Working in A's frame, calculate the difference in readings on A's and B's clocks when B reaches A.

(b) Calculate the difference again, now working in B's frame, and show that the result agrees (neglecting higher-order terms) with your result from part (a).

5.20. Opposite circular motion ****

A and B move at speed v (with $v \ll c$) in opposite directions around a circle of radius r (so they pass each other after every half-revolution). They both see their two clocks ticking, on *average*, at the same rate. That is, if they compare their clocks each time they pass each other, both clocks will show the same time elapsed. Demonstrate this in three ways. Work in:

(a) The lab frame (the inertial frame whose origin is the center of the circle).

(b) The frame whose origin is B and whose axes remain parallel to an inertial set of axes.

(c) The rotating frame that is centered at the origin and rotates along with B. This part is very tricky; the solution is given in Cranor *et al.* (2000), but don't peek too soon.

Section 5.3: Uniformly accelerating frame

5.21. Force and acceleration *

Using the v in Eq. (5.7), verify that $f = \gamma^3 ma$ in frame S, where $a \equiv dv/dt$. Recall that we defined g as f/m.

5.22. Various quantities **

A particle starts at rest in a given inertial frame and accelerates with proper acceleration g. Let τ be the time on the particle's clock. Starting with the v in Eq. (5.7), use time dilation to show that at time t in the original inertial frame, the various quantities are related by (dropping the c's)

$$gt = \sinh(g\tau), \qquad v = \tanh(g\tau), \qquad \gamma = \cosh(g\tau). \tag{5.29}$$

5.23. Using rapidity **

Another way to derive the v in Eq. (5.7) is to use the $v = \tanh(g\tau)$ rapidity result (where τ is the particle's proper time) from Section 2.6.2. Use time dilation to show that this implies $gt = \sinh(g\tau)$, which then implies Eq. (5.7).

5.24. Speed in an accelerating frame **

(a) In the setup in Problem 5.8, use Eq. (5.25) to find $|dx/d\tau|$ (which is the speed of the planet in the rocket's *accelerating* frame), as a function of τ. Then use Eqs. (5.25) and (5.29) to show that the speed can be rewritten as $(1 + gx)v$, where v is the speed of the rocket in the planet's frame (equivalently, the speed of the planet in the rocket's instantaneous *inertial* frame). This is the same result as in Eq. (5.24) in Problem 5.7.

(b) What is the maximum value of the speed $|dx/d\tau|$, in terms of g and the initial distance ℓ?

5.25. Redshift, blueshift **

We found in Section 5.2 that a clock at the rear of a rocket sees a clock at the front run fast by the factor $1 + gh/c^2$. However, we ignored effects of higher-order in $1/c^2$, so for all we know, we found only the first term in a Taylor series, and the factor is actually something like e^{gh/c^2} (as it is for a constant gravitational field; see Problem 5.2), or perhaps $1 + \ln(1 + gh/c^2)$.

(a) For the uniformly accelerating frame in Section 5.3.2, show that the factor is in fact exactly $1 + g_r h/c^2$, where g_r is the acceleration of the rear of the rocket. Show this by lining up a series of clocks and looking at the successive factors between them. (*Hint:* Take the log of the product of the factors, and use the approximation $\ln(1+\epsilon) \approx \epsilon$.) The overall factor takes a very nice form when written in terms of the a and b in Fig. 5.7; what is it?

(b) By the same reasoning as in part (a), it follows that the front clock sees the rear clock run slow by the factor $1 - g_f h/c^2$, where g_f is the acceleration of the front of the rocket. Show explicitly that $(1 + g_r h/c^2)(1 - g_f h/c^2) = 1$, as must be the case, because a clock can't gain time with respect to itself (and because the two clocks are at rest in the frame of the rocket; this reasoning wouldn't apply to two clocks flying past each other).

5.26. **Length contraction** $**$

A pencil points directly at you. You start at rest, some distance away from the pencil, and then accelerate toward it with acceleration a. After a time t, the pencil is length contracted in your frame by a factor $\sqrt{1 - v^2/c^2} \approx 1 - (at)^2/2c^2$, for small t (more precisely, for $at \ll c$). But the only way the pencil can shrink is for the back to move faster than the front, as measured by you in your accelerating frame. Using Eq. (5.24) in Problem 5.7, show that the contraction factor works out as it should, for small t.

5.27. **Accelerating stick's length** $**$

Consider the uniformly accelerating frame of a stick, the ends of which have worldlines given by the curves in Fig. 5.7. (So the stick has proper length $b - a$.) At time t in the original inertial frame (the lab frame), we know from Eqs. (5.8) and (5.9) that a point that undergoes acceleration g has position $\sqrt{1 + (gt)^2}/g$ relative to the point P in Fig. 5.7. An observer in the lab frame sees the stick being length-contracted by different factors along its length, because different points move with different speeds (at a given time in the lab frame). Show, by doing the appropriate integral, that the inertial observer concludes that the stick always has proper length $b - a$.

Section 5.4: Maximal-proper-time principle

5.28. **Maximum proper time** $**$

Show that the stationary value of the gravitational action in Eq. (5.19) is always a minimum (which means that the proper time is always a maximum). Do this by considering a function, $y(t) = y_0(t) + \xi(t)$, where y_0 is the path that yields the stationary value, and ξ is a small variation. (ξ is an arbitrary function, except for the condition that it vanishes at the boundary values of t.) *Hint*: In the action, integrate the $\dot{y}_0\dot{\xi}$ term by parts and invoke the fact that ξ is zero at the boundaries. Then group the terms according to their ξ dependence. Use the fact that the first-order dependence on ξ must be zero (why?).

Section 5.5: Twin paradox revisited

5.29. **Symmetric twin non-paradox** $**$

Two twins travel toward each other, both at speed v ($v \ll c$) with respect to an inertial observer. They synchronize their clocks when they pass each other. They travel to stars located at positions $\pm\ell$, and then decelerate and accelerate back up to speed v in the opposite direction (uniformly, and in a short time compared with the total journey time). In the frame of the inertial observer, it is clear (by symmetry) that both twins age the same amount by the time they pass each other again. Reproduce this result by working in the frame of one of the twins.

5.9 Solutions

5.1. **Clock on a tower**

Using the Taylor series $\sqrt{1-\epsilon} \approx 1 - \epsilon/2$, the SR time-dilation factor is $\sqrt{1-v^2/c^2} \approx 1 - v^2/2c^2$. The clock therefore loses a fraction $v^2/2c^2$ of the time elapsed during its motion up and down the tower. The upward and downward trips each take a time of h/v, so the time loss due to the SR effect is

$$\left(\frac{v^2}{2c^2}\right)\left(\frac{2h}{v}\right) = \frac{vh}{c^2}. \tag{5.30}$$

Our goal is to balance this time loss with the time gain due to the GR time-dilation effect. If the clock sits on top of the tower for a time T, then the time gain is $(gh/c^2)T$. However, we must not forget the increase in time due to the height gained while the clock is in motion. During its motion, the clock's average height is $h/2$. (We can use the average height here, because the GR effect is linear in h.) The total time in motion is $2h/v$, so the GR time gain while the clock is moving is

$$\left(\frac{g(h/2)}{c^2}\right)\left(\frac{2h}{v}\right) = \frac{gh^2}{c^2 v}. \tag{5.31}$$

Setting the total loss/gain in the clock's time equal to zero gives

$$-\frac{vh}{c^2} + \frac{gh}{c^2}T + \frac{gh^2}{c^2 v} = 0 \implies -v + gT + \frac{gh}{v} = 0 \implies T = \frac{v}{g} - \frac{h}{v}. \tag{5.32}$$

Remarks: This result implies that we must have $v \geq \sqrt{gh}$ in order for a nonnegative solution for T to exist. If $v < \sqrt{gh}$, then the SR effect is too small to cancel out the GR effect, even if the clock spends no time sitting at the top. If $v = \sqrt{gh}$, then $T = 0$, and we essentially have the same situation as in Exercise 5.15. Note that if v is very large compared with $\sqrt{gh}$ (but still small compared with c, so that our $\sqrt{1-v^2/c^2} \approx 1 - v^2/2c^2$ approximation is valid), then $T \approx v/g$, which is independent of h. The reason for this is that the time in Eq. (5.31) is negligible, and the other two contributions on the lefthand side of Eq. (5.32) are both proportional to h, so the h's cancel. ♣

5.2. **Exact time dilation**

(a) The time-dilation factor must satisfy $f(h_1 + h_2) = f(h_1)f(h_2)$, for the following reason. A clock at height $h_1 + h_2$ runs faster than a clock at height h_2 by a factor $f(h_1)$. In turn, a clock at height h_2 runs faster than a clock on the ground by a factor $f(h_2)$. The product of these two factors equals the factor by which a clock at height $h_1 + h_1$ runs faster than a clock on the ground, which is $f(h_1 + h_2)$, by definition. Therefore, $f(h_1 + h_2) = f(h_1)f(h_2)$, as desired. This reasoning is valid because all of the clocks are in the same frame. After a long time, they can therefore be slowly brought together without anything drastic happening to the readings. Since everyone must agree on the final readings, the above two expressions for the overall factor must be equal. See Question 47 in Appendix A for more on this issue.

Let's now determine the function f. For convenience, we'll change the notation and write the above relation as $f(x+y) = f(x)f(y)$. If we take the partial derivative of this equation with respect to x, and then set $x = 0$, we obtain (with a prime denoting differentiation) $f'(y) = f'(0)f(y)$. Since $f'(0)$ is a constant, we can separate variables and integrate:

$$\frac{df(y)}{dy} = f'(0)f(y) \implies \int \frac{df}{f} = f'(0)\int dy$$
$$\implies \ln f = f'(0)y + C \implies f(y) = e^C e^{f'(0)y}, \tag{5.33}$$

where C is a constant of integration. Since $f(0) = 1$, the e^C factor must equal 1, which means that $f(y) = e^{f'(0)y}$. We can determine $f'(0)$ by recalling that for small y, the $1 + gy/c^2$ result in Eq. (5.4) is valid. The derivative of this function has the constant value of g/c^2. So in particular we have $f'(0) = g/c^2$. The desired exact function f is therefore (switching from y back to h)

$$f(h) = e^{gh/c^2}. \tag{5.34}$$

In short, the exponential function is the only nontrivial function that satisfies $f(x + y) = f(x)f(y)$ for all x and y. The exceptions are the constant functions $f(x) \equiv 0$ and $f(x) \equiv 1$.

Remarks:

1. Alternatively, we can solve for $f(x)$ by instead solving for the log of f. If we let $g(x) \equiv \ln f(x)$, then $f(x) = e^{g(x)}$. So the $f(x + y) = f(x)f(y)$ relation becomes $e^{g(x+y)} = e^{g(x)}e^{g(y)}$, which implies $g(x + y) = g(x) + g(y)$. If we take the partial derivative of this equation with respect to x and then set $x = 0$, we obtain $g'(y) = g'(0)$. So the slope of $g(x)$ is constant, which implies that $g(y)$ is a linear function. Call it $Ay + C$. The original function f is then $f(y) = e^{g(y)} = e^C e^{Ay}$, in agreement with Eq. (5.33), with $A \leftrightarrow f'(0)$. The solution proceeds as above.
2. If we use the Taylor series $e^x \approx 1 + x + x^2/2$, we see that the exact result in Eq. (5.34) is approximately equal to $1 + gh/c^2 + (gh/c^2)^2/2$. The approximate result in Eq. (5.4) is therefore incorrect by a term of order $(gh/c^2)^2$, or equivalently by a term that is of order gh/c^2 smaller than the (already quite small) gh/c^2 term in Eq. (5.4). ♣

(b) Consider two clocks at heights y and $y + dy$, where dy is small. The lower clock sees the upper clock running at a rate $1 + g\,dy/c^2$. If we line up a series of clocks from $y = 0$ to $y = h$, separated by dy, then the factor by which the clock at $y = 0$ sees the clock at $y = h$ run fast equals the product of all the time-dilation factors between successive clocks. These factors all have the same value of $1 + g\,dy/c^2$ (and in the $dy \to 0$ limit, this approximate expression becomes exact). So if we divide the total distance h into a large number n of intervals (with $dy = h/n$), then the product of the factors is

$$f = \left(1 + \frac{g\,dy}{c^2}\right)^n. \tag{5.35}$$

Taking the log of both sides and using $\ln(1 + x) \approx x$, we obtain

$$\ln f = n \ln\left(1 + \frac{g\,dy}{c^2}\right) \approx n \cdot \frac{g\,dy}{c^2} = \frac{g(n\,dy)}{c^2} = \frac{gh}{c^2}. \tag{5.36}$$

Exponentiating this relation then gives $f(h) = e^{gh/c^2}$, in agreement with the result in part (a). In the $dy \to 0$ limit, the "$\approx$" sign in Eq. (5.36) becomes an equality.

Remark: This $f(h) = e^{gh/c^2}$ result holds only for a constant gravitational field. In the case of the earth's gravitational field, which falls off like $1/r^2$, Problem 5.3 shows that we again obtain an exponential factor, but with a somewhat more complicated exponent. For the gravitational field produced by the accelerating frame in Section 5.3, $1 + gh/c^2$ turns out to be the exact factor, where g is the acceleration at the location of the observer. There is no exponential function in this case. See Exercise 5.25. ♣

5.3. Varying gravitational field

Consider two clocks at radii r and $r + dr$, where dr is small. The gravitational acceleration at radius r is $g_r = GM/r^2$, so the lower clock sees the upper clock running at a rate $1 + g_r\,dr/c^2 = 1 + (GM/c^2r^2)\,dr$. (We're working in the approximation where g takes on the constant value of g_r over the entire dr interval. We'll eventually take the $dr \to 0$ limit, in which case this approximation becomes exact.) If we line up a series of clocks from r_{low} to r_{high}, separated by dr, then the factor by which the clock at r_{low} sees the clock at r_{high} run fast equals the product of all the time-dilation factors between successive clocks. With $r_1 \equiv r_{\text{low}}$ and $r_n \equiv r_{\text{high}}$, this product equals (using $\ln(1+x) \approx x$, and dropping the c's)[8]

$$
\begin{aligned}
f &= \left(1 + \frac{GM}{r_1^2}\,dr\right)\left(1 + \frac{GM}{r_2^2}\,dr\right)\cdots\left(1 + \frac{GM}{r_n^2}\,dr\right) \\
\Longrightarrow \quad \ln f &= \ln\left(1 + \frac{GM}{r_1^2}\,dr\right) + \ln\left(1 + \frac{GM}{r_2^2}\,dr\right) + \cdots + \ln\left(1 + \frac{GM}{r_n^2}\,dr\right) \\
&\approx \frac{GM}{r_1^2}\,dr + \frac{GM}{r_2^2}\,dr + \cdots + \frac{GM}{r_n^2}\,dr \\
\Longrightarrow \quad \ln f &\approx \int_{r_{\text{low}}}^{r_{\text{high}}} \frac{GM}{r^2}\,dr = \frac{GM}{r_{\text{low}}} - \frac{GM}{r_{\text{high}}}.
\end{aligned} \tag{5.37}
$$

We have used the fact that as dr goes to zero, the sum in the third line becomes the integral in the fourth line. Exponentiating this result and using $e^x \approx 1 + x$ gives (bringing the c's back in)

$$
f = \text{Exp}\left(\frac{GM}{c^2 r_{\text{low}}} - \frac{GM}{c^2 r_{\text{high}}}\right) \approx 1 + \frac{GM}{c^2}\left(\frac{1}{r_{\text{low}}} - \frac{1}{r_{\text{high}}}\right), \tag{5.38}
$$

as desired. The $e^x \approx 1 + x$ approximation is indeed valid here, because the x's we're concerned with are on the order of GM/c^2R (where R is the radius of the earth), which you can show takes on the value of $7 \cdot 10^{-10}$.

Remarks:

1. The gravitational *potential energy* of a mass m in the gravitational field of the earth is $-GMm/r$. (This is the negative integral of the force, $-GMm/r^2$.) The gravitational *potential* (without the word "energy") is defined as the potential energy per unit mass, and it is denoted by ϕ. So we have $\phi = -GM/r$. In terms of ϕ, the time-dilation factor in Eq. (5.38) can be written in the concise form, $f = 1 + \Delta\phi/c^2$. The original time-dilation factor in Eq. (5.4) can also be written in this form, with $\Delta\phi = gh$. This $\Delta\phi$ follows from the approximate mgh form of the gravitational potential energy, valid near the surface of the earth.

2. When r_{high} is very close to r_{low}, the result in Eq. (5.38) should reduce to our old result $1 + g_r h/c^2$, where $h \equiv r_{\text{high}} - r_{\text{low}}$ and $r \equiv r_{\text{low}}$. ($r \equiv r_{\text{high}}$ would work fine too.) Let's check that this is indeed the case. If $h \ll r$, we have (using $1/(1+x) \approx 1 - x$)

$$
\begin{aligned}
\frac{GM}{c^2}\left(\frac{1}{r_{\text{low}}} - \frac{1}{r_{\text{high}}}\right) &= \frac{GM}{c^2}\left(\frac{1}{r} - \frac{1}{r+h}\right) = \frac{GM}{c^2 r}\left(1 - \frac{1}{1+h/r}\right) \\
&\approx \frac{GM}{c^2 r}\big(1 - (1 - h/r)\big) = \frac{GM}{r^2}\frac{h}{c^2} \\
&= \frac{g_r h}{c^2},
\end{aligned} \tag{5.39}
$$

as desired.

3. The same type of reasoning that led to the exponential factor in Eq. (5.38) can be used in reverse to show that the high clock sees the low clock run slow by the factor

$$
\text{Exp}\left(\frac{GM}{c^2 r_{\text{high}}} - \frac{GM}{c^2 r_{\text{low}}}\right). \tag{5.40}
$$

The product of this factor with the exponential factor in Eq. (5.38) equals $e^0 = 1$, as it should, because a clock can't gain or lose time with itself. Note that this statement

[8] Technically, the product here should only go up to r_{n-1} instead of r_n, because r_{n-1} is the lower radius of the last infinitesimal interval, which ends at $r_n \equiv r_{\text{high}}$. But in the $dr \to 0$ limit, this distinction is inconsequential.

is relevant here because the high and low clocks are in the same frame. In the case of *special*-relativistic time dilation, where the two clocks are in different frames, both clocks see the other one running slow. So the product of the time-dilation factors is *not* equal to 1 (even though it is of course still true that a clock can't gain or lose time with itself).

4. The correction term in Eq. (5.38) can be written as (with $h \equiv r_{\text{high}} - r_{\text{low}}$)

$$\frac{GM}{c^2}\left(\frac{1}{r_{\text{low}}} - \frac{1}{r_{\text{high}}}\right) = \frac{GM}{c^2}\left(\frac{r_{\text{high}} - r_{\text{low}}}{r_{\text{low}} r_{\text{high}}}\right)$$
$$= \frac{1}{c^2}\left(\frac{GM}{r_{\text{low}}^2}\right) h \cdot \frac{r_{\text{low}}}{r_{\text{high}}} = \frac{g_{\text{low}} h}{c^2} \cdot \frac{r_{\text{low}}}{r_{\text{high}}}. \tag{5.41}$$

This way of writing the correction term tells us that if you are at radius r_{low}, and if you want to determine how fast a clock at r_{high} is running, then you just need to take the naive gh/c^2 result from Eq. (5.4) (with g being the g_{low} value at your radius) and then multiply by $r_{\text{low}}/r_{\text{high}}$. If h is small, then this multiplicative factor is essentially equal to 1, so we obtain $g_{\text{low}} h/c^2$, as expected. If $h \to \infty$, then $h/r_{\text{high}} \approx 1$, so we obtain $g_{\text{low}} r_{\text{low}}/c^2$. In other words, a clock at infinity runs at the rate you would obtain by incorrectly applying Eq. (5.4) to a clock at height r_{low} (that is, at radius $2r_{\text{low}}$). This fact also follows from letting $r_{\text{high}} = \infty$ in Eq. (5.38).

5. The exponential expression in Eq. (5.38) is an exact result, assuming that the GMm/r^2 Newtonian expression for the gravitational force holds. However, in reality this expression doesn't quite hold, although it is a very good approximation. It is replaced by the correct result from the full-fledged GR theory. So in the end, the second expression in Eq. (5.38) is an approximation for two reasons: (1) the approximate nature of the GMm/r^2 force, and (2) the $e^x \approx 1 + x$ approximation. But the result is still quite accurate, because the errors in both of these approximations are only of the order $(GM/c^2 r)^2$, which is a factor of $GM/c^2 r \approx 7 \cdot 10^{-10}$ (if we let r equal the radius R of the earth) smaller than the terms in Eq. (5.38).

6. The point of this problem is that if h isn't negligible compared with the radius R of the earth, then the result in Eq. (5.38) must be used. The original $1 + gh/c^2$ expression in Eq. (5.4) isn't adequate, because it involves an error on the order of h/R (which may be significant) times the second term in Eq. (5.38). (This follows from using the more accurate Taylor series $1/(1 + x) \approx 1 - x + x^2$ in Eq. (5.39).) This should be contrasted with the point of Problem 5.2, which was to deal with a tiny error on the order of gh/c^2 (which is extremely small) times the second term in Eq. (5.4). This error is academic, whereas the error of order h/R in this problem is quite relevant in the real world, as the following problem demonstrates. ♣

5.4. **Global positioning system**

Let's look at the SR effect first. Using the Taylor series $\sqrt{1-\epsilon} \approx 1 - \epsilon/2$, the time-dilation factor is $\sqrt{1 - v^2/c^2} \approx 1 - v^2/2c^2$. With $v = 3900\,\text{m/s}$, the $-v^2/2c^2$ correction factor equals

$$-\frac{v^2}{2c^2} = -\frac{(3900\,\text{m/s})^2}{2(3 \cdot 10^8\,\text{m/s})^2} = -8.5 \cdot 10^{-11}. \tag{5.42}$$

Each second, this is the fraction of a second that a satellite clock loses. Technically, this is the time loss relative to a stationary observer at, say, the center of the earth. We must remember that a clock on the earth's surface also loses time, because the clock has a nonzero speed due to the fact that it is rotating along with the earth. The speed of such a clock depends on its latitude. If the clock is at the equator, the speed is $2\pi R/(24\,\text{hr})$, which you can show is about $460\,\text{m/s}$. So this clock has a correction factor of

$$-\frac{v^2}{2c^2} = -\frac{(460\,\text{m/s})^2}{2(3 \cdot 10^8\,\text{m/s})^2} \approx -0.12 \cdot 10^{-11}. \tag{5.43}$$

This is small compared with the above effect of $-8.5 \cdot 10^{-11}$. But if we include it (see the remark below) then relative to a ground clock, a satellite clock loses about $-8.5 \cdot 10^{-11} - (-0.1 \cdot 10^{-11}) = -8.4 \cdot 10^{-11}$ of a second, for each second that passes.

Now for the GR effect. From Eq. (5.38) in Problem 5.3, the correction factor is

$$\begin{aligned}
&\frac{GM}{c^2}\left(\frac{1}{r_{\text{low}}} - \frac{1}{r_{\text{high}}}\right) \\
&= \frac{(6.67 \cdot 10^{-11}\,\text{m}^3/(\text{kg}\,\text{s}^2))(6 \cdot 10^{24}\,\text{kg})}{(3 \cdot 10^8\,\text{m/s})^2}\left(\frac{1}{6.4 \cdot 10^6\,\text{m}} - \frac{1}{26.6 \cdot 10^6\,\text{m}}\right) \\
&= 5.3 \cdot 10^{-10}.
\end{aligned} \tag{5.44}$$

Each second, this is the fraction of a second that a satellite clock gains. This is about six times as large as the SR effect (with the opposite sign). Putting the two effects together, we see that each second, a satellite clock gains

$$5.3 \cdot 10^{-10} - 0.84 \cdot 10^{-10} \approx 4.5 \cdot 10^{-10} \tag{5.45}$$

of a second. Since there are $24 \cdot 60 \cdot 60 = 86,400$ seconds in a day, the time gained by a satellite clock over the course of one day is

$$(4.5 \cdot 10^{-10})(86,400\,\text{s}) \approx 4 \cdot 10^{-5}\,\text{s}. \tag{5.46}$$

This equals $40 \cdot 10^{-6}\,\text{s} = 40$ microseconds. This might not seem like a large time, but when it is multiplied by the speed of light (which is effectively what is done when calculating the position of something on the earth), it produces a distance of $(3 \cdot 10^8\,\text{m/s})(4 \cdot 10^{-5}\,\text{s}) = 12,000\,\text{m} = 12\,\text{km}$. Therefore, if relativistic corrections weren't taken into account, then after only one day, the GPS system would give a location that is incorrect by more than 10 km, which would be far too large for the system to be of any use.

REMARK: We found above that the effect of the 460 m/s speed of a ground clock is only about 1% of the effect of the 3900 m/s speed of a satellite clock. However, even though it is small in comparison, it cannot be ignored, because it is still large in an absolute sense. This is true because it leads to a time loss in one day of $(0.12 \cdot 10^{-11})(86,400\,\text{s}) = 1 \cdot 10^{-7}\,\text{s}$. When this is multiplied by the speed of light, it produces a distance of $(3 \cdot 10^8\,\text{m/s})(1 \cdot 10^{-7}\,\text{s}) = 30\,\text{m}$, which is still large enough to make the system useless for most purposes.

To obtain the necessary accuracy for day-to-day applications, more precision is needed in the various parameters we used above. For example, the rounded-off satellite speed of 3900 m/s isn't precise enough, nor is the $c = 3 \cdot 10^8$ m/s value. And to obtain the correct version of the 460 m/s speed of a point on the equator, we must remember that there are 86,400 seconds in a *solar* day (defined as the time between moments when the sun is at its highest point in the sky), not a *sidereal* day (defined as the time for a complete 360° rotation of the earth). A sidereal day is about 4 minutes short of 24 hours (86,164.1 seconds, to be precise). It is during a *sidereal* day that a point on the equator covers a distance of $2\pi R$. There are also other more complicated effects that need to be considered, such as the nonspherical shape of the earth. This affects both the gravitational potential due to the earth and the speed of a ground clock. Additionally, the orbits of the satellites aren't perfectly circular; the (small) eccentricity must be taken into account. Suffice it to say, it is an extremely complicated problem to solve. But fortunately people have figured everything out and fine tuned the GPS system over the years, so that the rest of us can benefit from it! For further reading, see Ashby (2002). ♣

5.5. **Circular motion**

(a) In A's frame, there is only the SR time-dilation effect. A sees B move at speed v, so B's clock runs slow by a factor of $\sqrt{1 - v^2/c^2}$. And since $v \ll c$, we may use $\sqrt{1-\epsilon} \approx 1 - \epsilon/2$ to approximate this as $1 - v^2/2c^2$.

(b) In this frame, there are both SR and GR time-dilation effects. A moves at speed v with respect to B in this frame, so the SR effect is that A's clock runs slow by a factor $\sqrt{1 - v^2/c^2} \approx 1 - v^2/2c^2$. But B undergoes a (centripetal) acceleration of $a = v^2/r$ toward A. So for all B knows, he lives in a world where the acceleration due to gravity is v^2/r. A is "higher" (by a distance r) in the gravitational field, so the GR effect is that A's clock runs fast by a factor $1 + ar/c^2$. Using $a = v^2/r$, this becomes $1 + v^2/c^2$. Multiplying the $1 - v^2/2c^2$ SR factor by the $1 + v^2/c^2$ GR factor, we find (to order $1/c^2$) that A's clock runs fast by a factor $1 + v^2/2c^2$. This means (to order $1/c^2$, using $1/(1 + \epsilon) \approx 1 - \epsilon$) that B's clock runs slow by a factor $1 - v^2/2c^2$, in agreement with the answer to part (a).

(c) In this frame, there is no relative motion between A and B, so there is only the GR time-dilation effect. The gravitational field (that is, the centripetal acceleration) at a distance x from the center is $g_x = v_x^2/x = (x\omega)^2/x = x\omega^2$, where $\omega = v/r$ is B's angular frequency of rotation around A. Imagine lining up a series of clocks along a radius, with separation dx. Then the GR time-dilation result tells us that each clock loses $g_x\, dx/c^2 = x\omega^2\, dx/c^2$ of a second during each second, relative to the next clock inward. Integrating these fractions from $x = 0$ to $x = r$ tells us that B's clock loses a fractional time of

$$\int_0^r \frac{x\omega^2\, dx}{c^2} = \frac{r^2\omega^2}{2c^2} = \frac{v^2}{2c^2}, \tag{5.47}$$

compared with A's clock. This agrees with the results in parts (a) and (b).

REMARK: If you want to imagine an analogous line of clocks in part (b), you can imagine lining a stick with them, with B at one end of the stick. The sensible thing to do with the stick is to have it be motionless with respect to B's frame (as the line of clocks was in part (c)). But this means that the stick doesn't rotate with respect to the inertial axes, because B's axes don't rotate in part (b). So all of the clocks feel the same acceleration $r\omega^2 = v^2/r$, in contrast with the decreasing $x\omega^2$ accelerations in part (c). The integral of all the fractions therefore doesn't pick up the factor of $1/2$ as it did in part (c), and so we simply end up with a GR effect of v^2/c^2, as we found in part (b). A nonzero SR effect then arises because A is flying past the clock on the other end of the stick, whereas in part (c) A is at rest with respect to the end clock on the (rotating) stick. ♣

5.6. More circular motion

(a) In the lab frame, the situation is symmetric with respect to A and B. Therefore, if A and B are decelerated in a symmetric manner and brought together, their clocks must read the same time.

Let's assume (in the interest of obtaining a contradiction), that A sees B's clock run slow. Then after an arbitrarily long time, A will see B's clock an arbitrarily large time behind his. Now bring A and B to a stop. There is no possible way that the stopping motion can make B's clock gain an arbitrarily large amount of time, as seen by A. This is true because everything takes place in a finite region of space, so there is an upper bound on the GR time-dilation effect (because it behaves like gh/c^2, and h is bounded here). Therefore, A will end up seeing B's clock reading less. This contradicts the result in the previous paragraph. Likewise for the case where A sees B's clock run fast. A must therefore see B's clock run at the same rate as his own.

REMARK: Note how this problem differs from the problem where A and B move with equal constant speeds directly away from each other, and then reverse directions and head back at equal constant speeds to meet up again. For this new "linear" problem, the symmetry reasoning in the first paragraph above still holds, so A and B will again have the same clock readings when they meet up again. But the reasoning in the second paragraph does not hold. It better not, because each person does *not* see the other person's clock running at the same rate as his own, due to SR time dilation. The difference is that in this linear scenario, the experiment is not contained in a small region of space. So the turning-around effect of order gh/c^2 becomes arbitrarily large as the time of travel becomes arbitrarily large, because h grows with time. See Problem 5.12 and Exercise 5.29. ♣

(b) In this frame, there are both SR and GR time-dilation effects. Since A and B are always moving in opposite directions in the lab frame, A moves at speed $2v$ with respect to B in this frame. (We don't need to use the velocity-addition formula, because $v \ll c$.) So the SR effect is that A's clock runs slow by a factor $\sqrt{1-(2v)^2/c^2} \approx 1 - 2v^2/c^2$. But B undergoes an acceleration of $a = v^2/r$ toward A, so the GR effect is that A's clock runs fast by a factor $1 + a(2r)/c^2$, because they are separated by a distance $2r$. Using $a = v^2/r$, this factor becomes $= 1 + 2v^2/c^2$. Multiplying the $1 - 2v^2/c^2$ SR factor by the $1 + 2v^2/c^2$ GR factor, we find (to order $1/c^2$, although from part (a) we know that the result is actually exact) that the two clocks run at the same rate.

(c) In this frame, there is no relative motion between A and B, so there is only (at most) a GR effect. But A and B are both at the same gravitational potential, because they are at the same radius. Therefore, they both see the clocks running at the same rate. If you want, you can line up a series of clocks along the diameter between A and B, as we did along a radius in part (c) of Problem 5.5. The clocks will gain time as you march in toward the center, and then lose the same amount of time as you march back out to the diametrically opposite point.

5.7. **Time and speed in an accelerating frame**

Let S be the instantaneous inertial frame of the rocket at the given moment. Then after an infinitesimal time t as measured in S, the Minkowski diagram from the point of view of S is shown in Fig. 5.11. From Eq. (2.32) the angle θ shown is given by $\tan\theta = \beta = gt/c \to gt$, dropping the c's. Likewise, from Eq. (2.30) the angle α shown is given by $\tan\alpha = v$.

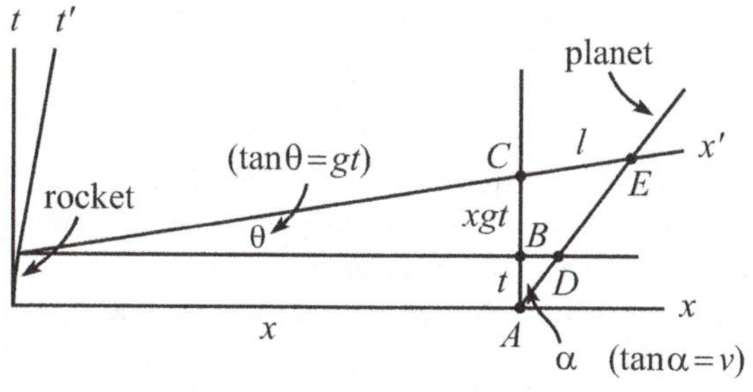

Figure 5.11

The x' and ct' axes are the axes of the new instantaneous inertial frame of the rocket at time t. Since BC has length $x\tan\theta = x(gt)$, we see that AC has length $t + xgt$. (Technically, x should be replaced by $x - gt^2/2$ here, since the rocket has moved a small distance to the right. But this correction is negligible for small t.) The similar triangles ABD and ACE (we are using the fact that the x' axis is essentially horizontal) therefore give

$$\frac{AE}{AD} = \frac{AC}{AB} = \frac{t + xgt}{t} = 1 + gx. \tag{5.48}$$

If the planet's speed is v, then standard time dilation says that the segment AD corresponds to $t\sqrt{1-v^2}$ units of time in the planet's frame. Combining this with Eq. (5.48) tells us that AE corresponds to $t(1+gx)\sqrt{1-v^2}$ units of time in the planet's frame. During the infinitesimal time t, the rocket's "now" axis sweeps from the x axis up to the x' axis. So a time t that elapses on the rocket's clock corresponds to a time $t(1+gx)\sqrt{1-v^2}$ on the planet's clock. Therefore,

$$dt_p = dt_r(1+gx)\sqrt{1-v^2}, \tag{5.49}$$

as desired. We have used the fact that the rocket's time is essentially equal to the time in the instantaneous inertial frame S. This is true because any time-dilation effects between the rocket and S are second order in the infinitesimal time t.

Now for the speed. Since the x' axis is essentially horizontal (for small t), the length of segment CE is essentially $\ell = (AC)\tan\alpha = t(1+gx)\cdot v$. To first order in t, the unit size on the x' axis is the same as on the x axis (the difference is second order in $\beta = gt$, from Eq. (2.33)), so the planet is now a distance ℓ farther away from the rocket. The speed of the planet in the rocket's accelerating frame is therefore

$$V = \frac{\ell}{t} = (1+gx)v, \tag{5.50}$$

as desired. Basically, due to the swinging up of the x' axis in the Minkowski diagram (which leads to the similar-triangle ratio in Eq. (5.48)), everything associated with the planet (the planet's speed in Eq. (5.50) and the rate of its clocks in Eq. (5.49)) happens faster by a factor $1+gx$ in the accelerating frame, compared with the rate in the instantaneous inertial frame S.

Remarks:

1. If $x = 0$ (that is, the rocket is passing right by the planet), the speed V in Eq. (5.50) is simply v. This makes sense, because the swinging up of the x' axis now has no effect, since there is zero distance between the rocket and the planet. The length CE in Fig. 5.11 is then essentially the same as the length BD. That is, $\ell = t\tan\alpha = tv$, so the speed is $\ell/t = v$.

2. If we combine Eqs. (5.49) and (5.50) and eliminate v, and if we then invoke the Equivalence Principle, we arrive at the result that a clock moving at speed V at height h in a gravitational field (with the structure of the accelerating frame in Section 5.3.2) is seen by someone on the ground to run at a rate (putting the c's back in),

$$\sqrt{\left(1+\frac{gh}{c^2}\right)^2 - \frac{V^2}{c^2}}. \tag{5.51}$$

   You should convince yourself that this result is consistent with the discussion in the second remark in the airplane example in Section 5.2. The speed V here corresponds to the speed as measured by the ground observer in the airplane example, which is different from how we defined v in that example. The v's in the two setups have the same meaning.

3. When dealing with *distances*, there is no need to distinguish between the rocket's instantaneous inertial frame S and its accelerating frame (call it S_{acc}), because the x axes of these two frames are the same at any instant. They both start out as the x axis in Fig. 5.11, and then a short time later they are both the x' axis, and so on. However, when dealing with *speeds*, we must distinguish between the frames, because the swinging up of the x' axis in Fig. 5.11 is critical for S_{acc} but irrelevant for S. This is true because when calculating a speed, we take the difference of two positions, $x_{\text{f}} - x_{\text{i}}$, and then divide by the corresponding time difference. In a given S frame, both x_{i} and x_{f} are measured using the *same* x axis. The planet starts at point A in Fig. 5.11 and ends up at point D. (The new inertial frame of the rocket involving the x' axis is irrelevant here, since we're just dealing with the original frame S.) But in S_{acc}, x_{i} and x_{f} are measured using *different* x axes (first the x axis, and then the x' axis). The planet starts at point A and ends up at point E (the intersection of the planet's worldline and the x' axis). The question, "What is the speed of a given object in the frame of an accelerating rocket?" is therefore ambiguous. We must state whether we are talking about the instantaneous inertial frame S or the accelerating frame S_{acc}. ♣

5.8. **Accelerator's point of view**

(a) First solution: Eq. (5.8) says that the distance traveled by the rocket (as measured in the original inertial frame), as a function of the time in the inertial frame, is

$$d = \frac{1}{g}\left(\sqrt{1+(gt)^2} - 1\right) = \frac{1}{g}\left(\cosh(g\tau) - 1\right), \tag{5.52}$$

where we have used the $gt = \sinh(g\tau)$ result from Exercise 5.22, along with the fact that $1 + \sinh^2(g\tau) = \cosh^2(g\tau)$. An inertial observer on the planet therefore measures the rocket-planet distance to be

$$\ell - \frac{1}{g}\left(\cosh(g\tau) - 1\right). \tag{5.53}$$

The rocket observer sees this length contracted by a factor γ, which equals $\cosh(g\tau)$ from Exercise 5.22. (Imagine a long stick connected to the planet. This stick is length contracted in the rocket frame.) So the rocket-planet distance x, as measured in the instantaneous inertial frame of the rocket, is

$$x = \frac{\ell - \frac{1}{g}\left(\cosh(g\tau) - 1\right)}{\cosh(g\tau)} \quad \Longrightarrow \quad 1 + gx = \frac{1+g\ell}{\cosh(g\tau)}. \tag{5.54}$$

As an exercise, you can show that in the nonrelativistic limit, this correctly reduces to the Newtonian expression, $x \approx \ell - g\tau^2/2$. You will need to write the cosh function in terms of exponentials, and then use the Taylor series for e^x.

Second solution: Eq. (5.24) in Problem 5.7 gives the speed of the planet in the *accelerating* frame of the rocket. Using the result from Exercise 5.22 to write v in terms of τ, the signed velocity $dx/d\tau$ is

$$\frac{dx}{d\tau} = -(1+gx)\tanh(g\tau). \tag{5.55}$$

Separating variables and integrating yields

$$\int \frac{dx}{1+gx} = -\int \tanh(g\tau)\,d\tau \quad \Longrightarrow \quad \ln(1+gx) = -\ln\left(\cosh(g\tau)\right) + C$$
$$\Longrightarrow \quad 1 + gx = \frac{A}{\cosh(g\tau)}, \tag{5.56}$$

where $A \equiv e^C$. Since the initial condition is $x = \ell$ when $\tau = 0$, we must have $A = 1 + g\ell$, in agreement with Eq. (5.54).

(b) Eq. (5.23) in Problem 5.7 says that the planet's clock runs fast (or slow) according to

$$dt = d\tau\,(1+gx)\sqrt{1-v^2}. \tag{5.57}$$

The speed v from Exercise 5.22 gives $\sqrt{1-v^2} = 1/\cosh(g\tau)$. Using this, along with the result for $1 + gx$ in Eq. (5.54), we can integrate Eq. (5.57) to obtain

$$\int dt = \int \frac{(1+g\ell)\,d\tau}{\cosh^2(g\tau)} \quad \Longrightarrow \quad gt = (1+g\ell)\tanh(g\tau). \tag{5.58}$$

The constant of integration is zero because $t = 0$ when $\tau = 0$. As an exercise, you can show that in the nonrelativistic limit, this result correctly reduces to the Newtonian expression, $t = \tau$.

5.9. **Getting way ahead**

The explanation of how the two clocks can show significantly different readings in the ground frame is the following. (Don't read beyond this sentence until you've thought qualitatively about what could possibly lead to a large difference in the readings, quantitative details aside.)

The rocket becomes increasingly length contracted in the ground frame, which means that the front end isn't traveling quite as fast as the back end. Therefore, the time-dilation factor for the front clock isn't as large as it is for the back clock. So the front clock loses less time relative to the ground, and hence ends up ahead of the back clock. Of course, it's not at all obvious that everything works out quantitatively and that the front clock eventually ends up an arbitrarily large time ahead of the back clock. In fact, it's quite surprising that this is the case, because the difference in speeds of the two clocks is very small. But let's now show that the above explanation does indeed correctly account for the difference in the clock readings given in Eq. (5.27).

At a general time t in the ground frame, let the back of the rocket be located at position x. Then the front is located at position $x + L\sqrt{1-v^2}$, due to length contraction. Taking the time derivatives of the two positions, we see that the speeds of the back and front are (with $v \equiv dx/dt$)[9]

$$v_{\text{b}} = v \qquad \text{and} \qquad v_{\text{f}} = v(1 - L\gamma\dot{v}), \tag{5.59}$$

as you can verify. If we assume that the back is the part that accelerates at g (it doesn't matter which point we pick, to leading order), then we can invoke the result in Eq. (5.7) to say that

$$v_{\text{b}} = v = \frac{gt}{\sqrt{1+(gt)^2}}, \tag{5.60}$$

where t is the time in the ground frame. Having written down v, we must now find the γ factors associated with the speeds of the front and back. From Eq. (5.13), the γ factor associated with the speed of the back, namely v, is

$$\gamma_{\text{b}} = \frac{1}{\sqrt{1-v^2}} = \sqrt{1+(gt)^2}\,. \tag{5.61}$$

The γ factor associated with the speed of the front, $v_{\text{f}} = v(1 - L\gamma\dot{v})$, is a bit more complicated. We must first calculate $\dot{v}$. From Eq. (5.60), we obtain $\dot{v} = g/(1+g^2t^2)^{3/2}$, which gives

$$v_{\text{f}} = v(1 - L\gamma\dot{v}) = \frac{gt}{\sqrt{1+(gt)^2}}\left(1 - \frac{gL}{1+g^2t^2}\right). \tag{5.62}$$

The γ factor (or rather $1/\gamma$, which is what we'll be concerned with) associated with this speed is calculated as follows. In the first line below, we ignore the higher-order $(gL)^2$ term, because it is really $(gL/c^2)^2$, and we are assuming that gL/c^2 is small. In obtaining the third line, we use the Taylor series $\sqrt{1+\epsilon} \approx 1 + \epsilon/2$.

$$\begin{aligned}
\frac{1}{\gamma_{\text{f}}} = \sqrt{1-v_{\text{f}}^2} &\approx \sqrt{1 - \frac{g^2t^2}{1+g^2t^2}\left(1 - \frac{2gL}{1+g^2t^2}\right)} \\
&= \frac{1}{\sqrt{1+g^2t^2}}\sqrt{1 + \frac{2g^3t^2L}{1+g^2t^2}} \\
&\approx \frac{1}{\sqrt{1+g^2t^2}}\left(1 + \frac{g^3t^2L}{1+g^2t^2}\right).
\end{aligned} \tag{5.63}$$

We can now calculate the time that each clock shows, at time t in the ground frame. Due to time dilation as observed from the ground frame, the time on the back clock changes according to $dt_{\text{b}} = dt/\gamma_{\text{b}}$. Using the γ_{b} from Eq. (5.61), we obtain

$$t_{\text{b}} = \int_0^t \frac{dt}{\sqrt{1+g^2t^2}}\,. \tag{5.64}$$

[9] Since these two speeds aren't equal, there is an ambiguity concerning which speed we should use in the length-contraction factor, $\sqrt{1-v^2}$. Equivalently, the rocket doesn't have a single inertial frame that describes all of it. But you can show that any differences arising from this ambiguity are of higher order in gL/c^2 than what we need to be concerned with.

The integral of $dx/\sqrt{1+x^2}$ is $\sinh^{-1} x$. (To derive this, make the substitution $x \equiv \sinh\theta$.) Letting $x \equiv gt$, this gives

$$gt_{\rm b} = \sinh^{-1}(gt). \tag{5.65}$$

The time on the front clock changes according to $dt_{\rm f} = dt/\gamma_{\rm f}$. Using the $\gamma_{\rm f}$ from Eq. (5.63), we obtain

$$t_{\rm f} = \int_0^t \frac{dt}{\sqrt{1+g^2t^2}} + \int_0^t \frac{g^3t^2L\,dt}{(1+g^2t^2)^{3/2}}. \tag{5.66}$$

The integral of $x^2\,dx/(1+x^2)^{3/2}$ is $\sinh^{-1} x - x/\sqrt{1+x^2}$. (To derive this, make the substitution $x \equiv \sinh\theta$, and use $\int d\theta/\cosh^2\theta = \tanh\theta$.) Letting $x \equiv gt$, this gives

$$gt_{\rm f} = \sinh^{-1}(gt) + (gL)\left(\sinh^{-1}(gt) - \frac{gt}{\sqrt{1+g^2t^2}}\right). \tag{5.67}$$

Using Eqs. (5.60) and (5.65), we can rewrite this as

$$gt_{\rm f} = gt_{\rm b}(1+gL) - gLv. \tag{5.68}$$

Dividing by g, and putting the c's in to make the units correct, we finally have

$$t_{\rm f} = t_{\rm b}\left(1 + \frac{gL}{c^2}\right) - \frac{Lv}{c^2}, \tag{5.69}$$

as we wanted to show. If we look at this calculation from the reverse point of view, we see that by using only special-relativity concepts, we have demonstrated that someone at the back of a rocket must see a clock at the front running fast by a factor $(1 + gL/c^2)$. There are, however, much easier ways of deriving this, as we saw in Section 5.2 and in Problem 2.12 ("Acceleration and redshift").

5.10. **Lv/c^2 revisited**

Consider first the case where the rocket accelerates while you sit there. Problem 5.9 is exactly relevant here, and it tells us that in your frame the clock readings are related by

$$t_{\rm f} = t_{\rm b}\left(1 + \frac{gL}{c^2}\right) - \frac{Lv}{c^2}. \tag{5.70}$$

You will eventually see the front clock an arbitrarily large time ahead of the back clock. But note that for small times (before things become relativistic), the standard Newtonian expression for the speed, $v \approx gt_{\rm b}$, is valid, so we have

$$t_{\rm f} \approx \left(t_{\rm b} + \frac{Lv}{c^2}\right) - \frac{Lv}{c^2} = t_{\rm b}. \tag{5.71}$$

So in this setup where the rocket is the thing that accelerates, both clocks show essentially the same time near the start. This makes sense; both clocks have essentially the same speed at the beginning, so to lowest order their γ factors are the same, which means that the clocks run at the same rate. But eventually the front clock will get ahead of the back clock.

Now consider the case where you accelerate while the rocket sits there. Problem 5.8 is relevant here, if we let the rocket in that problem now become you, and if we let two planets a distance L apart become the two ends of the rocket in this problem. The times you observe on the front and back clocks on the rocket are then, using Eq. (5.26) and assuming that you are accelerating toward the rocket (see Fig. 5.12),

$$gt_{\rm f} = (1 + g\ell)\tanh(g\tau) \qquad \text{and} \qquad gt_{\rm b} = (1 + g(\ell + L))\tanh(g\tau). \tag{5.72}$$

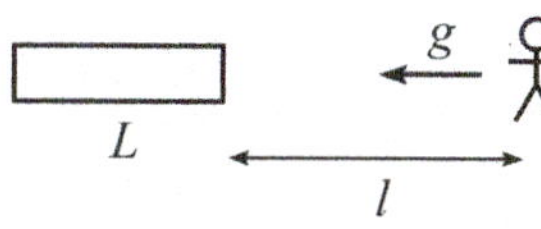

Figure 5.12

Combining these equations quickly yields $gt_{\rm b} = gt_{\rm f} + gL\tanh(g\tau)$. But from Exercise 5.22, we know that your speed relative to the rocket is $v = \tanh(g\tau)$. We therefore obtain $t_{\rm b} = t_{\rm f} + Lv$, or $t_{\rm b} = t_{\rm f} + Lv/c^2$ with the c^2. So in this case we arrive at the standard Lv/c^2 rear-clock-ahead result.

The point here is that in this second case, the clocks are synchronized in the rocket frame, and this is the assumption that went into our derivation of the Lv/c^2 result in Section 1.3.1. In the first case above where the rocket accelerates, the clocks are *not* synchronized in the rocket frame (except right at the start), due to the GR time dilation, so it isn't surprising that we don't obtain the Lv/c^2 result.

5.11. **Circling the earth**

This is one setup where we really do need to use the correct term, *stationary*-proper-time principle. It turns out that B's path yields a saddle point (that is, a stationary point that is neither a maximum nor a minimum) of the proper time. The value at this saddle point is less than A's proper time, but that is irrelevant, because we care only about local stationary points, not about global extrema. Let's show that B's path is in fact a saddle point of the proper time.

First, note that we do indeed have a stationary value of the proper time, due to the fact that B's path is a physical one, which from the discussion in Section 5.4 means that the action (and hence the proper time) is stationary.

To show that the stationary value is a saddle point (that is, not a maximum or a minimum), we must show that there exist slight tweaks to the motion that give both larger and smaller proper times. The proper time can indeed be made smaller by having B speed up and slow down (while keeping the same total time for each complete circle). This will cause a net increase in the time-dilation effect as viewed by A, thereby yielding a smaller proper time.[10] And the proper time can indeed be made larger by having B take a nearby path that doesn't quite form a great circle on the earth. (Imagine the curve traced out by a rubber band that has just begun to slip away from a great-circle position.) This path is shorter, so B won't have to travel as fast to get back in a given time, so the time-dilation effect will be smaller as viewed by A, thereby yielding a larger proper time.

The hypothetical modified paths just mentioned will of course require B to be acted on by some force in addition to gravity. That is fine. The hypothetical paths don't need to be physically relevant to our original setup involving only gravity; they most certainly won't be the path that B actually takes.

5.12. **Twin paradox**

(a) In the earth frame, the spaceship travels at speed v for essentially the whole time. Therefore, the traveler ages less by a factor $\sqrt{1-v^2/c^2} \approx 1 - v^2/2c^2$, where we have used the Taylor series $\sqrt{1-\epsilon} \approx 1 - \epsilon/2$. So the fractional time loss is $v^2/2c^2$. Since the total time in the earth frame is essentially $2\ell/v$, we see that the traveler ages less by a time of $(v^2/2c^2)(2\ell/v) = \ell v/c^2$. The time-dilation factor is different during the short turning-around period, but that is inconsequential in the earth frame.

(b) During the constant-speed part of the trip, the traveler sees the earth clock running slow by a factor $\sqrt{1-v^2/c^2} \approx 1 - v^2/2c^2$. The time (as measured by the traveler) for the constant-speed part is essentially $2\ell/v$, so relative to the traveler's clock, the earth clock loses a time of $(v^2/2c^2)(2\ell/v) = \ell v/c^2$. (We should technically be using the length-contracted distance ℓ/γ here, but this would only modify the time loss by a negligible term of order v^2/c^2 smaller than the $\ell v/c^2$ result.)

However, during the turnaround period, the spaceship is accelerating toward the earth, so the traveler sees the earth clock running fast, due to the GR time dilation. (The earth is high in the effective gravitational field.) If the turnaround takes a time T, then the magnitude of the acceleration is $a = 2v/T$, because the spaceship goes from velocity v to $-v$ in time T. The earth clock therefore runs fast by a factor $1 + a\ell/c^2 = 1 + 2\ell v/Tc^2$. This happens for a time T, so the earth clock gains a time of $(2\ell v/Tc^2)T = 2\ell v/c^2$.

[10]This is true for the same reason that a person who travels at constant speed in a straight line between two points shows a larger proper time than a second person who speeds up and slows down. This follows directly from SR time dilation, as viewed by the first person. If you want, you can imagine unrolling B's circular orbit into a straight line, and then invoke the result just mentioned. As far as SR time-dilation effects from A's point of view go, it doesn't matter if the circle is unrolled into a straight line.

Combining the results of the preceding two paragraphs, we conclude that the traveler sees the earth clock gain a net time of $2\ell v/c^2 - \ell v/c^2 = \ell v/c^2$. The traveler is therefore younger by $\ell v/c^2$, in agreement with the result in part (a).

5.13. **Twin paradox again**

(a) The only difference between this problem and the previous one is the nature of the turnaround, so all we need to show here is that the traveler still sees the earth clock gain a time of $2\ell v/c^2$ during the turnaround period.

Let the radius of the semicircle be r. Then the magnitude of the (centripetal) acceleration is $a = v^2/r$. Let θ be the angle shown in Fig. 5.13. For a given θ, the earth is at a height of essentially $h = \ell\cos\theta$ in the effective gravitational field felt by the spaceship. (This height is the projection of the full distance ℓ onto the line containing the radius at the given instant.) The fractional time that the earth gains while the traveler is at an angle θ is therefore $f = ah/c^2 = (v^2/r)(\ell\cos\theta)/c^2$. Integrating this over the time of the turnaround, and using $dt = r\,d\theta/v$, we see that the earth gains a time of

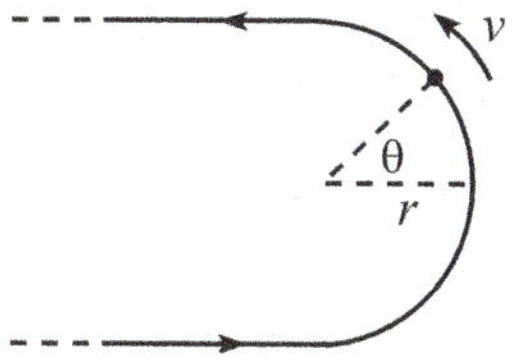

Figure 5.13

$$\Delta t = \int f\,dt = \int_{-\pi/2}^{\pi/2}\left(\frac{v^2\ell\cos\theta}{rc^2}\right)\left(\frac{r\,d\theta}{v}\right) = \frac{\ell v}{c^2}\sin\theta\bigg|_{-\pi/2}^{\pi/2} = \frac{2\ell v}{c^2}, \tag{5.73}$$

as desired.

(b) Let the acceleration vector at a given instant be $\mathbf{a}$, and let $\boldsymbol{\ell}$ be the vector from the spaceship to the earth. Since the turnaround is done in a small region of space, $\boldsymbol{\ell}$ is essentially constant here. Let θ be the angle between $\mathbf{a}$ and $\boldsymbol{\ell}$. This is the same θ as in part (a), because if the earth is far off to the left then $\boldsymbol{\ell}$ is horizontal, just like the horizontal radius in Fig. 5.13. So as in part (a), the earth is at a height $h = \ell\cos\theta$ in the effective gravitational field felt by the spaceship. The fractional time gain is therefore $ah/c^2 = a\ell\cos\theta/c^2$. But $a\ell\cos\theta$ equals the dot product $\mathbf{a}\cdot\boldsymbol{\ell}$. (The dot product of two vectors $\mathbf{A}$ and $\mathbf{B}$ is $\mathbf{A}\cdot\mathbf{B} = AB\cos\theta$.) So the fractional time gain is $\mathbf{a}\cdot\boldsymbol{\ell}/c^2$. Integrating this over the time of the turnaround, we see that the earth gains a time of

$$\begin{aligned}\Delta t = \int_{t_i}^{t_f}\frac{\mathbf{a}\cdot\boldsymbol{\ell}}{c^2}\,dt &= \frac{\boldsymbol{\ell}}{c^2}\cdot\int_{t_i}^{t_f}\mathbf{a}\,dt\\ &= \frac{\boldsymbol{\ell}}{c^2}\cdot(\mathbf{v}_f - \mathbf{v}_i)\\ &= \frac{\boldsymbol{\ell}\cdot(2\mathbf{v}_f)}{c^2}\\ &= \frac{2\ell v}{c^2},\end{aligned} \tag{5.74}$$

as we wanted to show. The point here is that no matter how complicated the motion is during the turnaround, the net effect is simply to change the velocity from $\mathbf{v}$ outward to $\mathbf{v}$ inward.

5.14. **Twin paradox times**

(a) As viewed by A, the relation between the twins' times is

$$dt_B = \sqrt{1 - v^2}\,dt_A. \tag{5.75}$$

The magnitude of B's acceleration is g, so assuming $v_0 \ll c$, we may say that B's velocity as a function of t_A is essentially given by $v = v_0 - gt_A$ (the standard Newtonian expression). The out and back parts of the trip therefore each take a time of essentially v_0/g in A's frame. The total elapsed time on B's clock is then (using the Taylor series $\sqrt{1-\epsilon} \approx 1 - \epsilon/2$)

$$
\begin{aligned}
T_B = \int dt_B &= 2\int_0^{v_0/g} \sqrt{1-v^2}\,dt_A \\
&\approx 2\int_0^{v_0/g} \left(1 - \frac{v^2}{2}\right) dt_A \\
&\approx 2\int_0^{v_0/g} \left(1 - \frac{1}{2}(v_0 - gt)^2\right) dt \\
&= 2\left(t + \frac{1}{6g}(v_0 - gt)^3\right)\Bigg|_0^{v_0/g} \\
&= \frac{2v_0}{g} - \frac{v_0^3}{3gc^2},
\end{aligned} \tag{5.76}
$$

where we have put the c's in to make the units right. The ratio of B's elapsed time to A's is therefore

$$
\frac{T_B}{T_A} \approx \frac{T_B}{2v_0/g} \approx 1 - \frac{v_0^2}{6c^2}. \tag{5.77}
$$

Note that if the trip were instead done at constant speed v_0 with an abrupt turnaround, then the $1 - v^2/2$ integrand in Eq. (5.76) would be constant, so we would simply obtain $T_B/T_A \approx 1 - v_0^2/2c^2$. The second term here is three times the second term in Eq. (5.77).

(b) As viewed by B, the relation between the twins' times is given by Eq. (5.17),

$$
dt_A = \sqrt{1 - \frac{v^2}{c^2}}\left(1 + \frac{gy}{c^2}\right) dt_B. \tag{5.78}
$$

Assuming $v_0 \ll c$, we may say that A's velocity as a function of t_B is given essentially by $v = v_0 - gt_B$. Integrating this gives A's height as $v_0 t_B - g t_B^2/2$ (the standard Newtonian expression). The up and down parts of the trip each take a time of essentially v_0/g in B's frame. Therefore, the total elapsed time on A's clock is (using the approximation in Eq. (5.18) and dropping the c's)

$$
\begin{aligned}
T_A = \int dt_A &\approx 2\int_0^{v_0/g} \left(1 - \frac{v^2}{2} + gy\right) dt_B \\
&\approx 2\int_0^{v_0/g} \left(1 - \frac{1}{2}(v_0 - gt)^2 + g\left(v_0 t - \frac{gt^2}{2}\right)\right) dt \\
&= 2\left(t + \frac{1}{6g}(v_0 - gt)^3 + g\left(\frac{v_0 t^2}{2} - \frac{gt^3}{6}\right)\right)\Bigg|_0^{v_0/g} \\
&= 2\left(\frac{v_0}{g} + \frac{v_0^3}{g}\left(\frac{1}{2} - \frac{1}{6}\right)\right) - 2\left(\frac{v_0^3}{6g}\right) \\
&= \frac{2v_0}{g} + \frac{v_0^3}{3gc^2},
\end{aligned} \tag{5.79}
$$

where we have put the c's in to make the units right. We therefore have

$$
\frac{T_A}{T_B} \approx \frac{T_A}{2v_0/g} \approx 1 + \frac{v_0^2}{6c^2} \implies \frac{T_B}{T_A} \approx 1 - \frac{v_0^2}{6c^2}, \tag{5.80}
$$

up to higher-order corrections, where we have used the Taylor series $1/(1+\epsilon) \approx 1-\epsilon$. This agrees with the result in part (a), as the Equivalence Principle requires.

Chapter 6

Appendices

6.1 Appendix A: Qualitative relativity questions

Basic principles

1. QUESTION: You are in a spaceship sailing along in outer space. Is there any way you can measure your speed without looking outside?

   ANSWER: There are two points to be made here. First, the question is meaningless, because absolute speed doesn't exist. The spaceship doesn't have a speed; it only has a speed relative to something else. Second, even if the question asked for the speed with respect to, say, a given piece of stellar dust, the answer would be "no." Uniform speed is not measurable from within the spaceship. Acceleration, on the other hand, is measurable (assuming there is no gravity around to confuse it with).

2. QUESTION: Two people, A and B, are moving with respect to each other in one dimension. Is the speed of B as viewed by A equal to the speed of A as viewed by B?

   ANSWER: Yes. The first postulate of relativity states that all inertial frames are equivalent, which implies that there is no preferred location or direction in space. If the relative speed measured by the left person were larger than the relative speed measured by the right person, then there would be a preferred direction in space. Apparently people on the left always measure a larger speed. This violates the first postulate. Likewise if the left person measures a smaller speed. The two speeds must therefore be equal.

3. QUESTION: Is the second postulate of relativity (that the speed of light is the same in all inertial frames) really necessary? Or is it already implied by the first postulate (that the laws of physics are the same in all inertial frames)?

   ANSWER: It is necessary. The speed-of-light postulate is not implied by the laws-of-physics postulate. The latter doesn't imply that baseballs have the same speed in all inertial frames, so it likewise doesn't imply that light has the same speed.

   It turns out that nearly all the results in special relativity can be deduced by using only the laws-of-physics postulate. What you can find (with some work) is that there is some limiting speed, which may be finite or infinite; see Section 2.7. The speed-of-light postulate fills in the last bit of information by telling us what the limiting speed is.

4. Question: Is the speed of light equal to c, under all circumstances?

   Answer: No. In Section 1.2 we stated the speed-of-light postulate as, "The speed of light in vacuum has the same value c in any inertial frame." There are two key words here: "vacuum" and "inertial." If we are dealing with a medium (such as glass or water) instead of vacuum, then the speed of light is smaller than c. And in an accelerating (noninertial) reference frame, the speed of light can be larger or smaller than c.

5. Question: In a given frame, a clock reads noon when a ball hits it. Do observers in all other frames agree that the clock reads noon when the ball hits is?

   Answer: Yes. Some statements, such as the one at hand, are frame independent. The relevant point is that everything happens at one location, so we don't have to worry about any Lv/c^2 loss-of-simultaneity effects. The L here is zero. Equivalently, the ball-hits-clock event and the clock-reads-noon event are actually the same event. They are described by the same space and time coordinates. Therefore, since we have only one event, there is nothing for different observers to disagree on.

6. Question: Can information travel faster than the speed of light?

   Answer: No. If it could, we would be able to generate not only causality-violating setups, but also genuine contradictions. See the discussions at the ends of Sections 2.3 and 2.4.

7. Question: Is there such a thing as a perfectly rigid object?

   Answer: No. From the discussions of causality in Sections 2.3 and 2.4, we know that information can't travel infinitely fast (in fact, it can't exceed the speed of light). So it takes time for the atoms in an object to communicate with each other. If you push on one end of a stick, the other end won't move right away.

8. Question: Can an object (with nonzero mass) move at the speed of light?

   Answer: No. Here are four reasons why. (1) Energy: $v = c$ implies $\gamma = \infty$, which implies that $E = \gamma mc^2$ is infinite. The object must therefore have an infinite amount of energy (unless $m = 0$, as for a photon). All the energy in the universe, let alone all the king's horses and all the king's men, can't accelerate something to speed c. (2) Momentum: again, $v = c$ implies $\gamma = \infty$, which implies that $p = \gamma mv$ is infinite. The object must therefore have an infinite amount of momentum (unless $m = 0$, as for a photon). (3) Force: since $F = \gamma^3 ma$, we have $a = F/m\gamma^3$. And since $\gamma \to \infty$ as $v \to c$, the acceleration a becomes smaller and smaller (for a given F) as v approaches c. It can be shown that with the γ^3 factor in the denominator, the acceleration drops off quickly enough so that the speed c is never reached in a finite time. (4) Velocity-addition formula: no matter what speed you give an object with respect to the frame it was just in (that is, no matter how you accelerate it), the velocity-addition formula always yields a speed that is less than c. The only way the resulting speed can equal c is if one of the two speeds in the formula is c.

9. Question: Imagine closing a very large pair of scissors. If arranged properly, it is possible for the point of intersection of the blades to move faster than the speed of light. Does this violate anything in relativity?

   Answer: No. If the angle between the blades is small enough, then the tips of the blades (and all the other atoms in the scissors) can move at a speed well below c, while the intersection point moves faster than c. But this doesn't violate anything in relativity. The intersection point isn't an actual object, so there is nothing wrong with it moving faster than c.

You might be worried that this result allows you to send a signal down the scissors at a speed faster than c. However, since there is no such thing as a perfectly rigid object, it is impossible to get the far end of the scissors to move right away, when you apply a force at the handle. The scissors would have to already be moving, in which case the motion is independent (at least for a little while) of any decision you make at the handle to change the motion of the blades.

10. QUESTION: A mirror moves toward you at speed v. You shine a light toward it, and the light beam bounces back at you. What is the speed of the reflected beam?

    ANSWER: The speed is c, as always. You will observe the light (which is a wave) having a higher wave frequency due to the Doppler effect. But the speed is still c.

11. QUESTION: Person A chases person B. As measured in the ground frame, they have speeds v_A and v_B. If they start a distance L apart (as measured in the ground frame), how much time will it take (as measured in the ground frame) for A to catch B?

    ANSWER: In the ground frame, the relative speed is $v_A - v_B$. Person A must close the initial gap of L, so the time it takes is $L/(v_A - v_B)$. There is no need to use any fancy velocity-addition or length-contraction formulas, because all quantities in this problem are measured with respect to the *same* frame. So it quickly reduces to a simple "(rate)(time) = (distance)" problem. Alternatively, the two positions in the ground frame are given by $x_A = v_A t$ and $x_B = L + v_B t$. Setting these positions equal to each other gives $t = L/(v_A - v_B)$.

    Note that no object in this setup moves with speed $v_A - v_B$. This is simply the rate at which the gap between A and B closes, and a gap isn't an actual thing.

12. QUESTION: How do you synchronize two clocks that are at rest with respect to each other?

    ANSWER: One way is to put a light source midway between the two clocks and send out signals, and then set the clocks to a certain value when the signals hit them. Another way is to put a watch right next to one of the clocks and synchronize it with that clock, and then move the watch very slowly over to the other clock and synchronize that clock with it. Any time-dilation effects can be made arbitrarily small by moving the watch sufficiently slowly, because the time-dilation effect is second order in v (and because the travel time is only first order in $1/v$).

The fundamental effects

13. QUESTION: Two clocks at the ends of a train are synchronized with respect to the train. If the train moves past you, which clock shows a higher time?

    ANSWER: The rear clock shows a higher time. It shows Lv/c^2 more than the front clock, where L is the proper length of the train.

14. QUESTION: Does the rear-clock-ahead effect imply that the rear clock runs faster than the front clock?

    ANSWER: No. Both clocks run at the same rate in the ground frame. It's just that the rear clock is always a fixed time of Lv/c^2 ahead of the front clock.

15. QUESTION: Moving clocks run slow. Does this result have anything to do with the time it takes light to travel from the clock to your eye?

    ANSWER: No. When we talk about how fast a clock is running in a given frame, we are referring to what the clock actually reads in that frame. It will certainly take time for the light from the clock to reach your eye, but it is understood that you subtract off this transit time in order to calculate the time (in your frame) at which the clock actually shows a particular reading. Likewise, other relativistic effects, such as length contraction and the loss of simultaneity, have nothing to do with the time it takes light to reach your eye. They deal only with what really *is*, in your frame. One way to avoid the complication of the travel time of light is to use the lattice of clocks and meter sticks described in Section 1.3.4.

16. QUESTION: A clock on a moving train reads T. Does the clock read γT or T/γ in the ground frame?

    ANSWER: Neither. It reads T. Clock *readings* don't get dilated. *Elapsed times* are what get dilated. If the clock advances from T_1 to T_2, then the time between these readings, as measured in the train frame, is just $T_2 - T_1$. But the time between the readings, as measured in the ground frame, is $\gamma(T_2 - T_1)$, due to time dilation.

17. QUESTION: Does time dilation depend on whether a clock is moving across your vision or directly away from you?

    ANSWER: No. A moving clock runs slow, no matter which way it is moving. This is clearer if you think in terms of the lattice of clocks and meter sticks in Section 1.3.4. If you imagine a million people standing at the points of the lattice, then they all observe the clock running slow. Time dilation is an effect that depends on the *frame* and the speed of a clock with respect to it. It doesn't matter where you are in the frame (as long as you're at rest in it), as you look at a moving clock.

18. QUESTION: Does special-relativistic time dilation depend on the acceleration of the moving clock you are looking at?

    ANSWER: No. The time-dilation factor is $\gamma = 1/\sqrt{1 - v^2/c^2}$, which doesn't depend on the acceleration a. The only relevant quantity is the v at a given instant; it doesn't matter if v is changing. As long as you represent an inertial frame, then the clock you are viewing can undergo whatever motion it wants, and you will observe it running slow by the simple factor of γ. See the third remark in the solution to Problem 2.12.

    However, if *you* are accelerating, then you can't naively apply the results of special relativity. To do things correctly, it is easiest to think in terms of general relativity (or at least the Equivalence Principle). This is discussed in Chapter 5.

19. QUESTION: Two twins travel away from each other at relativistic speed. The time-dilation result says that each twin sees the other twin's clock running slow. So each says the other has aged less. How would you reply to someone who asks, "But which twin really *is* younger?"

    ANSWER: It makes no sense to ask which twin really is younger, because the two twins aren't in the same reference frame; they are using different coordinates to measure time. It's as silly as having two people run away from each other into the distance (so that each person sees the other become small), and then asking: Who is really smaller?

20. QUESTION: A train moves at speed v. A ball is thrown from the back to the front. In the train frame, the time of flight is T. Is it correct to use time dilation to say that the time of flight in the ground frame is γT?

    ANSWER: No. The time-dilation result holds only for two events that happen at the *same place* in the relevant reference frame (the train, here). Equivalently, it holds if you are looking at a *single* moving clock. The given information tells us that the reading on the front clock (when the ball arrives) minus the reading on the back clock (when the ball is thrown) is T. It makes no sense to apply time dilation to the difference in these readings, because they come from two different clocks.

Another way of seeing why simple time dilation is incorrect is to use the Lorentz transformation. If the proper length of the train is L, then the correct time on the ground (between the ball-leaving-back and ball-hitting-front events) is given by $\Delta t_g = \gamma(\Delta t_t + v\Delta x_t/c^2) = \gamma T + \gamma v L/c^2$, which isn't equal to γT. Equivalently, if you look at a *single* clock on the train, for example the back clock, then it starts at zero but ends up at $T + Lv/c^2$ due to the rear-clock-ahead effect (because the front clock shows T when the ball arrives). Applying time dilation to this elapsed time on a *single* clock gives the correct time in the ground frame. This is the argument we used when deriving Δt_g in Section 2.1.2.

21. QUESTION: Someone says, "A stick that is length-contracted isn't *really* shorter, it just *looks* shorter." How do you respond?

    ANSWER: The stick really *is* shorter in your frame. Length contraction has nothing to do with how the stick looks, because light takes time to travel to your eye. It has to do with where the ends of the stick are at simultaneous times in your frame. This is, after all, how you measure the length of something. At a given instant in your frame, the distance between the ends of the stick is genuinely less than the proper length of the stick. If a green sheet of paper slides with a relativistic speed v over a purple sheet (of the same proper size), and if you take a photo when the centers coincide, then the photo will show some of the purple sheet (or essentially all of it in the $v \to c$ limit).

22. QUESTION: Consider a stick that moves in the direction in which it points. Does its length contraction depend on whether this direction is across your vision or directly away from you?

    ANSWER: No. The stick is length-contracted in both cases. Of course, if you look at the stick in the latter case, then all you see is the end, which is just a dot. But the stick is indeed shorter in your reference frame. As in Question 17 above concerning time dilation, length contraction depends on the frame, not where you are in it.

23. QUESTION: If you move at the speed of light, what shape does the universe take in your frame?

    ANSWER: The question is meaningless, because it's impossible for you to move at the speed of light. A meaningful question to ask is: What shape does the universe take if you move at a speed very close to c (with respect to, say, the average velocity of all the stars)? The answer is that in your frame everything will be squashed along the direction of your motion, due to length contraction. Any given region of the universe will be squashed down to a pancake.

24. QUESTION: Eq. (1.14) says that the time in the observer's frame is *longer* than the proper time, while Eq. (1.20) says that the length in the observer's frame is *shorter* than the proper length. Why does this asymmetry exist?

    ANSWER: The asymmetry arises from the different assumptions that lead to time dilation and length contraction. If a clock and a stick are at rest on a train moving with speed v relative to you, then time dilation is based on the assumption that $\Delta x_{\text{train}} = 0$ (this holds for two ticks on a train clock), while length contraction is based on the assumption that $\Delta t_{\text{you}} = 0$ (you measure a length by observing where the ends are at simultaneous times in your frame). These conditions deal with *different frames*, and this causes the asymmetry. Mathematically, time dilation follows from the second equation in Eq. (2.2):

$$\Delta t_{\text{you}} = \gamma(\Delta t_{\text{train}} + (v/c^2)\Delta x_{\text{train}}) \implies \Delta t_{\text{you}} = \gamma \Delta t_{\text{train}} \quad (\text{if } \Delta x_{\text{train}} = 0). \quad (6.1)$$

Length contraction follows from the first equation in Eq. (2.4):

$$\Delta x_{\text{train}} = \gamma(\Delta x_{\text{you}} - v\,\Delta t_{\text{you}}) \implies \Delta x_{\text{train}} = \gamma \Delta x_{\text{you}} \quad (\text{if } \Delta t_{\text{you}} = 0). \quad (6.2)$$

These equations *are* symmetric with respect to the γ *factors*. The γ goes on the side of the equation associated with the frame in which the given space or time interval is zero. However, the equations *aren't* symmetric with respect to *you*. You appear on different sides of the equations because for length contraction your frame is the one where the given interval is zero, whereas for time dilation it isn't. Since the end goal is to solve for the "you" quantity, we must divide Eq. (6.2) by γ to isolate Δx_{you}. This is why Eqs. (1.14) and (1.20) aren't symmetric with respect to the "observer" (that is, "you") label.

25. Question: When relating times via time dilation, or lengths via length contraction, how do you know where to put the γ factor?

    Answer: There are various answers to this, but probably the safest method is to (1) remember that moving clocks run slow and moving sticks are short, then (2) identify which times or lengths are larger or smaller, and then (3) put the γ factor where it needs to be so that the relative size of the times or lengths is correct.

Other kinematics topics

26. Question: Two objects fly toward you, one from the east with speed u, and the other from the west with speed v. Is it correct that their relative speed, as measured by you, is $u + v$? Or should you use the velocity-addition formula, $V = (u + v)/(1 + uv/c^2)$? Is it possible for their relative speed, as measured by you, to exceed c?

    Answer: Yes, no, yes, to the three questions. It is legal to simply add the two speeds to obtain $u + v$. There is no need to use the velocity-addition formula, because both speeds are measured with respect to the *same thing* (namely you), and because we are asking for the relative speed as measured by that thing. It is perfectly legal for the result to be greater than c, but it must be less than (or equal to, for photons) $2c$.

27. Question: In what situations is the velocity-addition formula relevant?

    Answer: The formula is relevant in the two scenarios in Fig. 1.41, assuming that the goal is to find the speed of A as viewed by C (or vice versa). The second scenario is the same as the first scenario, but from B's point of view. In the first scenario, the two speeds are measured with respect to different frames, so it isn't legal to simply add them. In the second scenario, although the speeds are measured with respect to the same frame (B's frame), the goal is to find the relative speed as viewed by someone else (A or C). So the simple sum $v_1 + v_2$ isn't relevant (as it was in Question 26).

28. Question: Two objects fly toward you, one from the east with speed u, and the other from the west with speed v. What is their relative speed?

    Answer: The question isn't answerable. It needs to be finished with, ". . . as viewed by so-and-so." The relative speed as viewed by you is $u + v$, and the relative speed as viewed by either object is $(u + v)/(1 + uv/c^2)$.

    However, if someone says, "A and B move with relative speed v," and if there is no mention of a third entity, then it is understood that v is the relative speed as viewed by either A or B.

29. QUESTION: A particular event has coordinates (x, t) in one frame. How do you use a Lorentz transformation (L.T.) to find the coordinates of this event in another frame?

    ANSWER: You don't. L.T.'s have nothing to do with single events. They deal only with *pairs* of events and the *separation* between them. As far as a single event goes, its coordinates in another frame can be anything you want, simply by defining your origin to be wherever and whenever you please. But for pairs of events, the separation is a well-defined quantity, independent of your choice of origin. It is therefore a meaningful question to ask how the separations in two different frames are related, and the L.T.'s answer this question.

    This question is similar to Question 16, where we noted that clock *readings* (that is, time *coordinates*) don't get dilated. Rather, *elapsed times* (the *separations* between time coordinates) are what get dilated.

30. QUESTION: When using the L.T.'s, how do you tell which frame is the moving "primed" frame?

    ANSWER: You don't. There is no preferred frame, so it doesn't make sense to ask which frame is moving. We used the primed/unprimed notation in the derivation in Section 2.1.1 for ease of notation, but don't take this to imply that there is a fundamental frame S and a less fundamental frame S'. In general, a better strategy is to use subscripts that describe the two frames, such as "g" for ground and "t" for train, as we did in Section 2.1.2. For example, if you know the values of Δt_{t} and Δx_{t} on a train (which we'll assume is moving in the positive x direction with respect to the ground), and if you want to find the values of Δt_{g} and Δx_{g} on the ground, then you can write down:

$$\begin{aligned} \Delta x_{\text{g}} &= \gamma(\Delta x_{\text{t}} + v\,\Delta t_{\text{t}}), \\ \Delta t_{\text{g}} &= \gamma(\Delta t_{\text{t}} + v\,\Delta x_{\text{t}}/c^2). \end{aligned} \tag{6.3}$$

    If instead you know the intervals on the ground and you want to find them on the train, then you just need to switch the subscripts "g" and "t" and change both signs to "−" (see the following question).

31. QUESTION: How do you determine the sign in the L.T.'s in Eq. (6.3)?

    ANSWER: The sign is a "+" if the frame associated with the left side of the equation (the ground, in Eq. (6.3)) sees the frame associated with the right side (the train) moving in the positive direction. The sign is a "−" if the motion is in the negative direction. This rule follows from looking at the motion of a specific point in the train frame. Since $\Delta x_{\text{t}} = 0$ for two events located at a specific point in the train, the L.T. for x becomes $\Delta x_{\text{g}} = \pm\gamma v\,\Delta t_{\text{t}}$. So if the point moves in the positive (or negative) direction in the ground frame, then the sign must be "+" (or "−") so that Δx_{g} is positive (or negative).

32. QUESTION: In relativity, the temporal order of two events in one frame may be reversed in another frame. Does this imply that there exists a frame in which I get off a bus before I get on it?

    ANSWER: No. The order of two events can be reversed in another frame only if the events are spacelike separated, that is, if $\Delta x > c\,\Delta t$ (which means that the events are too far apart for even light to go from one to the other). The two relevant events here (getting on the bus, and getting off the bus) are not spacelike separated, because the bus travels at a speed less than c, of course. They are timelike separated. Therefore, in all frames it is the case that I get off the bus after I get on it.

There would be causality problems if there existed a frame in which I got off the bus before I got on it. If I break my ankle getting off a bus, then I wouldn't be able to make the mad dash that I made to catch the bus in the first place, in which case I wouldn't have the opportunity to break my ankle getting off the bus, in which case I could have made the mad dash to catch the bus and get on, and, well, you get the idea.

33. QUESTION: Does the longitudinal Doppler effect depend on whether the source or the observer is the one that is moving in a given frame?

    ANSWER: No. Since there is no preferred reference frame, only the relative motion matters. If light needed an "ether" to propagate in, then there would be a preferred frame. But there is no ether. This should be contrasted with the everyday Doppler effect for sound. Sounds needs air (or some other medium) to propagate in. So in this case there *is* a preferred frame – the rest frame of the air.

Dynamics

34. QUESTION: How can you prove that $E = \gamma mc^2$ and $p = \gamma mv$ are conserved?

    ANSWER: You can't. Although there are strong theoretical reasons why the E and p given by these expressions should be conserved, in the end it comes down to experiment. And every experiment that has been done so far is consistent with these E and p being conserved. But this is no proof, of course. As is invariably the case, these expressions are undoubtedly just the limiting expressions of a more correct theory.

35. QUESTION: The energy of an object with mass m and speed v is $E = \gamma mc^2$. Is the statement, "A photon has zero mass, so it must have zero energy," correct or incorrect?

    ANSWER: It is incorrect. Although m is zero, the γ factor is infinite because $v = c$ for a photon. And infinity times zero is undefined. A photon does indeed have energy, and it happens to equal hf, where h is Planck's constant and f is the frequency of the light.

36. QUESTION: A particle has mass m. Is its relativistic mass equal to γm?

    ANSWER: Maybe. Or more precisely: if you want it to be. You can define the quantity γm to be whatever you want. But calling it "relativistic mass" isn't the most productive definition, because γm already goes by another name. It's just the energy, up to factors of c. The use of the word "mass" for this quantity, although quite permissible, is certainly not needed. See the discussion on page 129.

37. QUESTION: When using conservation of energy in a relativistic collision, do you need to worry about possible heat generated, as you do for nonrelativistic collisions?

    ANSWER: No. The energy γmc^2 is conserved in relativistic collisions, period. Any heat that is generated in a particle shows up as an increase in mass. Of course, energy is also conserved in nonrelativistic collisions, period. But if heat is generated, then the energy isn't all in the form of $mv^2/2$ kinetic energies of macroscopic particles.

38. QUESTION: How does the relativistic energy γmc^2 reduce to the nonrelativistic kinetic energy $mv^2/2$?

    ANSWER: The Taylor approximation $1/\sqrt{1 - v^2/c^2} \approx 1 + v^2/2c^2$ turns γmc^2 into $mc^2 + mv^2/2$. The first term is the rest energy. If we assume that a collision is elastic, which means that the masses don't change, then conservation of γmc^2 reduces to conservation of $mv^2/2$.

39. QUESTION: Given the energy E and momentum p of a particle, what is the quickest way to obtain the mass m?

ANSWER: The quickest way is to use the "Very Important Relation" in Eq. (3.12). Whenever you know two of the three quantities E, p, and m, this equation gives you the third. This isn't the only way to obtain m, of course. For example, you can use $v/c^2 = p/E$ to get v, and then plug the result into $E = \gamma mc^2$. But the nice thing about Eq. (3.12) is that you never have to deal with v.

40. QUESTION: For a collection of particles, why is the value of $E^2_{\text{total}} - p^2_{\text{total}}c^2$ invariant, as Eq. (3.26) states? What is the invariant value?

ANSWER: $E^2_{\text{total}} - p^2_{\text{total}}c^2$ is invariant because E_{total} and p_{total} transform according to the L.T.'s (see Eq. (3.25)), due to the fact that the single-particle transformations in Eq. (3.20) are *linear*. Given that E_{total} and p_{total} do indeed transform via the L.T.'s, the invariance of $E^2_{\text{total}} - p^2_{\text{total}}c^2$ follows from a calculation similar to the one for $c^2(\Delta t)^2 - (\Delta x)^2$ in Eq. (2.25).

In the center-of-mass (or really center-of-momentum) frame, the total momentum is zero. So the invariant value of $E^2_{\text{total}} - p^2_{\text{total}}c^2$ equals $(E^{\text{CM}}_{\text{total}})^2$. For a single particle, this is simply $(mc^2)^2 = m^2c^4$, as we know from Eq. (3.12).

41. QUESTION: Why do E and p transform the same way Δt and Δx do, via the L.T.'s?

ANSWER: In Section 3.2 we used the velocity-addition formula to show that E and p transform via the L.T.'s. This derivation, however, doesn't make it intuitively clear why E and p should transform like Δt and Δx. In contrast, the 4-vector approach in Section 4.2 makes it quite clear. To obtain the energy-momentum 4-vector $(E, \mathbf{p})$ from the displacement 4-vector $(dt, d\mathbf{r})$, we simply need to divide by the proper time $d\tau$ (which is an invariant) and then multiply by the mass m (which is again an invariant). The result is therefore still a 4-vector (which is by definition a 4-tuple that transforms according to the L.T.'s).

42. QUESTION: Why do the *differences* in the coordinates, Δx and Δt, transform via the L.T.'s, while it is the actual *values* of E and p that transform via the L.T.'s?

ANSWER: First, note that it wouldn't make any sense for the x and t coordinates themselves to transform via the L.T.'s, as we saw in Question 29. Second, as we noted in the preceding question, the 4-vector approach in Section 4.2 shows that E and p are proportional to the differences Δt and Δx. So E and p have these differences built into them. (Of course, due to the linearity of the L.T.'s, differences of E's and p's also transform via the L.T.'s.)

43. QUESTION: In a nutshell, why isn't F equal to ma (or even γma) in relativity?

ANSWER: F equals dp/dt in relativity, as it does in Newtonian physics. But the relativistic momentum is $p = \gamma mv$, and γ changes with time. So $dp/dt = m(\gamma\dot{v} + \dot{\gamma}v) = \gamma ma + \dot{\gamma}mv$. The second term here isn't present in the Newtonian case.

44. QUESTION: In a given frame, does the acceleration vector $\mathbf{a}$ necessarily point along the force vector $\mathbf{F}$?

ANSWER: No. We showed in Eq. (3.65) that $\mathbf{F} = m(\gamma^3 a_x, \gamma a_y)$. This isn't proportional to (a_x, a_y). The different powers of γ come from the facts that $\mathbf{F} = d\mathbf{p}/dt = d(\gamma m\mathbf{v})/dt$, and that γ has a first-order change if v_x changes, but not if v_y changes, assuming that v_y is initially zero. The particle therefore responds differently to forces in the x and y directions. It is easier to accelerate something in the transverse direction.

General relativity

45. QUESTION: How would the non equality (or non proportionality) of gravitational and inertial mass be inconsistent with the Equivalence Principle?

    ANSWER: In a box floating freely in space, two different masses that start at rest with respect to each other will remain that way. But in a freefalling box near the earth, two different masses that start at rest with respect to each other will remain that way if and only if their accelerations are equal. And since $F = ma \implies m_g g = m_i a \implies a = (m_g/m_i)g$, we see that the m_g/m_i ratios must be equal if the masses are to remain at rest with respect to each other. That is, m_g must be proportional to m_i. If this isn't the case, then the masses will diverge, which means that it is possible to distinguish between the two settings, in contradiction to the Equivalence Principle.

46. QUESTION: You are in either a large box accelerating at g in outer space or a large box on the surface of the earth. Is there any experiment you can do that will tell you which box you are in?

    ANSWER: Yes. The Equivalence Principle involves the word "local," or equivalently the words "small box." Under this assumption, you can't tell which box you are in. However, if the box is large, you can imagine letting go of two balls separated by a nonnegligible "vertical" distance. In the box in outer space, the balls will keep the same distance as they "fall." But near the surface of the earth, the gravitational force decreases with height, due to the $1/r^2$ dependence. The top ball will therefore fall slower, making the balls diverge. Alternatively, you can let go of two balls separated by a nonnegligible "horizontal" distance. In the box on the earth, the balls will head toward each other because the gravitational field lines converge to the center of the earth.

47. QUESTION: In a gravitational field, if a low clock sees a high clock run fast by a factor f_1, and if a high clock sees a low clock run slow by a factor f_2, then $f_1 f_2 = 1$. But in a special-relativistic setup, both clocks see the other clock running slow by the factor $f_1 = f_2 = 1/\gamma$. So we have $f_1 f_2 = 1/\gamma^2 \neq 1$. Why is the product $f_1 f_2$ equal to 1 in the GR case but not in the SR case?

    ANSWER: In the GR case, both clocks are in the same frame. After a long time, the two clocks can be slowly moved together without anything exciting or drastic happening to their readings. And when the clocks are finally sitting next to each other, it is certainly true that if B's clock reads, say, twice what A's reads, then A's must read half of what B's reads.

    In contrast, the clocks in the SR case are *not* in the same frame. If we want to finally compare the clocks by bringing them together, then something drastic *does* happen with the clocks. The necessary acceleration that must take place leads to the accelerating clock seeing the other clock whip ahead; see Exercise 1.30. It is still certainly true (as it was in the GR case) that when the clocks are finally sitting next to each other, if B's clock reads twice what A's reads, then A's reads half of what B's reads. But this fact implies nothing about the product $f_1 f_2$ while the clocks are sailing past each other, because the clock readings (or at least one of them) necessarily change in a drastic manner by the time the clocks end up sitting next to each other.

48. QUESTION: How is the maximal-proper-time principle consistent with the result of the standard twin paradox?

ANSWER: If twin A floats freely in outer space, and twin B travels to a distant star and back, then a simple time-dilation argument from A's point of view tells us that B is younger when he returns. This is consistent with the maximal-proper-time principle, because A is under the influence of only gravity (zero gravity, in fact), whereas B feels a normal force from the spaceship during the turning-around period. So A ends up older.

6.2 Appendix B: Derivations of the Lv/c^2 result

In the second half of Section 1.3.1, we showed that if a train with proper length L moves at speed v with respect to the ground, then as viewed in the ground frame, the rear clock reads Lv/c^2 more than the front clock, at any given instant (assuming, as usual, that the clocks are synchronized in the train frame). There are various other ways to derive this rear-clock-ahead result, so for the fun of it I've listed here all the derivations I can think of. The explanations are terse, but I refer you to the specific problem or section in the text where things are discussed in more detail. Many of these derivations are slight variations of each other, so perhaps they shouldn't all count as separate ones, but here's my list:

1. **Light source on train**: This is the original derivation in Section 1.3.1. Put a light source on a train, at distances $d_{\rm f} = L(c - v)/2c$ from the front and $d_{\rm b} = L(c + v)/2c$ from the back. You can show that the photons hit the ends of the train simultaneously in the ground frame. However, they hit the ends at different times in the train frame; the difference in the readings on clocks at the ends when the photons hit them is $(d_{\rm b} - d_{\rm f})/c = Lv/c^2$. Therefore, at a given instant in the ground frame (for example, the moment when the clocks at the ends are simultaneously illuminated by the photons), a person on the ground sees the rear clock read Lv/c^2 more than the front clock.

2. **Lorentz transformation**: The second of Eqs. (2.2) is $\Delta t_{\rm g} = \gamma(\Delta t_{\rm t} + v\Delta x_{\rm t}/c^2)$, where the subscripts refer to the frame (ground or train). If two events (for example, two clocks flashing their times) located at the ends of the train are simultaneous in the ground frame, then we have $\Delta t_{\rm g} = 0$. And $\Delta x_{\rm t} = L$, of course. The above Lorentz transformation therefore gives $0 = \Delta t_{\rm t} + vL/c^2 \implies \Delta t_{\rm t} = -Lv/c^2$. The minus sign here means that the event with the larger $x_{\rm t}$ value has the smaller $t_{\rm t}$ value. In other words, the front clock reads Lv/c^2 less than the rear clock, at a given instant in the ground frame.

3. **Invariant interval**: This is just a partial derivation, because it determines only the magnitude of the Lv/c^2 result, and not the sign. The invariant interval says that $c^2\Delta t_{\rm g}^2 - \Delta x_{\rm g}^2 = c^2\Delta t_{\rm t}^2 - \Delta x_{\rm t}^2$. If two events (for example, two clocks flashing their times) located at the ends of the train are simultaneous in the ground frame, then we have $\Delta t_{\rm g} = 0$. And $\Delta x_{\rm t} = L$, of course. And we also know from length contraction that $\Delta x_{\rm g} = L/\gamma$. The invariant interval then gives $c^2(0)^2 - (L/\gamma)^2 = c^2\Delta t_{\rm t}^2 - L^2$, which yields $c^2\Delta t_{\rm t}^2 = L^2(1 - 1/\gamma^2) = L^2v^2/c^2 \implies \Delta t_{\rm t} = \pm Lv/c^2$. As mentioned above, the sign isn't determined by this method. (The correct sign is "–" since the front clock is behind.)

4. **Minkowski diagram**: The task of Exercise 2.27 is to use a Minkowski diagram to derive the Lv/c^2 result. The basic goal is to determine how many ct units fit in the segment BC in Fig. 2.14, and also how many ct' units fit in the segment DF in Fig. 2.15.

5. **Walking slowly on a train**: In Problem 1.20 a person walks very slowly at speed u from the back of a train of proper length L to the front. In the frame of the train, the time-dilation effect is second order in u/c and therefore negligible (because the travel time is only first order in $1/u$). But in the frame of the ground, the time-dilation effect is (as you can show) *first* order in u/c. So there is a nonzero effect of the speed u; the γ factor for the person is different from the γ factor for a clock that is fixed on the train. An observer on the ground therefore sees the person's clock advance less than a clock that is fixed on the train.

   Now, the person's clock agrees with clocks at the rear and front at the start and finish, because of the negligible time dilation in the train frame. Therefore, since less time elapses on the person's clock than on the front clock (as measured in the ground frame), the person's clock must have started out reading more time than the front clock (in the ground frame). This then implies that the rear clock must show more time than the front clock, at any given instant in the ground frame. A quantitative analysis shows that this excess time is Lv/c^2.

6. **Consistency arguments**: There are many setups (for example, see the four problems in Section 1.4) where the Lv/c^2 result is an ingredient in explaining a result. Without it, you would encounter a contradiction, such as two different frames giving two different answers to a frame-independent question. So if you wanted to, you could work backwards (under the assumption that everything is consistent in relativity) and let the rear-clock-ahead effect be some unknown time T (which might be zero, for all you know), and then solve for the T that makes everything consistent. You would arrive at $T = Lv/c^2$.

7. **Gravitational time dilation**: This derivation holds only for small v. The task of Exercise 5.18 is to derive (for small v) the Lv/c^2 result by making use of the fact that Lv/c^2 looks a lot like the gh/c^2 term in the GR time-dilation result. If you stand on the ground near the front of a train of length L and then accelerate toward the back with acceleration g, you will see a clock at the back running faster by a factor $(1 + gL/c^2)$, which will cause it to read $(gL/c^2)t = Lv/c^2$ more than a clock at the front, after a time t. (Assume that you accelerate for a short period of time, so that $v \approx gt$, and so that the distance in the gL/c^2 term remains essentially L.)

8. **Accelerating rocket**: The task of Problem 5.9 is to show that if a rocket with proper length L accelerates at g and reaches a speed v, then in the ground frame the readings on the front and rear clocks are related by $t_f = t_b(1 + gL/c^2) - Lv/c^2$. In other words, the front clock reads $t_b(1 + gL/c^2) - Lv/c^2$ simultaneously with the rear clock reading t_b, in the ground frame. But in the rocket frame, gravitational time dilation tells us that the front clock reads $t_b(1 + gL/c^2)$ simultaneously with the rear clock reading t_b. The difference in clock readings (front minus rear) is therefore smaller in the ground frame than in the rocket frame, by an amount equal to Lv/c^2. This is the desired result. (Of course, the 4-star Problem 5.9 certainly isn't the quickest way to derive it!)

If the above derivations aren't sufficient to make you remember the rear-clock-ahead result, here's a little story that should do the trick:

Once upon a time there was a family with the surname of Rhee. This family was very wealthy, although it so happened that they had only one son. This son was destined to inherit the entire Rhee fortune, but despite the promise of future riches, his two prized possessions as a child were very modest ones. The first was a teddy bear he named Elvie, and the second was a clock in the shape of a large head. The face of the head served as the face of the clock, with the mouth at 6 o'clock and the eyes at 1:15 and 10:30. It was said to have been built in the likeness of a great-great-grandfather in the Rhee line, who was moderately disfigured.

When not playing with Elvie or watching the hands of the clock go round and round, the son spent a great deal of time making letters out of pieces of wire. His parents' view was that this was obviously the best way for him to learn the alphabet. As the saying goes, you never forget what you've built out of pieces of wire. However, the h's and y's caused great consternation, not to mention the f's, k's, t's and x's. (It is amusing to note that his parents created a personalized 24-letter alphabet for him, due to his repeated convulsions at the sight of i's and j's.)

His favorite letter was c, being the simplest letter, except perhaps for the l (which, however, requires at least an honest attempt at a straight line). He therefore made lots of c's. In fact, he made so many that he eventually got bored with them. So he started making variations. His favorite variation was bending his old c's into square ones with two corners, like block letters. That was fun. Lots of fun. He liked squaring the c's.

One day he was playing with his friend Albert in his room, tossing Elvie back and forth. An errant throw landed Elvie on top of a big c that he had squared, which happened to be right by the large clock head. Both children sat in silence for a few moments, contemplating the implications of what they were seeing. Albert smacked his hand on his forehead, wondering how he had never realized before what now seemed so obvious. "Of course!" he exclaimed, "The Rhee heir clock is a head by Elvie over c squared!"

6.3 Appendix C: Resolutions to the twin paradox

The twin paradox appeared in Chapters 1, 2, and 5, both in the text and in various problems. To summarize, the twin paradox deals with twin A who stays on the earth[1] and twin B who travels quickly to a distant star and back. When they meet up again, they discover that B is younger. This is true because A can use the standard special-relativistic time-dilation result to say that B's clock runs slow by a factor γ.

The "paradox" arises from the fact that the situation seems symmetrical. That is, it seems as though each twin should be able to consider herself to be at rest, so that she sees the other twin's clock running slow. So why does B turn out to be younger? The resolution to the paradox is that the setup is in fact *not* symmetrical, because B must turn around and thus undergo acceleration. She is therefore not always in an inertial frame, so she cannot always apply the simple special-relativistic time-dilation result.

While the above reasoning is sufficient to get rid of the paradox, it isn't quite complete, because (a) it doesn't explain how the result from B's point of view quantitatively

[1]We should actually have A floating in space, to avoid any general-relativistic time-dilation effects from the earth's gravity. But if B travels quickly enough, the special-relativistic effects will dominate the gravitational ones.

agrees with the result from A's point of view, and (b) the paradox can be formulated without any mention of acceleration, in which case a slightly different line of reasoning applies.

Below is a list of all the complete resolutions (explaining things from B's point of view) that I can think of. The descriptions are terse, but I refer you to the specific problem or section in the text where things are discussed in more detail. As with the Lv/c^2 derivations in Appendix B, many of these resolutions are slight variations of each other, so perhaps they shouldn't all count as separate ones, but here's my list:

1. **Rear-clock-ahead effect**: Let the distant star be labeled as C. Then on the outward part of the journey, B sees C's clock ahead of A's by Lv/c^2, because C is the rear clock in the universe as the universe flies by B. But after B turns around, A becomes the rear clock and is therefore now Lv/c^2 ahead of C. Since nothing unusual happens with C's clock, which is right next to B during the turnaround, A's clock must therefore jump forward very quickly (by an amount $2Lv/c^2$), from B's point of view. See Exercise 1.30.

2. **Looking out the portholes**: Imagine many clocks lined up between the earth and the star, all synchronized in the earth-star frame. And imagine looking out the portholes of the spaceship and making a movie of the clocks as you fly past them (or rather as the clocks fly past you, as viewed from the spaceship frame). Although you see each individual clock running slow, you see the "effective" clock in the movie (which is really many successive clocks) running fast. This is true because each successive clock is a "rear" clock relative to the previous one, which means that it is a little bit ahead of the previous one. This effect is just a series of small applications of the calculation in the first example in Section 1.4. The same reasoning applies during the return trip.

3. **Minkowski diagram**: Draw a Minkowski diagram with the axes in A's frame perpendicular. Then the lines of simultaneity (that is, the successive x axes) in B's frame are titled in different directions for the outward and return parts of the journey. The change in the tilt at the turnaround causes a large amount of time to advance on A's clock, as measured in B's frame. See Fig. 2.41.

4. **General-relativistic turnaround effect**: The acceleration that B feels when she turns around may equivalently be thought of as a gravitational field. Twin A on the earth is high up in the gravitational field, so B sees A's clock run very fast during the turnaround. This causes A's clock to show more time in the end. See Problem 5.12.

5. **Doppler effect**: By equating the total number of signals one twin sends out with the total number of signals the other twin receives, we can relate the total times on their clocks. See Exercise 2.32.

6.4 Appendix D: Lorentz transformations

In this Appendix, we will present an alternative derivation of the Lorentz transformations given in Eq. (2.2). The goal here will be to derive them from scratch, using only the two postulates of relativity. We will *not* use any of the fundamental effects from Section 1.3. Our strategy will be to use the relativity postulate ("all inertial frames are equivalent") to figure out as much as we can, and to then invoke the speed-of-light postulate at the end. The main reason for doing things in this order is that it will allow us to derive a very interesting result in Section 2.7.

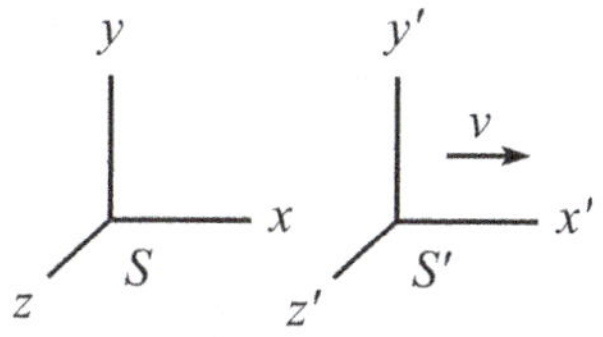

Figure 6.1

As in Section 2.1, consider a reference frame S' moving relative to another frame S; see Fig. 6.1. Let the constant relative speed between the frames be v. Let the corresponding axes of S and S' point in the same direction, and let the origin of S' move along the x axis of S. As in Section 2.1, we want to find the constants, A, B, C, and D, in the relations,

$$\begin{aligned} \Delta x &= A\,\Delta x' + B\,\Delta t', \\ \Delta t &= C\,\Delta t' + D\,\Delta x'. \end{aligned} \tag{6.4}$$

The four constants will end up depending on v (which is constant, given the two inertial frames). Since we have four unknowns, we need four facts. The facts we have at our disposal (using only the two postulates of relativity) are the following.

1. The physical setup: S' moves with velocity v with respect to S.

2. The principle of relativity: S should see things in S' in exactly the same way as S' sees things in S (except perhaps for a minus sign in some relative positions, but this just depends on our arbitrary choice of directional signs for the axes).

3. The speed-of-light postulate: A light pulse with speed c in S' also has speed c in S.

The second statement here contains two independent bits of information. (It contains at least two, because we will indeed be able to solve for our four unknowns. And it contains no more than two, because otherwise our four unknowns would be over-constrained.) The two bits that are used depend on personal preference. Three that are commonly used are: (a) the relative speed looks the same from either frame, (b) time dilation (if any) looks the same from either frame, and (c) length contraction (if any) looks the same from either frame. It is also common to recast the second statement in the form: The Lorentz transformations are the same as their inverse transformations (up to a minus sign). We'll choose to work with (a) and (b). Our four independent facts are then:

1. S' moves with velocity v with respect to S.

2. S moves with velocity $-v$ with respect to S'. The minus sign here is due to the fact that we picked the positive x axes of the two frames to point in the same direction.

3. Time dilation (if any) looks the same from either frame.

4. A light pulse with speed c in S' also has speed c in S.

Let's see what these imply, in the above order. In what follows, we could obtain the final result a little quicker if we invoked the speed-of-light fact prior the time-dilation one. But we'll do things in the above order so that we can easily carry over the results of this appendix to the discussion in Section 2.7.

- Fact 1 says that a given point in S' moves with velocity v with respect to S. Letting $x' = 0$ (which is understood to be $\Delta x' = 0$, but we'll drop the Δ's from here on) in Eqs. (6.4) and dividing them gives $x/t = B/C$. This must equal v. Therefore, $B = vC$, and the transformations become

$$\begin{aligned} x &= Ax' + vCt', \\ t &= Ct' + Dx'. \end{aligned} \tag{6.5}$$

- Fact 2 says that a given point in S moves with velocity $-v$ with respect to S'. Letting $x = 0$ in the first of Eqs. (6.5) gives $x'/t' = -vC/A$. This must equal $-v$. Therefore, $C = A$, and the transformations become

$$\begin{aligned} x &= Ax' + vAt', \\ t &= At' + Dx'. \end{aligned} \tag{6.6}$$

Note that these are consistent with the Galilean transformations, which have $A = 1$ and $D = 0$.

- Fact 3 can be used in the following way. How fast does a person in S see a clock in S' tick? (The clock is assumed to be at rest with respect to S'.) Let our two events be two successive ticks of the clock. Then $x' = 0$, and the second of Eqs. (6.6) gives

$$t = At'. \tag{6.7}$$

In other words, one second on S''s clock takes a time of A seconds in S's frame.

Consider the analogous situation from S''s point of view. How fast does a person in S' see a clock in S tick? (The clock is now assumed to be at rest with respect to S, in order to create the analogous setup. This is important.) If we invert Eqs. (6.6) to solve for x' and t' in terms of x and t, we find

$$\begin{aligned} x' &= \frac{x - vt}{A - Dv}, \\ t' &= \frac{At - Dx}{A(A - Dv)}. \end{aligned} \tag{6.8}$$

Two successive ticks of the clock in S satisfy $x = 0$, so the second of Eqs. (6.8) gives

$$t' = \frac{t}{A - Dv}. \tag{6.9}$$

In other words, one second on S's clock takes a time of $1/(A - Dv)$ seconds in S''s frame.

Eqs. (6.7) and (6.9) apply to the same situation (someone looking at a clock flying by). Therefore, the factors on the righthand sides must be equal, that is,

$$A = \frac{1}{A - Dv} \qquad \Longrightarrow \qquad D = \frac{1}{v}\left(A - \frac{1}{A}\right). \tag{6.10}$$

Our transformations in Eq. (6.6) therefore take the form

$$\begin{aligned} x &= A(x' + vt'), \\ t &= A\left(t' + \frac{1}{v}\left(1 - \frac{1}{A^2}\right)x'\right). \end{aligned} \tag{6.11}$$

These are consistent with the Galilean transformations, which have $A = 1$.

- Fact 4 may now be used to say that if $x' = ct'$, then $x = ct$. In other words, if $x' = ct'$, then

$$c = \frac{x}{t} = \frac{A((ct') + vt')}{A\left(t' + \frac{1}{v}\left(1 - \frac{1}{A^2}\right)(ct')\right)} = \frac{c + v}{1 + \frac{c}{v}\left(1 - \frac{1}{A^2}\right)}. \tag{6.12}$$

Solving for A gives

$$A = \frac{1}{\sqrt{1 - v^2/c^2}}. \tag{6.13}$$

We have chosen the positive square root so that the coefficient of t' is positive; if t' increases, then t should also increase. (The forward direction of time should be the same in the two frames.) The transformations are now no longer consistent with the Galilean transformations, because c is not infinite, which means that A is not equal to 1.

The constant A is commonly denoted by γ, so we may finally write our Lorentz transformations, Eqs. (6.11), in the form (using $1 - 1/\gamma^2 = v^2/c^2$),

$$\begin{aligned} x &= \gamma(x' + vt'), \\ t &= \gamma(t' + vx'/c^2), \end{aligned} \tag{6.14}$$

where

$$\gamma \equiv \frac{1}{\sqrt{1 - v^2/c^2}}, \tag{6.15}$$

in agreement with Eq. (2.2).

6.5 Appendix E: Nonrelativistic dynamics

Chapter 3 covers relativistic dynamics, that is, relativistic momentum, energy, force, etc. If you haven't taken a standard mechanics course, the present appendix will get you up to speed on Newtonian (nonrelativistic) dynamics. The discussion here will necessarily be brief. The goal is to cover the basics of the relevant topics, as opposed to presenting a comprehensive introduction to dynamics.

Newton's laws

The subject of dynamics is governed by Newton's three laws. The *first law* states that an object continues to move with constant velocity (which may be zero) unless acted on by a force. Of course, we haven't defined what a force is yet, so this law might seem a little circular. But what the first law does is define an *inertial frame*, which is simply a frame in which the first law holds. If an object isn't interacting with any other objects and is moving with constant velocity in a given frame, then that frame is an inertial one. If, on the other hand, the object's velocity is changing (still assuming no interaction with any other objects), then the frame isn't inertial. For example, if an object is floating freely in space, and if it is enclosed in a box and the box is accelerating, then with respect to the box, the object's velocity changes. The box therefore represents a *noninertial* reference frame. In Newtonian physics, inertial frames are important because they are the frames in which the second law (discussed below) holds. In relativistic physics, inertial frames are important because they are the frames in which Einstein's two postulates hold.

Newton's *second law* is often written as

$$\mathbf{F} = m\mathbf{a}, \tag{6.16}$$

where m is the mass of an object and $\mathbf{a}$ is its acceleration. ($\mathbf{F} = m\mathbf{a}$ is a vector equation, so it is really shorthand for the three equations, $F_x = ma_x$, $F_y = ma_y$, and $F_z = ma_z$.) However, what the law actually states is

$$\mathbf{F} = \frac{d\mathbf{p}}{dt} \qquad \text{(Newton's second law)} \tag{6.17}$$

where the momentum $\mathbf{p}$ of an object is defined as

$$\mathbf{p} = m\mathbf{v}, \tag{6.18}$$

with $\mathbf{v}$ being the object's velocity. In the common case where the mass m is constant, the time derivative in Eq. (6.17) acts only on the $\mathbf{v}$, so the second law simplifies to the form in Eq. (6.16):

$$\mathbf{F} = \frac{d\mathbf{p}}{dt} = \frac{d(m\mathbf{v})}{dt} = m\frac{d\mathbf{v}}{dt} \implies \mathbf{F} = m\mathbf{a}. \tag{6.19}$$

Since most mechanics problems involve objects with constant mass, the $\mathbf{F} = m\mathbf{a}$ form of the law usually suffices.

Newton's *third law* states that the forces that two objects exert on each other are equal and opposite. If we use the notation $\mathbf{F}_{ij}$ to denote the force on object i due to object j, then we have

$$\mathbf{F}_{12} = -\mathbf{F}_{21} \qquad \text{(Newton's third law)} \tag{6.20}$$

Conservation of momentum

In an isolated system of particles, the combination of the second and third laws implies that the total momentum is conserved (constant in time). Consider first the simple case of two isolated particles. Since the second law tells us that $\mathbf{F} = d\mathbf{p}/dt$, we have $\mathbf{F}_{12} = d\mathbf{p}_1/dt$ and $\mathbf{F}_{21} = d\mathbf{p}_2/dt$. The third law, $\mathbf{F}_{12} = -\mathbf{F}_{21}$, then becomes

$$\frac{d\mathbf{p}_1}{dt} = -\frac{d\mathbf{p}_2}{dt} \implies \frac{d(\mathbf{p}_1 + \mathbf{p}_2)}{dt} = 0 \implies \mathbf{p}_1 + \mathbf{p}_2 = \text{constant}. \tag{6.21}$$

In other words, the total momentum is conserved. Whatever momentum particle 2 gives to particle 1, particle 1 gives an equal and opposite momentum to particle 2. We can alternatively show that the total momentum is conserved by writing the second law in integral form:

$$\mathbf{F} = \frac{d\mathbf{p}}{dt} \implies \int \mathbf{F}\,dt = \int d\mathbf{p} \implies \int \mathbf{F}\,dt = \Delta\mathbf{p}. \tag{6.22}$$

This implies

$$\Delta\mathbf{p}_{\text{total}} = \Delta\mathbf{p}_1 + \Delta\mathbf{p}_2 = \int \mathbf{F}_{12}\,dt + \int \mathbf{F}_{21}\,dt = 0, \tag{6.23}$$

because $\mathbf{F}_{12} = -\mathbf{F}_{21}$, by the third law.

Consider now the case of three isolated particles. The rate of change of the total momentum of the system equals

$$\begin{aligned} \frac{d(\mathbf{p}_1 + \mathbf{p}_2 + \mathbf{p}_3)}{dt} &= \frac{d\mathbf{p}_1}{dt} + \frac{d\mathbf{p}_2}{dt} + \frac{d\mathbf{p}_3}{dt} \\ &= \mathbf{F}_1 + \mathbf{F}_2 + \mathbf{F}_3 \\ &= (\cancel{\mathbf{F}_{12}} + \cancel{\mathbf{F}_{13}}) + (\cancel{\mathbf{F}_{21}} + \cancel{\mathbf{F}_{23}}) + (\cancel{\mathbf{F}_{31}} + \cancel{\mathbf{F}_{32}}) \\ &= 0. \end{aligned} \tag{6.24}$$

We have used the fact that Newton's third law, $\mathbf{F}_{ij} = -\mathbf{F}_{ji}$, tells us that the forces cancel in pairs, as indicated. This reasoning easily extends to a general number N of particles. For every term $\mathbf{F}_{ij}$ in the generalization of the third line above, there is also a term $\mathbf{F}_{ji}$, and the sum of these two terms is zero by Newton's third law.

Work

The (nonrelativistic) kinetic energy of an object is defined as

$$K = \frac{mv^2}{2}. \tag{6.25}$$

The reason why this quantity is important (and why it is conserved) is that it appears in the *work-energy theorem*, which we will prove below. In 1-D, the *work* done by a constant force acting on an object (directed parallel to the line of motion of the object) is defined to be the force times the displacement:

$$W = F\,\Delta x. \tag{6.26}$$

Work is a signed quantity; if the force points opposite to the direction of motion (so that F and Δx have opposite signs), then W is negative. For a force that isn't constant, the work is defined to be the integral of the force:

$$W = \int F\,dx. \tag{6.27}$$

The *work-energy theorem* states that the work done on a particle equals the change in kinetic energy of the particle. That is,

$$W = \Delta K \qquad \text{(work-energy theorem)} \tag{6.28}$$

This theorem is consistent with our intuition. From $F = ma$, we know that if the force points in the same direction as the velocity of a particle, then the speed increases, so the kinetic energy increases. That is, ΔK is positive. This is consistent with the fact that the work W is positive if the force points in the same direction as the velocity. Conversely, if the force points opposite to the velocity of the particle, then the speed decreases, so the kinetic energy decreases. This is consistent with the fact that the work is negative.

To prove the work-energy theorem, we will need to use the fact that the acceleration a of an object can be written as $a = v\,dv/dx$. This is true because if we write a as dv/dt and then multiply by 1 in the form of dx/dx, we obtain

$$a = \frac{dv}{dt} = \frac{dx}{dt}\frac{dv}{dx} = v\frac{dv}{dx}, \tag{6.29}$$

as desired. Plugging $a = v\,dv/dx$ into $F = ma$ and then multiplying by dx and integrating from x_1 (and the corresponding v_1) to x_2 (and the corresponding v_2) gives

$$F = ma \implies F = mv\frac{dv}{dx} \implies \int_{x_1}^{x_2} F\,dx = \int_{v_1}^{v_2} mv\,dv. \tag{6.30}$$

The integral on the lefthand side is the work done, by definition. The integral of v on the righthand side is $v^2/2$, so we obtain

$$W = \frac{mv^2}{2}\bigg|_{v_1}^{v_2} = \frac{mv_2^2}{2} - \frac{mv_1^2}{2} = \Delta K, \tag{6.31}$$

as desired.

Conservation of energy

Consider the simple case of an isolated system of two particles (with no internal structure; see below). If these particles collide and interact only via contact forces (as opposed to, say, the gravitational force, which acts over a distance), then we claim that the sum of the kinetic energies of the particles is conserved. This follows from the work-energy theorem and Newton's third law:

$$\Delta K_{\text{total}} = \Delta K_1 + \Delta K_2 = W_1 + W_2 = \int \mathbf{F}_{12}\,dx + \int \mathbf{F}_{21}\,dx = 0, \tag{6.32}$$

because $\mathbf{F}_{12} = -\mathbf{F}_{21}$, by the third law. Whatever work particle 2 does on particle 1, particle 1 does an equal and opposite amount of work on particle 2. We have used the fact that the dx's in the two integrals in Eq. (6.32) are the same. This is true because the force is a *contact* force, which means that the point of application on particle 1 is the same point in space as the point of application on particle 2. As with conservation of momentum, the above conservation-of-energy result generalizes to more than two particles. The forces (and hence works) cancel in pairs due to Newton's third law, so the total kinetic energy of any isolated system of particles (with no internal structure) is conserved.

Note the parallel between the above derivations of conservation of momentum and conservation of energy. Conservation of momentum follows from integrating $\mathbf{F}$ with respect to *time* (that is, $\int \mathbf{F}\,dt$) and applying Newton's third law; see Eq. (6.23). Similarly, conservation of energy follows from integrating $\mathbf{F}$ with respect to *displacement* (that is, $\int \mathbf{F}\,dx$) and applying Newton's third law; see Eq. (6.32). The only difference is replacing t with x. The critical similarity is that the forces $\mathbf{F}_{ij}$ and $\mathbf{F}_{ji}$ (a third-law pair) between two objects act not only for the same *time*, but also for the same *displacement* (assuming that the force is a contact force).

In the collisions we are concerned with, we are assuming that each particle is the same before and after the collision. Equivalently, as mentioned above, we are assuming that the particles have no internal structure where other forms of energy might be hiding. The most common form of "hidden" energy is heat. This is simply the kinetic energy (and potential energy, too) of molecules on a microscopic scale, as opposed to the macroscopic kinetic energy associated with the motion of the particle as a whole. If no heat is generated, we call a collision *elastic*; whereas if heat is generated, we call it *inelastic*. So in summary, in an isolated collision,

1. Momentum is always conserved.
2. Energy is also always conserved, although *mechanical energy* (by which we mean the $mv^2/2$ energies of the macroscopic particles involved; that is, excluding heat) is conserved only if the collision is elastic (by definition).

In the more general case where there are non-contact forces (such as gravitational or electric) or where an object deforms (as with a spring), the work done by these types of forces may be (under certain circumstances) associated with a *potential energy*. The general conservation-of-energy statement is then that the total kinetic plus potential energy of a system (including heat) is conserved. But for the collisions we discuss in Chapter 3, we don't need to worry about potential energy.

Collision examples

Figure 6.2

Consider the 1-D setup shown in Fig. 6.2. Masses m_1 and m_2 have initial velocities v_1 and v_2. These are signed quantities, so they may be positive or negative (rightward is positive, leftward is negative). The masses collide elastically in 1-D and acquire final velocities u_1 and u_2. To determine u_1 and u_2, we simply need to write down the conservation of momentum and energy equations and then do some algebra. Conservation of p gives

$$\begin{aligned} p_{\text{before}} &= p_{\text{after}} \\ \Longrightarrow\quad m_1 v_1 + m_2 v_2 &= m_1 u_1 + m_2 u_2. \end{aligned} \tag{6.33}$$

And conservation of E gives

$$\begin{aligned} E_{\text{before}} &= E_{\text{after}} \\ \Longrightarrow\quad \frac{1}{2}m_1 v_1^2 + \frac{1}{2}m_2 v_2^2 &= \frac{1}{2}m_1 u_1^2 + \frac{1}{2}m_2 u_2^2. \end{aligned} \tag{6.34}$$

Eqs. (6.33) and (6.34) are two equations in the two unknowns u_1 and u_2. Solving them involves solving a quadratic equation (although there are some sneakier and less time consuming ways). The result is fairly messy and not terribly enlightening, so we won't bother writing it down. The point we want to make here is simply that to solve for the final motion of the particles, you just need to write down the conservation of p and E equations and then do some algebra.

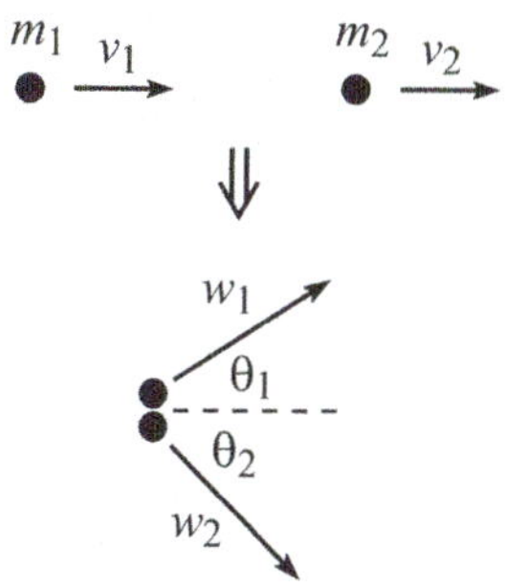

Figure 6.3

Consider now the 2-D setup shown in Fig. 6.3. The initial motion is the same as in Fig. 6.2, but now the masses are free to scatter elastically in two dimensions. The final velocities are described by the final *speeds* w_1 and w_2 (these are positive quantities, by definition) and the angles θ_1 and θ_2. Since momentum is a vector, it is conserved in both the x and y directions. So conservation of $\mathbf{p}$ gives us *two* equations:

$$\begin{aligned} p_x: &\quad m_1v_1 + m_2v_2 = m_1w_1\cos\theta_1 + m_2w_2\cos\theta_2, \\ p_y: &\quad 0 = m_1w_1\sin\theta_1 - m_2w_2\sin\theta_2. \end{aligned} \tag{6.35}$$

(There is technically also the conservation-of-p_z equation, but that is just the trivial statement $0 = 0$ in the present setup.) The conservation-of-E equation is exactly the same as in Eq. (6.34), because kinetic energy depends only on the speed (the magnitude of the velocity); the direction is irrelevant. So we have

$$\frac{1}{2}m_1v_1^2 + \frac{1}{2}m_2v_2^2 = \frac{1}{2}m_1w_1^2 + \frac{1}{2}m_2w_2^2. \tag{6.36}$$

The preceding three equations are three equations in four unknowns (w_1, w_2, θ_1, and θ_2), so it is impossible to solve for all of the unknowns. (The physical reason for this is that we would need to be told exactly how the masses glance off each other, to be able to figure out what the final angles are.) But if someone gives us the additional information of what one of the four unknowns is, then we can solve for the other three (with some tedious algebra).

The nice thing about conservation laws is that we don't need to worry about the messy/intractible specifics of what goes on during a collision. We simply have to write down the initial and final expressions for $\mathbf{p}$ and E and then solve for whatever variable(s) we're trying to determine.

6.6 Appendix F: Problem-solving strategies

Given the large number of problems that appear in this book, it's a good idea to provide you with some strategies for solving them. The strategies discussed below are certainly not specific to relativity; they can be applied in any subject in physics (along with many other sciences). The examples we'll use are taken from basic mechanics. If you haven't studied mechanics yet, rest assured, you'll still be able to understand the problem-solving strategy being discussed, even without knowing the relevant physics. The strategies we'll cover here are (1) solving problems symbolically (with letters instead of numbers), (2) checking units, and (3) checking limiting cases.

6.6.1 Solving problems symbolically

If you are solving a problem where the given quantities are specified numerically, it is highly advantageous to immediately change the numbers to letters and then solve the problem in terms of the letters. After you obtain a symbolic answer in terms of these letters, you can plug in the actual numerical values to obtain a numerical answer. There are many advantages to using letters:

- It is quicker. It's much easier to multiply a g by an ℓ by writing them down on a piece of paper next to each other, than it is to multiply their numerical values on a calculator. If solving a problem involves five or ten such operations, the time would add up if you performed all the operations on a calculator.

- You are less likely to make a mistake. It's very easy to mistype an 8 for a 9 in a calculator, but you're probably not going to miswrite a q for an a on a piece of paper. But even if you do, you'll quickly realize that it should be an a. You certainly won't just give up on the problem and deem it unsolvable because no one gave you the value of q!

- You can do the problem once and for all. If someone comes along and says, oops, the value of ℓ is actually 2.4 m instead of 2.3 m, then you won't have to do the whole problem again. You can simply plug the new value of ℓ into your symbolic answer.

- You can see the general dependence of your answer on the various given quantities. For example, you can see that it grows with quantities a and b, decreases with c, and doesn't depend on d. There is *much* more information contained in a symbolic answer than in a numerical one. And besides, symbolic answers nearly always look nice and pretty.

- You can check units and special cases. These checks go hand-in-hand with the preceding "general dependence" advantage. We'll discuss these very important checks below.

Two caveats to all this: First, occasionally there are times when things get messy when working with letters. For example, solving a system of three equations in three unknowns might be rather cumbersome unless you plug in the actual numbers. But in the vast majority of problems, it is advantageous to work entirely with letters. Second, if you solve a problem that was posed with letters instead of numbers, it's always a good idea to pick some values for the various parameters to see what kinds of numbers pop out, just to get a general sense of the size of things.

6.6.2 Checking units/dimensions

The words *dimensions* and *units* are often used interchangeably, but there is technically a difference: dimensions refer to the general qualities of mass, length, time, etc., whereas units refer to the specific way we quantify these qualities. For example, in the standard meters-kilogram-second (mks) system of units we use in this book, the meter is the unit associated with the dimension of length, the joule is the unit associated with the dimension of energy, and so on. However, we'll often be sloppy and ignore the difference between units and dimensions.

The consideration of units offers two main benefits when solving problems:

- Consider units at the start. Considering the units of the relevant quantities before you start solving a problem can tell you roughly what the answer has to look like, up to numerical factors. This practice is called *dimensional analysis*.

- Check units at the end. Checking units at the end of a calculation (which is something you should *always* do) can tell you if your answer has a chance at being correct. It won't tell you that your answer is definitely correct, but it might tell you that your answer is definitely *incorrect*. For example, if your goal in a problem is to find a length, and if you end up with a mass, then you know that it's time to look back over your work.

In the mks system of units, the three fundamental mechanical units are the meter (m), kilogram (kg), and second (s). All other units in mechanics, for example the joule (J) or the newton (N), can be built up from these fundamental three. If you want to work with dimensions instead of units, then you can write everything in terms of length (L), mass (M), and time (T). The difference is only cosmetic.

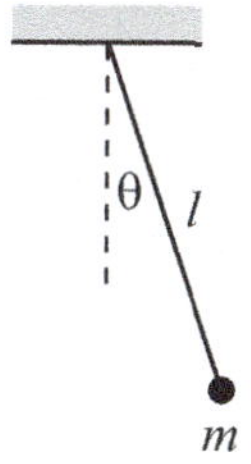

Figure 6.4

As an example of the above two benefits of considering units, consider a pendulum consisting of a mass m hanging from a massless string with length ℓ; see Fig. 6.4. Assume that the pendulum swings back and forth with an angular amplitude θ_0 that is small; that is, the string doesn't deviate far from vertical. What is the period, call it T_0, of the oscillatory motion? (The period is the time of a full back-and-forth cycle.)

With regard to the first of the above benefits, what can we say about the period T_0 (which has units of seconds), by looking only at units and not doing any calculations? Well, we must first make a list of all the quantities the period can possibly depend on. The given quantities are the mass m (with units of kg), the length ℓ (with units of m), and the angular amplitude θ_0 (which is unitless). Additionally, there might be dependence on g (the acceleration due to gravity, with units of m/s^2), If you think for a little while, you'll come to the conclusion that there isn't anything else the period can depend on (assuming that we ignore air resistance).

So the question becomes: How does T_0 depend on m, ℓ, θ_0, and g? Or equivalently: How can we produce a quantity with units of seconds from four quantities with units of kg, m, 1, and m/s^2? (The 1 signifies no units.) We quickly see that the answer can't involve the mass m, because there would be no way to get rid of the units of kg. We then see that if we want to end up with units of seconds, the answer must be proportional to $\sqrt{\ell/g}$, because this gets rid of the meters and leaves one power of seconds in the numerator. Therefore, by looking only at the units involved, we have shown that $T_0 \propto \sqrt{\ell/g}$.[2]

This is all we can say by considering units. For all we know, there might be a numerical factor out front, and also an arbitrary function of θ_0 (which won't mess up the units, because θ_0 is unitless). The correct answer happens to be $T_0 = 2\pi\sqrt{\ell/g}$, but there is no way to know this without solving the problem for real.[3] However, even though we haven't produced an exact result, there is still a great deal of information contained in our $T_0 \propto \sqrt{\ell/g}$ statement. For example, we see that the period is independent of m; a small mass and a large mass swing back and forth at the same frequency. Similarly, the period is independent of θ_0 (as long as θ_0 is small). We also see that if we quadruple the length of the string, then the period gets doubled. And if we place the same pendulum on the moon, where the g factor is $1/6$ of that on the earth, then the period increases by a factor of $\sqrt{6} \approx 2.4$; the pendulum swings back and forth more slowly. Not bad for only considering units!

While this is all quite interesting, the second of the above two benefits (checking the units of an answer) is actually the one that you will get the most mileage out of when solving problems, mainly because you should make use of it *every* time you solve a problem. It only takes a second. In the present example with the pendulum, let's say that you solved the problem correctly and ended up with $T_0 = 2\pi\sqrt{\ell/g}$. You should immediately check the units, which do indeed correctly come out to be seconds. If you

---

[2]In this setup it was easy to determine the correct combination of the given parameters. But in more complicated setups, you might find it simpler to write down a general product of the given dimensionful quantities raised to arbitrary powers, and then solve a system of equations to determine the correct powers that yield the desired units.

[3]This $T_0 = 2\pi\sqrt{\ell/g}$ result, which is independent of θ_0, holds in the approximation where the amplitude θ_0 is small. For a general value of θ_0, the period actually *does* involve a function of θ_0. This function can't be written in closed form, but it starts off as $1 + \theta_0^2/16 + \cdots$. It takes a lot of work to show this, though. See Exercise 4.23 in Morin (2008).

made a mistake in your solution, such as flipping the square root upside down (so that you instead had $\sqrt{g/\ell}$), then your units check would yield the incorrect units of s^{-1}. You would then know to go back and check over your work. Throughout this book, we often won't bother to explicitly write down the units check if the check is a simple one (as with the above pendulum). But you should of course always do the check in your head.

Having said all this, we should note that it is common practice in relativity to drop the c's in calculations. On one hand, this goes against everything we've just said, because dropping the c's (which have units) means that the units of your answer will in general be incorrect. (You can put the c's back in at the end, of course. But you've lost the ability to catch a mistake by checking units.) On the other hand, the calculations are often *much* simpler without the c's, and this simplicity decreases the probability of making an algebraic mistake. In the end, it comes down to which scenario you prefer – messier calculations with a units check at the end, or simpler calculations with no units check. My view is that in nontrivial calculations, the simplicity usually wins.

6.6.3 Checking limiting/special cases

As with units, the consideration of limiting cases (or perhaps we should say more generally special cases) offers two main benefits. First, it can help you get started on a problem. If you are having trouble figuring out how a given system behaves, you can imagine making, for example, a certain length become very large or very small, and then you can see what happens to the behavior. Having convinced yourself that the length actually affects the system in extreme cases (or perhaps you will discover that the length doesn't affect things at all), it will then be easier to understand how it affects the system in general. This will in turn make it easier to write down the relevant quantitative equations (conservation laws, $F = ma$ equations, etc.), which will allow you to fully solve the problem. In short, modifying the various parameters and seeing the effects on the system can lead to an enormous amount of information.

Second, as with checking units, checking limiting cases (or special cases) is something you should *always* do at the end of a calculation. As with units, checking limiting cases won't tell you that your answer is definitely correct, but it might tell you that your answer is definitely incorrect. Your intuition about limiting cases is invariably *much* better than your intuition about generic values of the parameters. You should use this to your advantage.

As an example, consider the trigonometric formula for $\tan(\theta/2)$. The formula can be written in various ways. Let's say that you're trying to derive it, but you keep making mistakes and getting different answers. However, let's assume that you're fairly sure it takes the form of $\tan(\theta/2) = A(1 \pm \cos\theta)/\sin\theta$, where A is a numerical coefficient. Can you determine the correct form of the answer by checking special cases? Indeed you can, because you know what $\tan(\theta/2)$ equals for a few special values of θ:

- $\theta = 0$: We know that $\tan(0/2) = 0$, so this immediately rules out the $A(1 + \cos\theta)/\sin\theta$ form, because this isn't zero when $\theta = 0$; it actually goes to infinity at $\theta = 0$. The answer must therefore take the $A(1 - \cos\theta)/\sin\theta$ form. This appears to be $0/0$ when $\theta = 0$, but it does indeed go to zero, as you can check by using the Taylor series for $\sin\theta$ and $\cos\theta$; see the subsection on Taylor series below.

- $\theta = 90°$: We know that $\tan(90°/2) = 1$, which quickly gives $A = 1$. So the correct answer must be $\tan(\theta/2) = (1 - \cos\theta)/\sin\theta$.

- $\theta = 180°$: If you want to feel better about this $(1 - \cos\theta)/\sin\theta$ result, you can note that it gives the correct answer for another special value of θ; it correctly goes to infinity when $\theta = 180°$.

Of course, none of what we've done here demonstrates that $(1 - \cos\theta)/\sin\theta$ actually *is* the right answer. But checking the above special cases does two things: it rules out some incorrect answers, and it makes us feel better about the correct answer.

A type of approximation that frequently comes up involves expressions of the form $ab/(a+b)$, that is, a product over a sum. For example, let's say we're trying to determine the value of a mass, and it comes out to be

$$M = \frac{m_1 m_2}{m_1 + m_2}. \tag{6.37}$$

What does M look like in the limit where m_1 is much smaller than m_2? In this limit we can ignore the m_1 in the denominator, but we *can't* ignore it in the numerator. So we obtain $M \approx m_1 m_2/(0 + m_2) = m_1$. Why can we can ignore one of the m_1's but not the other? We can ignore the m_1 in the denominator because it appears there as an *additive* term. If m_1 is small, then erasing it essentially doesn't change the value of the denominator. However, in the numerator m_1 appears as a *multiplicative* term. Even if m_1 is small, its value certainly affects the value of the numerator. Decreasing m_1 by a factor of 10 decreases the numerator by the same factor of 10. So we certainly can't just erase it. (That would change the units of M, anyway.)

Alternatively, you can obtain the $M \approx m_1$ result in the limit of small m_1 by applying a Taylor series (discussed below) to M. But this would be overkill. It's much easier to just erase the m_1 in the denominator. In any case, if you're ever unsure about which terms you should keep and which terms you can ignore, just plug some very small numbers (or very large numbers, depending on what limit you're dealing with) into a calculator to see how the expression depends on the various parameters.

It should be noted that there is no need to wait until the end of a solution to check limiting cases (or units, too). Whenever you arrive at an intermediate result that lends itself to checking limiting cases, you should check them. If you find that something is amiss, this will prevent you from wasting time carrying onward with incorrect results.

6.6.4 Taylor series

A tool that is often useful when checking limiting cases is the Taylor series. A Taylor series expresses a function $f(x)$ as a series expansion in x (that is, a sum of terms involving different powers of x). Perhaps the most well-known Taylor series is the one for the function $f(x) = e^x$:

$$e^x = 1 + x + \frac{x^2}{2!} + \frac{x^3}{3!} + \frac{x^4}{4!} + \cdots. \tag{6.38}$$

A number of other Taylor series are listed near the beginning of Appendix G. The rest of Appendix G contains a discussion of Taylor series and various issues that arise when using them. In the present section, we'll just take the above expression for e^x as given and see where it leads us.[4]

The classic example of the usefulness of a Taylor series in relativity is the demonstration of how the relativistic energy γmc^2 leads to the nonrelativistic energy $mv^2/2$; see Eq. (3.10). However, for the present purposes of illustrating the utility of Taylor

[4]Calculus is required if you want to *derive* a Taylor series. However, if you just want to *use* a Taylor series (which is what we do in this book), then algebra is all you need. So although some Taylor-series manipulations might look a bit scary, there's nothing more than algebra involved.

series, let's consider a mechanics setup in which a beach ball is dropped from rest. It can be shown that if air drag is taken into account, and if the drag force is proportional to the velocity (so that it takes the form $F_{\text{d}} = -bv$, where b is the drag coefficient), then the ball's velocity (with upward taken as positive) as a function of time equals

$$v(t) = -\frac{mg}{b}\left(1 - e^{-bt/m}\right). \tag{6.39}$$

This is a somewhat complicated expression, so you might be a little doubtful of its validity. Let's therefore look at some limiting cases. If these limiting cases yield expected results, then we can feel more comfortable that the expression is actually correct.

If t is very small (more precisely, if $bt/m \ll 1$; see the discussion in Section 6.7.3), then we can use the Taylor series in Eq. (6.38) to make an approximation to $v(t)$, to leading order in t. (The leading-order term is the smallest power of t with a nonzero coefficient.) To first order in x, Eq. (6.38) gives $e^x \approx 1 + x$. If we let x be $-bt/m$, then we see that Eq. (6.39) can be written as

$$\begin{aligned} v(t) &\approx -\frac{mg}{b}\left(1 - \left(1 - \frac{bt}{m}\right)\right) \\ &\approx -gt. \end{aligned} \tag{6.40}$$

This answer makes sense, because the drag force is negligible at the start (because v, and hence bv, is very small), so we essentially have a freely falling body with acceleration g downward. And $v(t) = -gt$ is the correct expression in that case (a familiar result from introductory mechanics). This successful check of a limiting case makes us have a little more faith that Eq. (6.39) is actually correct.

If we mistakenly had, say, $-2mg/b$ as the coefficient in Eq. (6.39), then we would have obtained $v(t) \approx -2gt$ in the small-t limit, which is incorrect. So we would know that we needed to go back and check over our work. Although it isn't obvious that an extra factor of 2 in Eq. (6.39) is incorrect, an extra 2 *is* obviously incorrect in the limiting $v(t) \approx -2gt$ result. As mentioned above, your intuition about limiting cases is generally much better than your intuition about generic values of the parameters.

We can also consider the limit of large t (or rather, large bt/m). In this limit, $e^{-bt/m}$ is essentially zero, so the $v(t)$ in Eq. (6.39) becomes (there's no need for a Taylor series in this case)

$$v(t) \approx -\frac{mg}{b}. \tag{6.41}$$

This is the "terminal velocity" that the ball approaches as time goes on. Its value makes sense, because it is the velocity for which the total force (gravitational plus air drag), $-mg - bv$, equals zero. And zero force means constant velocity. Mathematically, the velocity never quite reaches the value of $-mg/b$, but it gets extremely close as t becomes large.

Whenever you derive approximate answers as we just did, you gain something and you lose something. You lose some truth, of course, because your new answer is an approximation and therefore technically not correct (although the error becomes arbitrarily small in the appropriate limit). But you gain some aesthetics. Your new answer is invariably much cleaner (often involving only one term), and that makes it a lot easier to see what's going on.

In the above beach-ball example, we checked limiting cases of an answer that was correct. This whole process is more useful (and a bit more fun) when you check limiting cases of an answer that is *incorrect* (as in the case of the erroneous coefficient of $-2mg/b$ we mentioned above). When this happens, you gain the unequivocal information that your answer is wrong (assuming that your incorrect answer doesn't just happen to give the correct result in a certain limit, by pure luck). However, rather than leading you into despair, this information is something you should be quite happy about, considering that the alternative is to carry on in a state of blissful ignorance. Once you know that your answer is wrong, you can go back through your work and figure out where the error is (perhaps by checking limiting cases at various intermediate stages to narrow down where the error could be). Personally, if there's any way I'd like to discover that my answer is garbage, this is it. So you shouldn't check limiting cases (and units) because you're being told to, but rather because you *want* to.

6.7 Appendix G: Taylor series

6.7.1 Basics

We saw in Section 6.6.4 that Taylor series can be extremely useful for checking limiting cases, in particular in situations where a given parameter is small. In this appendix we'll discuss Taylor series in more detail.

A Taylor series expresses a given function of x as a series expansion in powers of x. The general form of a Taylor series is (the primes here denote differentiation)

$$f(x_0 + x) = f(x_0) + f'(x_0)x + \frac{f''(x_0)}{2!}x^2 + \frac{f'''(x_0)}{3!}x^3 + \cdots . \tag{6.42}$$

This equality can be verified by taking successive derivatives of both sides of the equation and then setting $x = 0$. For example, taking the first derivative and then setting $x = 0$ yields $f'(x_0)$ on the left. And this operation also yields $f'(x_0)$ on the right, because the first term is a constant and gives zero when differentiated, the second term gives $f'(x_0)$, and all of the rest of the terms give zero once we set $x = 0$, because they all contain at least one power of x. Likewise, if we take the second derivative of each side and then set $x = 0$, we obtain $f''(x_0)$ on both sides. And so on for all derivatives. Therefore, since the two functions on each side of Eq. (6.42) are equal at $x = 0$ and also have their nth derivatives equal at $x = 0$ for all n, they must in fact be the same function (assuming that they're nicely behaved functions, as we generally assume in physics).

Some specific Taylor series that often come up are listed below. They are all expanded around $x = 0$. That is, $x_0 = 0$ in Eq. (6.42). They are all derivable via Eq. (6.42), but sometimes there are quicker ways of obtaining them. For example, Eq. (6.44) is most easily obtained by taking the derivative of Eq. (6.43), which itself is just the sum of a geometric series.

$$\frac{1}{1+x} = 1 - x + x^2 - x^3 + \cdots \tag{6.43}$$

$$\frac{1}{(1+x)^2} = 1 - 2x + 3x^2 - 4x^3 + \cdots \tag{6.44}$$

$$\ln(1+x) = x - \frac{x^2}{2} + \frac{x^3}{3} - \cdots \tag{6.45}$$

$$e^x = 1 + x + \frac{x^2}{2!} + \frac{x^3}{3!} + \cdots \tag{6.46}$$

$$\cos x = 1 - \frac{x^2}{2!} + \frac{x^4}{4!} - \cdots \tag{6.47}$$

$$\sin x = x - \frac{x^3}{3!} + \frac{x^5}{5!} - \cdots \tag{6.48}$$

$$\sqrt{1+x} = 1 + \frac{x}{2} - \frac{x^2}{8} + \cdots \tag{6.49}$$

$$\frac{1}{\sqrt{1+x}} = 1 - \frac{x}{2} + \frac{3x^2}{8} + \cdots \tag{6.50}$$

$$(1+x)^n = 1 + nx + \binom{n}{2}x^2 + \binom{n}{3}x^3 + \cdots \tag{6.51}$$

Each of these series has a range of validity, that is, a "radius of convergence." For example, the series for e^x is valid for all x, while the series for $1/(1+x)$ is valid for $|x| < 1$. The various ranges won't particularly concern us, because whenever we use a Taylor series, we will assume that x is small (much smaller than 1). In this case, all of the series are valid.

The above series might look a little scary, but in most situations there is no need to include terms beyond the first-order term in x. For example, $\sqrt{1+x} \approx 1 + x/2$ is usually a good enough approximation. The smaller x is, the better the approximation is, because any term in the expansion is smaller than the preceding term by a factor of order x. Note that you can quickly verify that the $\sqrt{1+x} \approx 1 + x/2$ expression is valid to first order in x, by squaring both sides to obtain $1 + x \approx 1 + x + x^2/4$. Similar reasoning at second order shows that $-x^2/8$ is correctly the next term in the expansion.

As mentioned in Footnote 4 in Appendix F, we won't worry about taking derivatives to rigorously derive the above Taylor series. We'll just take them as given, which means that if you haven't studied calculus yet, that's no excuse for not using Taylor series! Instead of deriving them, let's just check that they're believable. This can easily be done with a calculator. For example, consider what e^x looks like if x is a very small number, say, $x = 0.0001$. Your calculator (or a computer, if your want more digits) will tell you that

$$e^{0.0001} = 1.000\,100\,005\,000\,166\,6\ldots \tag{6.52}$$

This can be written more informatively as

$$\begin{aligned} e^{0.0001} &= \quad 1.0 \\ &\quad + 0.000\,1 \\ &\quad + 0.000\,000\,005 \\ &\quad + 0.000\,000\,000\,000\,166\,6\ldots \\ &= 1 + (0.0001) + \frac{(0.0001)^2}{2!} + \frac{(0.0001)^3}{3!} + \cdots . \end{aligned} \tag{6.53}$$

This last line agrees with the form of the Taylor series for e^x in Eq. (6.46). If you made x smaller (say, 0.000001), then the same pattern would appear, but just with more zeros between the numbers than in Eq. (6.52). If you kept more digits in Eq. (6.52), you could verify the $x^4/4!$ and $x^5/5!$, etc., terms in the e^x Taylor series. But things aren't quite as obvious for these terms, because we don't have all the nice zeros as we do in the first 12 digits in Eq. (6.52).

Note that the lefthand sides of all of the Taylor series listed above involve 1's and x's. So how do we make an approximation to an expression of the form, say, $\sqrt{N+x}$, where x is small? We could of course use the general Taylor-series expression in Eq. (6.42) to generate the series from scratch by taking derivatives. But we can save ourselves some time by making use of the similar-looking $\sqrt{1+x}$ series in Eq. (6.49). We can turn the N into a 1 by factoring out an N from the square root, which gives $\sqrt{N}\sqrt{1+x/N}$. Having generated a 1, we can now apply Eq. (6.49), with the only modification being that the small quantity x that appears in that equation is replaced by the small quantity

x/N. This gives (to first order in x)

$$\sqrt{N+x} = \sqrt{N}\sqrt{1+\frac{x}{N}} \approx \sqrt{N}\left(1+\frac{1}{2}\frac{x}{N}\right) = \sqrt{N} + \frac{x}{2\sqrt{N}}. \tag{6.54}$$

You can quickly verify that this expression is valid to first order in x by squaring both sides. As a numerical example, if $N = 100$ and $x = 1$, then this approximation gives $\sqrt{101} \approx 10 + 1/20 = 10.05$, which is very close to the actual value of $\sqrt{101} \approx 10.0499$.

Example (Calculating a square root): Use the Taylor series $\sqrt{1+x} \approx 1 + x/2 - x^2/8$ to produce an approximate value of $\sqrt{5}$. How much does your answer differ from the actual value?

Solution: We'll first write 5 as 4 + 1, because we know what the square root of 4 is. However, we can't immediately apply the given Taylor series with $x = 4$, because we need x to be small. We must first factor out a 4 from the square root, so that we have an expression of the form $\sqrt{1+x}$, where x is small. Using $\sqrt{1+x} \approx 1 + x/2 - x^2/8$ with $x = 1/4$ (not 4!), we obtain

$$\begin{aligned}\sqrt{5} = \sqrt{4+1} = 2\sqrt{1+1/4} &\approx 2\left(1+\frac{1/4}{2}-\frac{(1/4)^2}{8}\right) \\ &= 2\left(1+\frac{1}{8}-\frac{1}{128}\right) \approx 2.2344.\end{aligned} \tag{6.55}$$

The actual value of $\sqrt{5}$ is about 2.2361. The approximate result is only 0.0017 less than this, so the approximation is quite good (the percentage difference is only 0.08%). Equivalently, the square of the approximate value is 4.9924, which is very close to 5. If you include the next term in the series, which happens to be $+x^3/16$, the result is $\sqrt{5} \approx 2.2363$, with an error of only 0.01%. By keeping a sufficient number of terms, you can produce any desired accuracy.

When trying to determine the square root of a number that isn't a perfect square, you could of course just guess and check, improving your guess on each iteration. But a Taylor series (calculated relative to the closest perfect square) provides a systematic method that doesn't involve guessing.

6.7.2 How many terms to keep?

When making a Taylor-series approximation, how do you know how many terms in the series to keep? For example, if the exact answer to a given problem takes the form of $e^x - 1$, then the Taylor series $e^x \approx 1 + x$ tells us that our answer is approximately equal to x. You can check this by picking a small value for x (say, 0.01) and plugging it in your calculator. This approximate form makes the dependence on x (for small x) much more transparent than the original expression $e^x - 1$ does.

But what if our exact answer had instead been $e^x - 1 - x$? The Taylor series $e^x \approx 1 + x$ would then yield an approximate answer of zero. And indeed, the answer *is* approximately zero. However, when making approximations, it is generally understood that we are looking for the *leading-order* term in the answer (that is, the smallest power of x with a nonzero coefficient). If our approximate answer comes out to be zero, then that means we need to go (at least) one term further in the Taylor series, which means $e^x \approx 1 + x + x^2/2$ in the present case. Our approximate answer is then $x^2/2$. (You should check this by letting $x = 0.01$.) Similarly, if the exact answer had instead been

$e^x - 1 - x - x^2/2$, then we would need to go out to the $x^3/6$ term in the Taylor series for e^x.

You should be careful to be consistent in the powers of x you deal with. If the exact answer is, say, $e^x - 1 - x - x^2/3$, and if you use the Taylor series $e^x \approx 1 + x$, then you will obtain an approximate answer of $-x^2/3$. This is incorrect, because it is inconsistent to pay attention to the $-x^2/3$ term in the exact answer while ignoring the corresponding $x^2/2$ term in the Taylor series for e^x. Including both terms gives the correct approximate answer as $x^2/6$.

So what is the answer to the above question: How do you know how many terms in the series to keep? Well, the answer is that before you do the calculation, there's really no way of knowing how many terms to keep. The optimal strategy is probably to just hope for the best and start by keeping only the term of order x. This will often be sufficient. But if you end up with a result of zero, then you can go to order x^2, and so on. Of course, you could play it safe and always keep terms up to, say, fourth order. But that is invariably a poor strategy, because you will probably never need to go out that far in a series.

6.7.3 Dimensionless quantities

Whenever you use a Taylor series from the above list to make an approximation in a physics problem, the parameter x must be *dimensionless*. If it weren't dimensionless, then the terms with the various powers of x in the series would all have different units, and it makes no sense to add terms with different units.

As an example of an expansion involving a properly dimensionless quantity, consider the approximation we made in going from Eq. (6.39) to Eq. (6.40) in the beach-ball example in Appendix F. In this setup, the small dimensionless quantity x is the bt/m term that appears in the exponent in Eq. (6.39). This quantity is indeed dimensionless, because from the original expression for the drag force, $F_d = -bv$, we see that b has units of N/(m/s), or equivalently kg/s. Hence bt/m is dimensionless.

We can restate the above dimensionless requirement in a more physical way. Consider the question, "What is the velocity $v(t)$ in Eq. (6.39), in the limit of small t?" This question is meaningless, because t has dimensions. Is a year a large or a small time? How about a hundredth of a second? There is no way to answer this without knowing what situation we're dealing with. A year is short on the time scale of galactic evolution, but a hundredth of a second is long on the time scale of an elementary-particle process. It makes sense only to look at the limit of a large or small *dimensionless* quantity x. And by "large or small," we mean compared with the number 1.

Equivalently, in the beach-ball example the quantity m/b has dimensions of time, so the value of m/b is a time that is inherent to the system. It therefore *does* make sense to look at the limit where $t \ll m/b$ (that is, $bt/m \ll 1$), because we are now comparing two things, namely t and m/b, that have the same dimensions. We will sometimes be sloppy and say things like, "In the limit of small t." But you know that we really mean, "In the limit of a small dimensionless quantity that has a t in the numerator (like bt/m)," or, "In the limit where t is much smaller than a certain quantity that has dimensions of time (like m/b)." Similarly, throughout this book the phrase "small v" is always understood to mean "v much smaller than c."

After you make an approximation, how do you know if it is a "good" one? Well, just as it makes no sense to ask if a dimensionful quantity is large or small without comparing it to another quantity with the same dimensions, it makes no sense to ask if an approximation is "good" or "bad" without stating what accuracy you want. In the beach-ball example, let's say that we're looking at a value of t for which $bt/m = 1/100$.

In Eq. (6.40) we kept the bt/m term in the Taylor series for $e^{-bt/m}$, and this directly led to our answer of $-gt$. We ignored the $(bt/m)^2/2$ term in the series. This is smaller than the bt/m term that we kept, by a factor of $(bt/m)/2 = 1/200$. So the error is roughly half a percent. (The corrections from the higher-order terms are even smaller.) If this is enough accuracy for whatever purpose you have in mind, then the approximation is a good one. If not, then it's a bad one, and you need to add more terms in the series until you get your desired accuracy.

6.8 Appendix H: Useful formulas

The first formula here can be quickly proved by showing that the Taylor series for both sides are equal.

$$e^{i\theta} = \cos\theta + i\sin\theta \tag{6.56}$$

$$\cos\theta = \frac{e^{i\theta} + e^{-i\theta}}{2} \qquad \sin\theta = \frac{e^{i\theta} - e^{-i\theta}}{2i} \tag{6.57}$$

$$\cos\frac{\theta}{2} = \pm\sqrt{\frac{1+\cos\theta}{2}} \qquad \sin\frac{\theta}{2} = \pm\sqrt{\frac{1-\cos\theta}{2}} \tag{6.58}$$

$$\tan\frac{\theta}{2} = \pm\sqrt{\frac{1-\cos\theta}{1+\cos\theta}} = \frac{1-\cos\theta}{\sin\theta} = \frac{\sin\theta}{1+\cos\theta} \tag{6.59}$$

$$\sin 2\theta = 2\sin\theta\cos\theta \qquad \cos 2\theta = \cos^2\theta - \sin^2\theta \tag{6.60}$$

$$\sin(\alpha+\beta) = \sin\alpha\cos\beta + \cos\alpha\sin\beta \tag{6.61}$$

$$\cos(\alpha+\beta) = \cos\alpha\cos\beta - \sin\alpha\sin\beta \tag{6.62}$$

$$\tan(\alpha+\beta) = \frac{\tan\alpha + \tan\beta}{1 - \tan\alpha\tan\beta} \tag{6.63}$$

$$\cosh x = \frac{e^x + e^{-x}}{2} \qquad \sinh x = \frac{e^x - e^{-x}}{2} \tag{6.64}$$

$$\tanh x = \frac{\sinh x}{\cosh x} = \frac{e^x - e^{-x}}{e^x + e^{-x}} \tag{6.65}$$

$$\cosh^2 x - \sinh^2 x = 1 \tag{6.66}$$

$$\frac{d}{dx}\cosh x = \sinh x \qquad \frac{d}{dx}\sinh x = \cosh x \tag{6.67}$$

$$\sinh(x+y) = \sinh x\cosh y + \cosh x\sinh y \tag{6.68}$$

$$\cosh(x+y) = \cosh x\cosh y + \sinh x\sinh y \tag{6.69}$$

$$\tanh(x+y) = \frac{\tanh x + \tanh y}{1 + \tanh x\tanh y} \tag{6.70}$$

6.9 References

Ashby, N. (2002). Relativity and the global positioning system. *Physics Today,* **55**(5), 41-47.

Chandrasekhar, S. (1979). Einstein and general relativity: Historical perspectives. *American Journal of Physics,* **47**, 212-217.

Costella, J. P., McKellar, B. H. J., Rawlinson, A. A., and Stephenson, G. J. (2001). The Thomas rotation. *American Journal of Physics,* **69**, 837-847.

Cranor, M. B., Heider, E. M., and Price R. H. (2000). A circular twin paradox. *American Journal of Physics,* **68**, 1016-1020.

Eisner, E. (1967). Aberration of light from binary stars – a paradox? *American Journal of Physics,* **35**, 817-819.

Fadner, W. L. (1988). Did Einstein really discover "$E = mc^2$"? *American Journal of Physics,* **56**, 114-122.

Handschy, M. A. (1982). Re-examination of the 1887 Michelson–Morley experiment. *American Journal of Physics,* **50**, 987-990.

Hollenbach, D. (1976). Appearance of a rapidly moving sphere: A problem for undergraduates. *American Journal of Physics,* **44**, 91-93.

Holton, G. (1988). Einstein, Michelson, and the "Crucial" Experiment. In *Thematic Origins of Scientific Thought, Kepler to Einstein,* Cambridge: Harvard University Press.

Lee, A. R. and Kalotas, T. M. (1975). Lorentz transformations from the first postulate. *American Journal of Physics,* **43**, 434-437.

Lee, A. R. and Kalotas, T. M. (1977). Causality and the Lorentz transformation. *American Journal of Physics,* **45**, 870.

Medicus, H. A. (1984). A comment on the relations between Einstein and Hilbert. *American Journal of Physics,* **52**, 206-208.

Mermin, N. D. (1983). Relativistic addition of velocities directly from the constancy of the velocity of light. *American Journal of Physics,* **51**, 1130-1131.

Morin, D. (2008). *Introduction to Classical Mechanics, With Problems and Solutions,* Cambridge: Cambridge University Press.

Muller, R. A. (1992). Thomas precession: Where is the torque? *American Journal of Physics,* **60**, 313-317.

Pound, R. V. and Rebka, G. A. (1960). Apparent weight of photons. *Physical Review Letters,* **4**, 337-341.

Rebilas, K. (2002). Comment on "The Thomas rotation," by John P. Costella *et al.* *American Journal of Physics,* **70**, 1163-1165.

Rindler, W. (1994). General relativity before special relativity: An unconventional overview of relativity theory. *American Journal of Physics,* **62**, 887-893.

Index

accelerating frame, 197–200
action, 201, 202

baseballs, 14
billiards, 136
bouncing stick, 45
break or not break, 106

causality, 85–87, 90–92, 101, 232
centripetal acceleration, 186
checking limits, 248–249
checking units, 246–248
collisions
 nonrelativistic, 244–245
 relativistic, 134–137
Compton scattering, 153
conservation of
 energy, 125–128, 133–134, 243
 momentum, 124–125, 133–134, 242
contact force, 244
cookies, relativistic, 46
correspondence between (ct, x), (E, pc), 132
correspondence principle, 128

decay of muon, 23, 26
decays, 137–138
dimensional analysis, 246
dimensionless parameter, 254
dimensions, 246
Doppler effect
 classical, 194
 longitudinal, 50, 92–93, 132, 146, 151, 189, 232
 transverse, 94–96
drawing pictures, 30
dropping the c's, 130

eccentricity, 59
effective mass, 153, 158
Einstein, 2–5, 8, 192
elapsed times vs. readings, 22, 228
elastic collision, 128, 244
electron-volt, 139
energy
 nonrelativistic, 243
 relativistic, 125–128
energy-momentum 4-vector, 135, 180, 183
equivalence principle, 192–193
ether, 5–10
event, 2

Fizeau experiment, 47
force
 4-vector, 180
 in 1-D, 140–142
 in 2-D, 142–143
 relativistic, 140–145
 transformation of, 143–145, 184
4-momentum, 135, 180
4-vector
 acceleration, 180, 185–186
 definition, 177
 dislacement, 179
 energy-momentum, 135, 180, 183
 force, 180, 184
 velocity, 179
frame dragging, 9
frame independence, 29, 226
frequency
 Doppler, longitudinal, 93
 Doppler, transverse #1, 95
 Doppler, transverse #2, 95
 GR red/blue-shift, 195

Galilean invariance, 3
Galilean transformations, 2–5, 73, 74, 101
γ factor, 19
gap-closing speed, 13, 227
global positioning system (GPS), 205
golden ratio, 51, 157, 175
gravitational mass, 193, 234
gravitational redshift, 196, 209

harmonic oscillator, 153
heat, 128, 232
Heron's formula, 161
Higgs boson, 157

hyperbolic functions, 96, 255
hyperbolic worldline, 198

index of refraction, 47
inelastic collision, 128, 244
inertial frame, 10, 11, 241
inertial mass, 193, 234
inner product, 135, 181
interferometer, 7
invariance
- Galilean, 3
- of $(\Delta s)^2$, 82
- of $E^2 - p^2$, 132
- of inner product, 181–182
- of mass, 129–130
- of speed of light, 11

invariant interval, 82–85

kilowatt-hour, 129
kinetic energy
- nonrelativistic, 127, 242
- relativistic, 128

lattice of clocks, meter sticks, 28
least action, principle of, 201
length contraction, 23–27, 32, 45, 79, 89–90, 210, 229
- no transverse, 27

length, how to measure, 25
light, 1, 4–11, 151
light clock, 19
lightlike separation, 84
limits, 248
linearity, 133, 181
local, assumption of, 193, 234
Lorentz transformations, 4, 71–79, 97, 177, 231, 238–241
- matrix form, 74, 97
- sign in, 75, 231
- symmetry of, 74

Lorentz–FitzGerald contraction, 8, 24

magnetic field, 63
magnetic force, 46
mass
- gravitational, 193, 234
- inertial, 193, 234
- invariance of, 129–130
- relativistic, 129–130, 232

maximal proper time, 200–202
Maxwell's equations, 4–5
mechanical energy, 127
MeV (mega-electron-volt), 140
Michelson–Morley experiment, 5–10
Minkowski diagram, 87–90
- gridlines, 89

momentum
- nonrelativistic, 242
- relativistic, 124–125

Mona Lisa, 25
muon, 23, 26, 140

Newton's laws, 241
nonrelativistic dynamics, 241–245
norm, 182, 183

particle-physics units, 139–140
pendulum, 247
period of relativistic oscillator, 153
photon, 129, 151, 153, 232
physical laws, form of, 186–187
Planck's constant, 151
pole in barn, 45, 50, 108
postulates of relativity, 10–12
potential energy, 244
principle of
- correspondence, 128
- equivalence, 192–193
- Galilean invariance, 3
- least action, 201
- maximal proper time, 200–202
- relativity, 10

problem-solving strategies, 42, 245–251
proper acceleration, 98
proper length, 24, 84
proper time, 21, 83, 179
- maximal, 200–202

Pythagorean triple, 47

rapidity, 96
readings vs. elapsed times, 22, 228
rear clock ahead, 14–17, 78, 108, 207, 208, 235–237
redshift, 93, 106, 146
- gravitational, 196, 209

relative speed, 39, 225, 230
relativistic mass, 129–130, 232
relativistic strings, 149–150
relativity postulate, 10
relativity without c, 99–102
rigidity, 226
Rindler space, 200
rocket motion, 145–149
rotation matrix, 97

saddle point, 222
seeing things, 14, 30, 227
setting $c = 1$, 130
sidereal day, 215
simultaneity
  line of, 89, 117–118
  loss of, 12–17
space station, 208
spacelike separation, 83–84, 231
spacetime diagram, 87
spacetime interval, 83
spacetime separation, 83
special cases, 248
speed c, cannot travel at, 142, 226
speed of light, 1, 5, 8, 11, 102, 225
speed-of-light postulate, 11
stationary action, 201
stationary proper time, 200
stellar aberration, 9
sticking to a frame, 30
strategies for solving problems, 42, 245–251
superluminal signal, 86
symbolic answer, 245
synchronizing clocks, 29, 44, 227

Taylor series, 249–255
  dimensionless parameter in, 254
  general expression, 251
  list of, 251
threshold energy, 152
time dilation, 31, 108
  GR, 194–196
  GR, for $1/r^2$ field, 205
  SR, 17–23, 78, 84, 228
  SR, with acceleration, 21, 120, 228
timelike separation, 83, 231
transformation
  Galilean, 2–5, 73, 74, 101
  Lorentz, 71–79, 97, 231, 238–241
  of E and p, 131–134, 183
  of acceleration, 144, 185–186
  of force, 143–145, 184
  of 4-vector, 177
twin paradox, 22, 48, 50, 105, 109, 202–203, 207, 237–238

uniform acceleration, 197
uniformly accelerating frame, 199–200
units, 246

velocity addition
  longitudinal, 35–42, 48, 53, 54, 79–80, 105, 188
  transverse, 80–82
very important relation, 129

work, 141, 242
work-energy theorem, 141, 243
worldline, 88
  hyperbolic, 198